AF368532

Free Radical Research in Cancer

Free Radical Research in Cancer

Special Issue Editor

Ana Čipak Gašparović

MDPI • Basel • Beijing • Wuhan • Barcelona • Belgrade

Special Issue Editor
Ana Čipak Gašparović
Institute Ruder Boskovic
Croatia

Editorial Office
MDPI
St. Alban-Anlage 66
4052 Basel, Switzerland

This is a reprint of articles from the Special Issue published online in the open access journal *Antioxidants* (ISSN 2076-3921) from 2018 to 2019 (available at: https://www.mdpi.com/journal/antioxidants/special_issues/free_radical_cancer).

For citation purposes, cite each article independently as indicated on the article page online and as indicated below:

LastName, A.A.; LastName, B.B.; LastName, C.C. Article Title. *Journal Name* **Year**, *Article Number*, Page Range.

ISBN 978-3-03936-088-8 (Pbk)
ISBN 978-3-03936-089-5 (PDF)

Contents

About the Special Issue Editor

Ana Čipak Gašparović is a senior research associate in the Laboratory for Oxidative stress, Division of Molecular Medicine, Ruđer Bošković Institute. She obtained her diploma in biology (molecular biology) at Faculty of Science, University of Zagreb in 2001. She obtained her Ph.D. in biochemistry and molecular biology at Faculty of Science, University of Zagreb in 2009. In 2013 she obtained the title of project management specialist on joint study of the Ruđer Bošković Institute and the College for Business and Management "Baltazar Adam Krcelic". Since 2001 She worked in Laboratory for Oxidative stress, Ruđer Bošković Institute since 2001. In 2010 when she moved to the position of research associate and since 2016 to the position of senior research associate. She was on several study visits lasting from 1 week to 2 month at the University of Graz, University of Salzburg, and University of New South Wales, Sidney. Since 2011 she is a lecturer of "Oxidative stress in carcinogenesis—models and methods", a part of Molecular Biosciences at joint Ph.D. Program of University of Osijek, Ruđer Bošković Institute and University of Dubrovnik, and on doctoral studies of School of Medicine, University of Zagreb. During her work she was mentor to diploma student and Ph.D. student She is a member of Croatian Association for Cancer Research (CACR), Croatian Society for Biochemistry and Molecular Biology (CSBMB), Society for Free Radical Research (SFRR), The International HNE-Club. She participates in several COST Actions related to cancer. She has worked on 2 Croatian MSES projects and was a researcher on the following bilateral projects: 2 Croatian-Austrian, Croatian-French, Croatian-Hungarian and a Croatian-Slovenian. She was principal investigator of Croatian-Austrian bilateral project She was also a researcher on FP7 project Thymistem. In the focus of her research are mechanisms of cancer cell adaptation to chronic stress with emphasis on the role of specific aquaporins, as well as cancer stem cells and their interactions with the microenvironment.

Editorial

Free Radical Research in Cancer

Ana Čipak Gašparović

Division of Molecular Medicine, Ruđer Bošković Institute, HR-10000 Zagreb, Croatia; acipak@irb.hr

Received: 7 February 2020; Accepted: 12 February 2020; Published: 15 February 2020

It can be challenging to find efficient therapy for cancer due to its biological diversity. One of the factors that contribute to its biological diversity are free radicals. Evolutionary, aerobic organisms evolve in an oxygen atmosphere, improving the energy production system by using oxygen. Oxygen is beneficial, but it can also be detrimental if free radicals are formed [1]. Free radicals, as well as some non-radical species that have oxygen, are reactive oxygen species (ROS). ROS can damage DNA leading to mutations, single, or double-strand breaks [2]. These events, if the cell is unable to repair the damage, are deleterious. If not fatal, these changes in genetic material result in tumor development by losing cell cycle control. Further, these mutations create genetic instability that result in tumor heterogeneity, and thereby increase the possibility of surviving stress conditions. In addition to direct interaction with DNA, proteins, and lipids, ROS are also signaling molecules that take an active part in regulating cellular processes [3,4]. It was previously thought that ROS only damage cells, but we now know that some enzymes primarily produce ROS, and they are not by-products [3]. These are NAD(P)H oxidases (NOX) and they produce ROS in response to inflammatory signaling. This planned production of ROS may play a role in proliferation, as ROS are able to activate signaling pathways, such as mitogen activated-protein kinase (MAPK)/extracellular-regulated kinase 1/2 (ERK1/2), phosphoinositide-3-kinase (PI3K)/protein kinase (Akt), and more, thoroughly reviewed in [3]. An important factor in surviving ROS is the antioxidant system of the cell. The main role of this complex system is to remove the excess ROS. As there are many ROS, there are many different parts of this system acting in a similar or unique way in removing ROS, such as the glutathione system, superoxide dismutase-catalase catalase, thioredoxin system, and small molecules (e.g. vitamin C, vitamin E). In order to ensure the right levels of these enzymes and small molecules, ROS activate several antioxidative transcription factors, such as Nrf2 and the FoxO family. These transcription factors are responsible for activating the majority of antioxidative genes [5–7]. Generally, cancer cells have increased amounts of ROS; consequently, they adapt by increasing the antioxidative defense system [8], thereby, strongly linking ROS and antioxidative research.

Nevertheless, ROS were at first considered detrimental, and this was used as a therapeutic strategy in fighting cancer. Most of the conventional types of chemotherapy, as well as radiotherapy, are based on ROS production. Unfortunately, this strategy has to eradicate the tumor completely, otherwise the surviving cells adapt and build up their antioxidant systems, as well as other mechanisms (e.g., drug transporters) making themselves resistant. Strategies involving activation/inhibition of signaling pathways (and here, Nrf2 was certainly an attractive target) turned out to be a double-edged sword [9].

This Special Issue aims to provide different approaches to study the role of free radicals in cancer. Recent findings are presented within eight original papers and four review papers, spanning from cancer therapy and resistance development to side effects of cancer therapy, with its effects on human health, in a process governed by free radicals.

The focus of the review papers is on free radicals, ROS, and cancer therapy. As mentioned above, ROS modulate cellular signaling pathways and are therefore important to maintain redox homeostasis. A review by Kim et al. [10] provides an overview of cellular ROS production, both controlled and uncontrolled, as well as ROS elimination (keeping in mind the importance of this homeostasis). Further, redox changes in cancer are described, with emphasis on chemotherapy based on ROS production. The paradox of chemotherapy is discussed: the chemotherapy resistance can be acquired through either increased proliferation (leading to resistance) or by changing to a cancer stem-like cell phenotype, with a low proliferation rate. In hand with this review is the work of Mendes and Serpa [11], which discusses metabolic remodeling of lung cancer. These metabolic changes occur via several mechanisms, which include mutations, as well as responses to oxidative or alkylating treatments. These events lead to chemotherapy resistance that occur because of changes in drug transporters, as well as in antioxidants. Metabolic remodeling is therefore a challenge in cancer therapy, and can be used—if the changes are well monitored and defined—to adapt to clinical therapy, in order to avoid recurrence.

The review papers by Clavo et al. [12], and Prasad and Srivastava [13], discuss adjuvant cancer therapy by reduction of ROS. Natural compounds, such as Triphala and Ayurvedic medicine, have antioxidative properties, and prevent free radical formation and lipid peroxidation. In addition to antioxidative properties, the authors also discuss the chemopreventive and chemotherapeutic effects of Triphala, which are encouraged by the results of three clinical studies. Another strategy in fighting cancer, described by Clavo et al. [12], is the use of ozone as an adjuvant therapy to conventional chemotherapy. The authors present evidence of beneficial effects of ozone therapy on animal models and describe possible mechanisms by which these effect may occur.

Mechanisms, by which cellular processes are changed in cancer, spread on different molecules (such as enzymes, transcription factors, or ion channels). An example of an ion channel is the transient receptor potential melastatin 2 (TRPM2), a Ca^{2+} channel that can be activated by H_2O_2 [14]. A study presented by Hack et al. [14] showed parallel expression of NOX4 and TRPM2 in human granulosa cell tumor samples, suggesting that induction of oxidative stress could be beneficial for the therapy, as activation of this channel by H_2O_2 increased Ca^{2+} levels and apoptotic cell death.

Acquired resistance was a model in two papers and was achieved through growth of cells under conditions of chronic oxidative stress. Both models used breast cancer cell lines in their study. In a study by Glorieux and Calderon [15], NQO1 affected cancer redox homeostasis and sensitivity to drugs. Consequently, NQO1 polymorphism may be used as an important factor if quinone-based chemotherapeutic drugs are considered as cancer therapy. Interestingly, NQO1 is a target gene for NRF2, an antioxidative transcription factor. Using breast cancer cell lines stimulated for cancer-stem-like phenotypes under chronic oxidative stress, we showed an increase in NRF2, but also in some epithelial-mesenchymal transition markers, indicating that NRF2 can play a role in breast cancer resistance [16].

In addition to breast cancer, ROS and NRF2 were studied in regards to the androgen receptor and its splice-variant AR-V7 [17]. As therapy for prostate cancer, a triterpenoid antioxidant drug was tested for its ability to regulate androgen receptor expression. This drug proved to enhance efficacy of clinically approved anti-androgen, but also decreased ROS and increased NRF2, indicating possible mechanisms of action. There are numerous consequences of prostate cancer therapy due to ROS production, but effects on sperm are not fully investigated. Takeshima et al. [18] show evidence that cancer chemotherapy has similar effects on semen as idiopathic infertility, suggesting antioxidant therapy to reduce ROS.

As mentioned, many conventional cancer therapies are based on free radical/ROS production. Photodynamic therapy is also a cancer therapy that uses chemosensitizers to generate free radicals, which then act against the tumor. Such a photosensitizer, a tailored boron-dipyrromethene (BODIPY) derivative, was used on A375 and SKMEL28 cancer cell lines [19]. Authors show positive effects of this compound by inducing singlet oxygen and NO to cause cell death.

Finally, Rodríguez-García et al. [20] studied protein carbonylation in patients with myelodysplastic syndromes. These patients had increased protein carbonyls, but levels decreased after treatment with an iron chelator (deferasirox). Analysis of the p21 gene expression in bone marrow cells revealed correlation between high protein carbonyls and increased expression, and vice versa. The paper suggests that the fine-tuning of oxidative stress levels in bone marrow can determine the disease progression in these patients.

Conflicts of Interest: The authors declare no conflict of interest.

References

1. Cadenas, E.; Sies, H. Oxidative stress: excited oxygen species and enzyme activity. *Adv. Enzyme Regul.* **1985**, *23*, 217–237. [CrossRef]
2. Barzilai, A.; Yamamoto, K.I. DNA damage responses to oxidative stress. *DNA Repair (Amst).* **2004**, *3*, 1109–1115. [CrossRef] [PubMed]
3. Moloney, J.N.; Cotter, T.G. ROS signalling in the biology of cancer. *Semin. Cell Dev. Biol.* **2018**, *80*, 50–64. [CrossRef] [PubMed]
4. Zhang, J.; Wang, X.; Vikash, V.; Ye, Q.; Wu, D.; Liu, Y.; Dong, W. ROS and ROS-Mediated Cellular Signaling. *Oxid. Med. Cell. Longev.* **2016**, *2016*, 4350965. [CrossRef] [PubMed]
5. Raghunath, A.; Sundarraj, K.; Nagarajan, R.; Arfuso, F.; Bian, J.; Kumar, A.P.; Sethi, G.; Perumal, E. Antioxidant response elements: Discovery, classes, regulation and potential applications. *Redox Biol.* **2018**, *17*, 297–314. [CrossRef] [PubMed]
6. Saito, R.; Suzuki, T.; Hiramoto, K.; Asami, S.; Naganuma, E.; Suda, H.; Iso, T.; Yamamoto, H.; Morita, M.; Baird, L.; et al. Characterizations of Three Major Cysteine Sensors of Keap1 in Stress Response. *Mol. Cell. Biol.* **2016**, *36*, 271–284. [CrossRef] [PubMed]
7. Laissue, P. The forkhead-box family of transcription factors: key molecular players in colorectal cancer pathogenesis. *Mol. Cancer* **2019**, *18*, 5. [CrossRef] [PubMed]
8. Kim, S.J.; Kim, H.S.; Seo, Y.R. Understanding of ROS-Inducing Strategy in Anticancer Therapy. *Oxid. Med. Cell. Longev.* **2019**, *2019*, 5381692. [CrossRef] [PubMed]
9. Milkovic, L.; Zarkovic, N.; Saso, L. Controversy about pharmacological modulation of Nrf2 for cancer therapy. *Redox Biol.* **2017**, *12*, 727–732. [CrossRef] [PubMed]
10. Kim, E.K.; Jang, M.; Song, M.J.; Kim, D.; Kim, Y.; Jang, H.H. Redox-mediated mechanism of chemoresistance in cancer cells. *Antioxidants* **2019**, *8*, 471. [CrossRef] [PubMed]
11. Mendes, C.; Serpa, J. Metabolic remodelling: An accomplice for new therapeutic strategies to fight lung cancer. *Antioxidants* **2019**, *8*, 603. [CrossRef] [PubMed]
12. Clavo, B.; Rodríguez-Esparragón, F.; Rodríguez-Abreu, D.; Martínez-Sánchez, G.; Llontop, P.; Aguiar-Bujanda, D.; Fernández-Pérez, L.; Santana-Rodríguez, N. Modulation of Oxidative Stress by Ozone Therapy in the Prevention and Treatment of Chemotherapy-Induced Toxicity: Review and Prospects. *Antioxidants* **2019**, *8*, 588. [CrossRef] [PubMed]
13. Prasad, S.; Srivastava, S.K. Oxidative Stress and Cancer: Chemopreventive and Therapeutic Role of Triphala. *Antioxidants* **2020**, *9*, 72. [CrossRef] [PubMed]
14. Hack, C.T.; Buck, T.; Bagnjuk, K.; Eubler, K.; Kunz, L.; Mayr, D.; Mayerhofer, A. A Role for H2O2 and TRPM2 in the Induction of Cell Death: Studies in KGN Cells. *Antioxidants* **2019**, *8*, 518. [CrossRef] [PubMed]
15. Glorieux, C.; Calderon, P.B. Cancer Cell Sensitivity to Redox-Cycling Quinones is Influenced by NAD(P)H: Quinone Oxidoreductase 1 Polymorphism. *Antioxidants* **2019**, *8*, 369. [CrossRef] [PubMed]
16. Čipak Gašparović, A.; Milković, L.; Dandachi, N.; Stanzer, S.; Pezdirc, I.; Vrančić, J.; Šitić, S.; Suppan, C.; Balic, M. Chronic Oxidative Stress Promotes Molecular Changes Associated with Epithelial Mesenchymal Transition, NRF2, and Breast Cancer Stem Cell Phenotype. *Antioxidants* **2019**, *8*, 633. [CrossRef] [PubMed]
17. Khurana, N.; Chandra, P.K.; Kim, H.; Abdel-Mageed, A.B.; Mondal, D.; Sikka, S.C. Bardoxolone-Methyl (CDDO-Me) Suppresses Androgen Receptor and Its Splice-Variant AR-V7 and Enhances Efficacy of Enzalutamide in Prostate Cancer Cells. *Antioxidants* **2020**, *9*, 68. [CrossRef] [PubMed]
18. Takeshima, T.; Kuroda, S.; Yumura, Y. Cancer Chemotherapy and Chemiluminescence Detection of Reactive Oxygen Species in Human Semen. *Antioxidants* **2019**, *8*, 449. [CrossRef] [PubMed]

19. Lazzarato, L.; Gazzano, E.; Blangetti, M.; Fraix, A.; Sodano, F.; Picone, G.M.; Fruttero, R.; Gasco, A.; Riganti, C.; Sortino, S. Combination of PDT and NOPDT with a Tailored BODIPY Derivative. *Antioxidants* **2019**, *8*, 531. [CrossRef] [PubMed]
20. Rodríguez-García, A.; Morales, M.L.; Garrido-García, V.; García-Baquero, I.; Leivas, A.; Carreño-Tarragona, G.; Sánchez, R.; Arenas, A.; Cedena, T.; Ayala, R.M.; et al. Protein Carbonylation in Patients with Myelodysplastic Syndrome: An Opportunity for Deferasirox Therapy. *Antioxidants* **2019**, *8*, 508. [CrossRef] [PubMed]

 *antioxidants*

Article

Bardoxolone-Methyl (CDDO-Me) Suppresses Androgen Receptor and Its Splice-Variant AR-V7 and Enhances Efficacy of Enzalutamide in Prostate Cancer Cells

Namrata Khurana [1,2,3], Partha K. Chandra [2], Hogyoung Kim [1], Asim B. Abdel-Mageed [1], Debasis Mondal [2,4,*] and Suresh C. Sikka [1,*]

[1] Department of Urology, Tulane University School of Medicine, 1430 Tulane Avenue, New Orleans, LA 70112, USA; nkhurana@wustl.edu (N.K.); hkim8@tulane.edu (H.K.); amageed@tulane.edu (A.B.A.-M.)

[2] Department of Pharmacology, Tulane University School of Medicine, 1430 Tulane Avenue, New Orleans, LA 70112, USA; pchandr1@tulane.edu

[3] Department of Internal Medicine-Medical Oncology, Washington University in St. Louis Medical Campus, 660 S Euclid Ave, St. Louis, MO 63110-1010, USA

[4] Department of Microbiology, Lincoln Memorial University—Debusk College of Osteopathic Medicine, 9737 Coghill Drive, Knoxville, TN 37932, USA

* Correspondence: debasis.mondal@lmunet.edu (D.M.); ssikka@tulane.edu (S.C.S.); Tel.: +865-338-5715 (D.M.); +504-988-5179 (S.C.S.)

Received: 4 December 2019; Accepted: 7 January 2020; Published: 12 January 2020

Abstract: Androgen receptor (AR) signaling is fundamental to prostate cancer (PC) progression, and hence, androgen deprivation therapy (ADT) remains a mainstay of treatment. However, augmented AR signaling via both full length AR (AR-FL) and constitutively active AR splice variants, especially AR-V7, is associated with the recurrence of castration resistant prostate cancer (CRPC). Oxidative stress also plays a crucial role in anti-androgen resistance and CRPC outgrowth. We examined whether a triterpenoid antioxidant drug, Bardoxolone-methyl, known as CDDO-Me or RTA 402, can decrease AR-FL and AR-V7 expression in PC cells. Nanomolar (nM) concentrations of CDDO-Me rapidly downregulated AR-FL in LNCaP and C4-2B cells, and both AR-FL and AR-V7 in CWR22Rv1 (22Rv1) cells. The AR-suppressive effect of CDDO-Me was evident at both the mRNA and protein levels. Mechanistically, acute exposure (2 h) to CDDO-Me increased and long-term exposure (24 h) decreased reactive oxygen species (ROS) levels in cells. This was concomitant with an increase in the anti-oxidant transcription factor, Nrf2. The anti-oxidant N-acetyl cysteine (NAC) could overcome this AR-suppressive effect of CDDO-Me. Co-exposure of PC cells to CDDO-Me enhanced the efficacy of a clinically approved anti-androgen, enzalutamide (ENZ), as evident by decreased cell-viability along with migration and colony forming ability of PC cells. Thus, CDDO-Me which is in several late-stage clinical trials, may be used as an adjunct to ADT in PC patients.

Keywords: bardoxolone methyl; prostate cancer; castration-resistant prostate cancer; androgen receptor (AR), AR-V7; anti-androgen; enzalutamide; androgen deprivation therapy

---

1. Introduction

Prostate cancer (PC) is the second leading cause of cancer-related mortality in men in the United States [1]. Notwithstanding the initial efficacy of androgen deprivation therapy (ADT), outgrowth of castration-resistant prostate cancer (CRPC) is the primary cause of death among patients [2]. The development of CRPC is linked with continuous androgen receptor (AR) signaling even in the absence of androgens [3–7]. Several mechanisms responsible for the constitutive AR signaling in

CRPC cells include AR gene amplification, ligand-independent AR activation by cytokines or kinases, both intracrine and/or intratumoral androgen production, overexpression of AR co-activators, and most importantly, the expression of constitutively active AR splice variants (AR-Vs) [8,9]. Despite the castrated levels of androgens, these spliced forms of AR lacking the C-terminal ligand binding domain (LBD), promote the transcriptional activation of AR target genes as they still retain the transactivating N-terminal domain (NTD) [8–10].

AR-V7 (also known as AR3) is the most significant functional protein encoding AR splice variant [11–19]. Augmented levels of AR-V7 were identified in CRPC tumor specimens [18] and circulating tumor cells [13]. Elevated AR-V7 expression was found after the development of CRPC tumors when primary tumor tissues were examined before and after the development of castration resistance [11,14–19]. Moreover, overexpression of AR-V7 is one of the key factors in the development of resistance to the potent second-generation anti-androgens, e.g., enzalutamide (ENZ) and abiraterone acetate (ABI) [20,21]. Studies have also shown a critical role of full-length AR (AR-FL) in dimerizing and transactivating AR-V7 [22], which is involved in castration-resistant cell growth [23]. Therefore, there is a critical requisite for potential therapeutic strategies which can efficiently reduce AR-FL and AR-V7 linked constitutive tumor promoting signaling in the CRPC cells.

Bardoxolone-methyl, the C-28 methyl ester of 2-cyano-3,12-dioxoolean-1,9-dien-28-oic acid (CDDO) known as CDDO-Me or RTA 402 is one of the synthetic triterpenoids that has been shown to have anti-inflammatory, as well as anticarcinogenic activities [24,25]. Studies with CDDO-Me have been conducted in various kinds of cancers such as prostate [26], breast [27], ovary [28], lung [29], leukemia [30], pancreatic [31], and osteosarcoma [32]. CDDO-Me activates Keap1/Nrf2/ARE pathway [33,34], inhibits nuclear factor kappa-B (NF-kB) [35] and Janus-activated kinase (JAK)/STAT (signal transducer and activator of transcription) pathway [36], and is effective at low nanomolar concentrations [24]. The α, β-unsaturated carbonyl groups present on its rings form reversible adducts with the thiol groups of critical cysteine residues in target proteins such as Keap-1 and inhibitor of kappa-B (IkB) kinase (IKKβ). The binding of CDDO-Me to Keap1 releases Nrf2 impeding its ubiquitination, thus leading to the stabilization and nuclear import of this potent transcription factor [24]. Activated Nrf2 reduces intracellular reactive oxygen species (ROS) levels via the transcriptional induction of numerous antioxidant proteins, e.g., superoxide dismutase (SOD) and glutathione peroxidase (GPX) leading to a synchronized antioxidant and anti-inflammatory response [33]. Similarly, when CDDO-Me binds to IKKβ, it prevents NF-kB dissociation from its bound complex with IkB in the cytosol, thus resulting in the suppression of NF-kB activation and the downstream cascade of pro-inflammatory signaling pathways [35]. Various other mechanisms responsible for the anticancer action of CDDO-Me involve inhibition of proliferation of cancer cells, induction of apoptosis, and arrest of cancer cells in the G_2/M phase [30,37–39]. In vivo studies have also reported potent inhibitory effects of CDDO-Me on tumor growth, metastasis and angiogenesis [40,41]. CDDO-Me has demonstrated promising anticancer effects in Phase I clinical trials against multiple solid tumors [42]. Although multiple studies with CDDO-Me have been conducted in PC [26,37,41,43,44], its efficacy to suppress the expression of both AR-FL and AR-V7 in PC cells has not been investigated before.

In the current study, we have shown that at physiologically achievable plasma concentrations (i.e., nanomolar doses) [24], CDDO-Me suppresses gene expression and protein levels of both AR-FL and AR-V7 in the LNCaP, C4-2B, and CWR22Rv1 (22Rv1) cells. Pre-exposure to the antioxidant N-acetyl cysteine (NAC) was able to abrogate the AR suppressive effect of CDDO-Me. Most importantly, co-treatment with physiologically achievable doses of CDDO-Me could sensitize PC cells to the cytotoxic effects of a clinically approved anti-androgen drug ENZ. Our findings thus implicate the potential of CDDO-Me as an adjunct therapy in patients with CRPC tumors; especially those overexpressing AR-FL and AR-V7.

2. Materials and Methods

2.1. Cell Culture

LNCaP (an androgen-dependent PC cell line expressing only AR-FL) and 22Rv1 (an androgen-independent PC cell line expressing both AR-FL and AR-V7) were purchased from American Type Culture Collection (ATCC; Rockville, MD, USA). The C4-2B (an androgen-independent PC cell line expressing only AR-FL) cell line was obtained from Dr. Leland Chung's lab in Cedar Sinai Medical Center (Los Angeles, CA, USA) [45]. All the three cell lines were cultured in Rosewell Park Memorial Institute (RPMI)—1640 media supplemented with 10% fetal bovine serum (FBS) (Atlanta Biologicals; Lawrenceville, GA, USA) and 1% antibiotic–antimycotic (Thermo Scientific; Waltham, MA, USA) in a humidified incubator containing 5% CO_2 at 37 °C. The experiments were performed in a phenol-red free RPMI media supplemented with 10% charcoal-stripped FBS (CS-FBS) from Atlanta Biologicals to simulate androgen depleted conditions.

2.2. Reagents

MTT [3-(4,5-dimethylthiazol-2-yl)-2,5-diphenyltetrazoliumbromide] was obtained from Sigma-Aldrich (St. Louis, MO, USA). Enzalutamide (ENZ) was purchased from ApexBio (Houston, TX, USA). Cycloheximide (CHX) was bought from Cayman chemicals (Ann Arbor, MI, USA). CDDO-Me was purchased from Selleckchem (Houston, TX, USA). The drugs were dissolved in 100% DMSO and the final DMSO concentration which was used in the experiments was less than 0.1%. N-acetyl cysteine (NAC) was obtained from Santa Cruz Biotechnology (Santa Cruz, CA, USA), dissolved in water and diluted in media immediately before use. The primary antibodies including rabbit polyclonal anti-AR (N-20) (sc-816), mouse monoclonal anti-Nrf2 (437C2a) (sc-81342), and anti-GAPDH (sc-47724) were obtained from Santa Cruz Biotechnology. The horseradish peroxidase (HRP)-conjugated goat anti-rabbit (A0545) and goat anti-mouse (A9044) secondary antibodies were bought from Sigma-Aldrich (St. Louis, MO, USA). The goat antirabbit secondary antibody tagged with Texas red (T-2767) was bought from Thermo Scientific.

2.3. MTT Assay

MTT assays were carried out to determine the cell viability post treatment with the drug(s). Briefly, ~5000 cells were cultured in 96-well plates followed by synchronization in a serum free medium overnight. The viability of the cells was determined at 72 h post exposure to drug(s) with the MTT solution (5 mg/mL for 3–4 h at 37 °C). DMSO was used to solubilize the formazan crystals and the optical density (O.D.) was measured at 540 nm with μQuant spectrophotometric plate reader (Bio-Tek; Seattle, WA, USA).

2.4. Western Blot Analysis

The radioimmunoprecipitation assay (RIPA) lysis buffer (Santa Cruz Biotechnology) was used to harvest the whole cell lysates post exposure to drug(s). Quantification of the total protein was done using the bicinchoninic acid (BCA) protein assay reagent (Thermo Scientific). In brief, 10 μg of protein was electrophoresed in SDS-PAGE gels (10%) and transferred onto nitrocellulose membranes using semi-dry electro-transfer. The membranes were incubated overnight with the primary antibodies against AR (1:500 dilution), Nrf2 (1:500 dilution), and GAPDH (1:3000 dilution) at 4 °C after blocking with 5% casein in TBS-T buffer (tris buffer saline with 0.1% tween-20). The membranes were then incubated with the corresponding HRP-conjugated secondary antibodies (1:2000 dilution) for 1 h and developed using the Supersignal west femto substrate (Thermo Scientific). The scanning of the immunoblots was done using the ImageQuant LAS 500 scanner (GE Healthcare; Princeton, NJ, USA). Image J software (NIH; Bethesda, MD, USA) was used to quantify the band intensities. The densitometric values for AR proteins (AR-FL and AR-V7) were normalized to the GAPDH values for calculating the fold change.

2.5. ROS Assay

DCFDA/H2DCFDA—a cellular ROS assay kit (Abcam; Cambridge, MA, USA; Cat # ab113851) was used to measure reactive oxygen species (ROS). Cells were harvested and seeded in a dark, clear bottom 96-well microplate with 25,000 cells per well. The cells were stained with 2′,7′-dichlorofluorescin diacetate (DCFDA) and treated with different agents for the specified period of time. The DCFDA fluorescence (Ex/Em = 485/535 nm) was measured immediately using a microplate reader (Bio-Tek).

2.6. Wound-Heal Assay

Wound-heal assay was performed to monitor the migratory phenotype of PC cells post exposure to drug(s) [46]. In brief, cells were cultured in 6-well plates (1×10^6 cells per well) until a confluent monolayer was formed. A 200 µl pipette tip was used to scratch the monolayer. The wells were then washed with PBS and images (10× magnification) were captured of the wound at 0 time point with a Leica Microsystems microscope (Buffalo Grove, IL, USA). Images of the wound were then captured at 72 h post exposure to the drug(s). The cell migration (wound closure) was measured by calculating the distance between 4–5 random points within the wound edges.

2.7. Colony Forming Units Assay

Cells (500/dish) were cultured in 60 mm petri dishes in three replicates in 2% FBS containing media and exposed to the drug(s) after 48 h. The drug(s) were replenished in the second week. After two weeks, the cell colonies were stained with 0.2% crystal violet in 20% methanol post fixation with 100% ethanol. The colony forming units (CFU) were counted with the Image J software. The total number of CFUs were then compared in untreated (control) and drug-treated cultures.

2.8. Immunofluorescence Microscopy

Immunofluorescence microscopy (IFM) was used to visualize subcellular localization of AR post exposure to CDDO-Me. In brief, cells (3×10^4) were cultured in chamber slides (EMD Millipore; Billerica, MA, USA) and then fixed in ice cold methanol after treatment. After permeabilization of the cells with 0.1% Triton-X 100, blocking was done in 10% goat serum. The cells were then incubated with the primary antibody (1:300 dilution) overnight at 4 °C. This was followed by incubation with the corresponding secondary antibody tagged with texas red (1:1000 dilution) for 1 h. The cover slips were mounted after nuclear stain diamino-2-phenylindole (DAPI) containing vectashield mounting media (Burlingame, CA, USA) was added to the slides. The images (60× magnification) were captured with the fluorescent microscope (Leica Microsystems; Buffalo Grove, IL, USA).

2.9. Quantitative RT-PCR

The mRNA levels for both AR-FL and AR-V7 were measured using the quantitative reverse transcriptase polymerase chain reaction (qRT-PCR). In brief, after treatment, total mRNA was extracted using the RNeasy mini plus kit (Qiagen; Valencia, CA, USA) in accordance with the manufacturer's instructions. The iScript cDNA synthesis kit (Bio-Rad, Hercules, CA, USA) was used to prepare complementary DNA (cDNA) according to the manufacturer's instructions. The following primer sequences were used: (1) AR-FL:—forward: 5′-CAGCCTATTGCGAGAGAGCTG-3′ and reverse: 5′-GAAAGGATCTTGGGCACTTGC-3′; (2) AR-V7:—forward: 5′-CCATCTTGTCGTCTTCGGAAAT GTTATGAAGC-3′ and reverse: 5′-TTTGAATGAGGCAAGTCAGCCTTTCT-3′; and (3) β-actin:—forward: 5′TGAGACCTTCAACACCCCAGCCATG-3′ and reverse: 5′-GTAGATGGGCACAGTGTGGGTG-3′. The iQ™ SYBR green supermix (Bio-Rad) was used to measure the AR transcript levels and C1000™ Thermocycler (CFX96; Bio-Rad) was used to carry out the amplification reactions. The following amplification conditions were used: Priming at 95 °C for 5 min, and then 35 cycles of 95 °C for 30 s, 55 °C for 30 s, and 72 °C for 30 s. The data (ΔCt values) for AR (AR-FL and AR-V7) transcript levels were normalized to the corresponding β-actin values for calculating the fold change.

2.10. Statistical Analysis

The graphPad Prism (version-6) Software (San Diego, CA, USA) was used for the statistical analyses. Results were expressed as the standard error of the mean (± SEM). A two-tailed student's *t*-test was used to determine significant changes from controls and *p*-values of < 0.05 were considered significant. The CompuSyn software (ComboSyn, Inc., Paramus, NJ, USA) was used for synergy determination and combination index (CI) was calculated on the basis of Chou–Talalay method, which determines additive (CI = 1), synergistic (CI < 1), or antagonistic (CI > 1) effects quantitatively [47].

3. Results

3.1. Exposure to Low-Dose CDDO-Me Decreases AR-FL and AR-V7 Protein Levels in PC Cells, in a Time- and Dose-Dependent Manner

The PC cells (LNCaP, C4-2B, and 22Rv1) were exposed to increasing concentrations of CDDO-Me (0–500 nM) and AR protein levels were measured at different time points (Figure 1A–C). Immunoblot analysis depicted that exposure to CDDO-Me causes time- and dose-dependent decreases in AR-FL in both LNCaP and C4-2B cells. Most interestingly, in 22Rv1 cells nanomolar (nM) doses of CDDO-Me were able to decrease both AR-FL and AR-V7 protein levels. In all the three cell lines, the reduction in the AR-FL and AR-V7 protein was apparent within 6 h of exposure to CDDO-Me, and was evident even with the lowest dose of CDDO-Me used (100 nM). At 24 h post exposure to the highest dose of CDDO-Me (500 nM), the AR-FL, and AR-V7 proteins were abrogated in all three cell lines. Indeed, these results were further corroborated by our IFM data, which showed clearly reduced levels of both cytoplasmic and nuclear AR immunofluorescence in the 22Rv1 cells post 24 h exposure to increasing doses of CDDO-Me (100–500 nM) (Figure 1D).

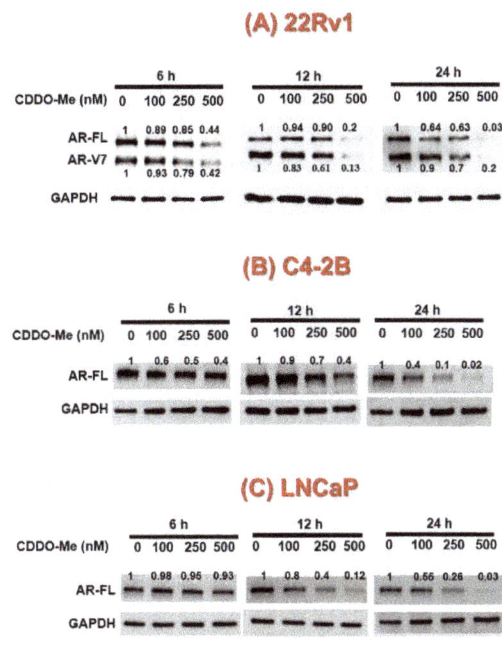

Figure 1. *Cont.*

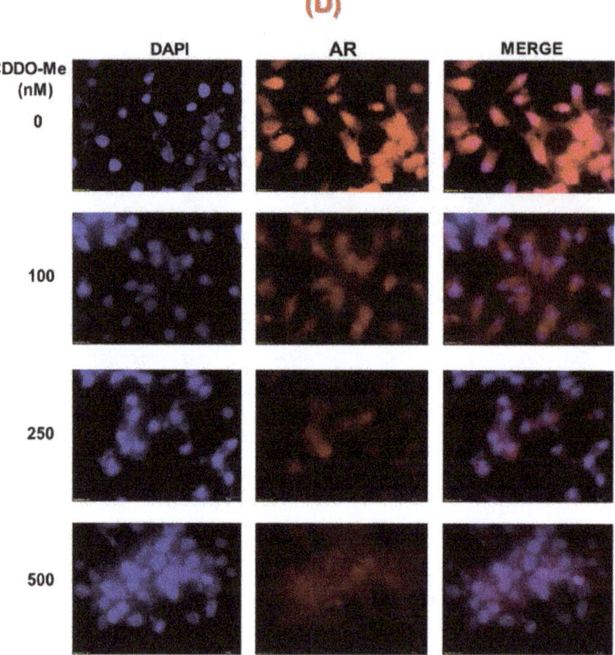

Figure 1. Effect of Bardoxolone-methyl (CDDO-Me) on androgen receptor (AR) levels in prostate cancer (PC) cells. 22Rv1, C4-2B, and LNCaP cells were treated with increasing concentrations of CDDO-Me (0–500 nM) and cell lysates were harvested at 6–24 h post treatment. A representative immunoblot of AR and GAPDH protein levels are shown for (**A**) 22Rv1 (**B**) C4-2B, and (**C**) LNCaP cells. (**D**) Immunofluorescence microscopy (IFM) images (60× magnification) of subcellular AR localization in PC cells. 22Rv1 cells were treated with increasing doses of CDDO-Me (100, 250, and 500 nM) for 24 h prior to fixation and immunolabeling. Left panels show DAPI stained nuclei (blue), middle panel shows AR immunoreactivity (red), and merged images are shown in the third panel.

3.2. CDDO-Me-Mediated Suppression of AR-FL and AR-V7 is Regulated at both mRNA and Protein Levels

To determine whether the CDDO-Me-mediated suppression of AR is regulated at the level of transcription or protein synthesis, we quantified AR-FL and AR-V7 specific mRNA by qRT-PCR and AR protein levels in the presence of the protein synthesis inhibitor, cycloheximide (CHX). Immunoblot analysis revealed that combined exposure to CHX and CDDO-Me reduced the half-life of AR-FL and AR-V7 protein more significantly as compared to the CHX treatment alone(Figure 2A,B). This suggested that CDDO-Me regulates both AR-FL and AR-V7 levels at the translational level, possibly by promoting protein degradation (data not shown). Interestingly, the qRT-PCR data also showed that CDDO-Me treatment can significantly suppress the mRNA levels of both AR-FL and AR-V7 in a time dependent manner (3, 6, and 9 h) in 22Rv1 cells (Figure 2C,D). As much as 8–10-fold decrease in AR specific message was clearly observed in cells that were exposed to CDDO-Me (500 nM) for 9 h. Thus, the potent AR-suppressive effect of low-dose CDDO-Me is regulated at both transcriptional and translational levels.

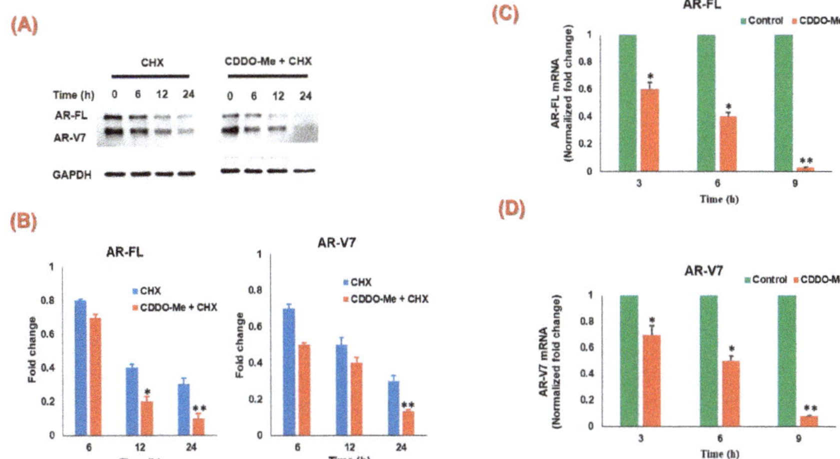

Figure 2. Transcriptional and post translational regulation of AR by CDDO-Me. (**A**) 22Rv1 cells were pretreated (2 h) with 5 µg/mL cycloheximide (CHX) followed by exposure to CDDO-Me (500 nM) for 0–24 h and AR levels were monitored by western immunoblot. A representative immunoblot of AR and GAPDH protein levels in 22Rv1 cells. (**B**) The normalized data are expressed as fold changes (mean ± SEM) in two independent experiments and significant differences between groups are shown as *$p < 0.05$. (**C,D**) 22Rv1 cells were treated with CDDO-Me (500 nM), total RNA extracted after 3, 6, and 9 h and quantitative RT-PCR (qRT-PCR) was performed. The normalized fold change in (**C**) AR-FL and (**D**) AR-V7 gene expression from two independent experiments is expressed as the mean ± SEM. Significant differences between groups are shown as p-values (*$p < 0.05$; **$p < 0.005$).

3.3. The Suppression of AR-FL and AR-V7 by CDDO-Me is Primarily Mediated via Oxidative Stress in both C4-2B and 22Rv1 Cells

Several studies have shown that oxidative stress signaling can regulate AR expression and CRPC progression [48,49]. Antioxidant agents have also been reported to activate the Nrf2 transcription factor by transient induction of ROS [50,51]. Therefore, we wanted to determine if CDDO-Me, which is a potent antioxidant agent and a well-known inducer of Nrf2 [24], can similarly induce oxidative stress and Nrf2 in PC cells. Exposure to CDDO-Me exerted a biphasic effect on ROS levels in the 22Rv1 cells. Acute exposure to CDDO-Me (2 h) was found to increase ROS in a dose-dependent manner, which could be blocked by co-exposure of cells with the antioxidant agent, N-acetyl cysteine (NAC) (Figure 3A). Interestingly, however at 6, 12, and 24 h post exposure to CDDO-Me, even the basal ROS levels were found to decrease considerably (Figure 3B), possibly due to the activation of the Nrf2 pathway. This hypothesis was corroborated by an increase in the total levels of Nrf2 protein in the C4-2B cells, where the dose-dependent increase in Nrf2 was evident post 24 h exposure to CDDO-Me (Figure 3C).

To determine whether transient induction of ROS was important for the AR-suppressive effect of CDDO-Me, the 22Rv1 cells were exposed to NAC both pre and post treatment with CDDO-Me for 24 h (Figure 3D). Pretreatment with NAC (5 mM) for overnight or even 2 h before CDDO-Me addition was able to abrogate the AR-suppressive effects of CDDO-Me (500 nM). Interestingly however, exposure to NAC at 6 h post treatment with CDDO-Me was not able to abolish its AR suppressive effects at 24 h. These findings suggested that the acute induction of ROS, observed within 2 h post exposure to CDDO-Me, was critical in decreasing the levels of AR-FL and AR-V7 in 22Rv1 cells. Similar results could be seen in C4-2B cells as well, where only NAC pretreatment, but not post treatment, was able to nullify the AR-suppression by CDDO-Me (Figure 3E).

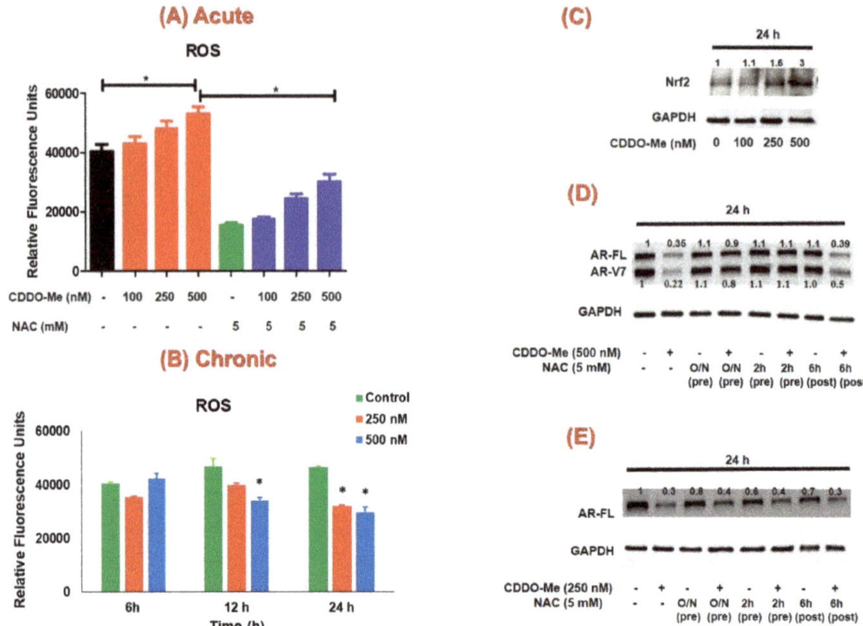

Figure 3. Effect of CDDO-Me mediated reactive oxygen species (ROS) on AR levels in PC cells. (**A**) Acute effect of CDDO-Me on ROS levels in 22Rv1 cells. 22Rv1 cells were exposed to CDDO-Me (100, 250, and 500 nM) for 2 h with and without 5 mM N-acetyl cysteine (NAC) (2 h pretreatment) and ROS levels were measured. (**B**) Chronic effect of CDDO-Me on ROS levels in 22Rv1 cells. 22Rv1 cells were treated with CDDO-Me (250 and 500 nM) and ROS levels were detected at 6, 12, and 24 h. The data (% of control) are expressed as the mean ± SEM of three independent experiments ($n = 3$) and significant differences between groups are shown as p-values (*$p < 0.05$) (**C**) Effect of CDDO-Me on Nrf2 protein levels. C4-2B cells were treated with increasing doses of CDDO-Me (100, 250, and 500 nM) for 24 h and total Nrf2 and GAPDH levels were detected by immunoblot. In (**D**) and (**E**), CDDO-Me exposure was carried out in cells that were either pretreated (2 h or overnight (O/N)) or posttreated (6 h) with NAC. Cell lysates were obtained at 24 h post CDDO-Me treatment of (**D**) 22Rv1 or (**E**) C4-2B cells. A representative immunoblot of AR and GAPDH protein levels is shown.

3.4. Co-Exposure to CDDO-Me Increases the Anticancer Efficacy of ENZ

To determine whether the AR-suppression by CDDO-Me enhances the efficacy of clinically approved anti-androgens, cytotoxicity was measured in LNCaP, C4-2B, and 22Rv1 cells co-exposed to CDDO-Me and ENZ, using the MTT cell viability assay (Figure 4). Cell viability data at 72 h clearly showed that the 22Rv1 cells, which expresses both AR-FL and AR-V7, were more sensitive to CDDO-Me than the C4-2B cells, which expresses AR-FL only (Figure 4A,B). The LNCaP cells, which express AR-FL and are androgen responsive, were least sensitive to CDDO-Me even at the highest dose tested (500 nM) (Figure 4C). Most importantly, we observed that co-exposure to CDDO-Me was able to increase the therapeutic efficacy of ENZ in the 22Rv1 cells. These CRPC cells are resistant to anti-androgens, and ENZ treatment alone showed negligible decreases in cell viability. However, CDDO-Me alone was shown to decrease cell growth and co-exposure to CDDO-Me was able to further augment the anticancer efficacy of ENZ in 22Rv1 cells. Similar enhancements in cytotoxicity following CDDO-Me and ENZ co-exposure were also evident in the LNCaP and C4-2B cells.

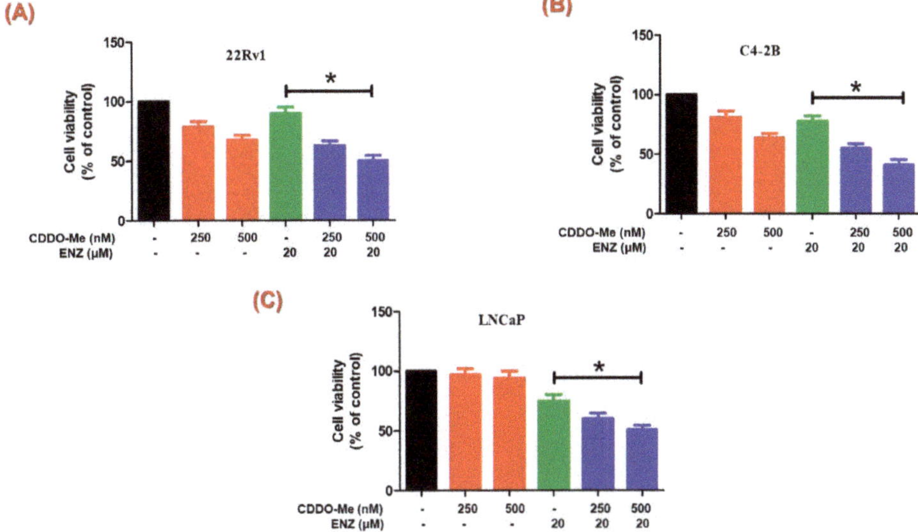

Figure 4. Effect of CDDO-Me and enzalutamide (ENZ) combination on cell viability of PC cells. Cytotoxic effect of CDDO-Me (250 and 500 nM) alone and in combination with ENZ (20 µM) at 72 h post treatment in (**A**) 22Rv1, (**B**) C4-2B, and (**C**) LNCaP cells is shown. PC cells were treated with CDDO-Me alone and in combination with ENZ and cell viability was measured at 72 h using MTT. The data (% of control) are expressed as the mean ± SEM of three independent experiments ($n = 3$) and significant differences between groups are shown as *p*-values (*$p < 0.05$).

3.5. Co-Exposure to CDDO-Me and ENZ Abrogates the Migratory Potential of CRPC Cells

Increased metastatic ability of CRPC cells has been linked to increased AR signaling [52] and can be tested in vitro by using different migration and invasion assays [46]. We carried out wound-heal assays to determine the effect of CDDO-Me, alone, or in combination with ENZ, on the migratory behavior of 22Rv1 cells (Figure 5A,B). The vigorous migratory ability of 22Rv1 cells was clearly evident by as much as a 50% reduction in wound-width within 72 h post wounding. Exposure to ENZ alone failed to show significant inhibition in migration of these aggressive CRPC cells. However, exposure to CDDO-Me alone significantly reduced cell migration and the combined treatment with CDDO-Me and ENZ almost totally abrogated the migratory potential of 22Rv1 cells.

3.6. CDDO-Me Increases ENZ Efficacy by Inhibiting the Clonogenic Ability of PC Cells

The clonogenic ability of CRPC cells is a major determinant of metastatic growth [53]. We carried out the colony forming unit (CFU) assay to investigate the chronic effects (14 days) of low-dose CDDO-Me, alone and in combination with ENZ, on the clonogenic ability of 22Rv1 cells (Figure 6A,B). Interestingly, treatment with ENZ alone (0.2 µM) did not cause any significant inhibition in the total number of CFUs. However, chronic exposure to even low-dose CDDO-Me (50 nM) caused approximately 50% reduction in the number of CFUs. Most importantly, co-exposure to CDDO-Me and ENZ enabled almost a total abrogation (~80%–90%) in the number of CFUs. The CFU suppressive effect of CDDO-Me was synergistically increased (CI < 1) when combined with low-dose ENZ, which alone did not significantly alter the CFUs. A remarkable suppression in CFUs generated by the 22Rv1 cells was evident following chronic exposure to very low-doses CDDO-Me and ENZ thus underscoring the therapeutic potential of this novel combination.

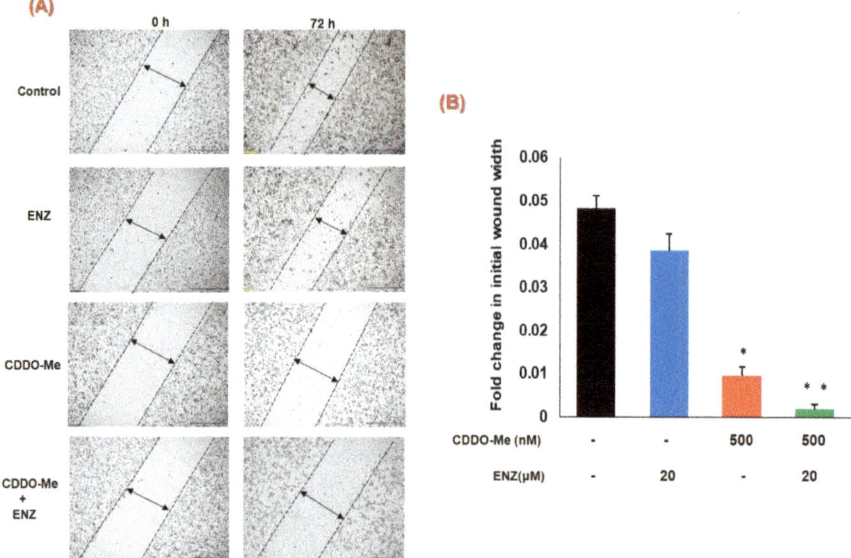

Figure 5. Effect of combination of CDDO-Me and ENZ on cell migration in 22Rv1 cells. Quantification of cell migration was examined by wound-healing assay. (**A**) A representative light microscope image of the wound at the 0 and 72 h time-points is shown in 22Rv1 cells. Images show wound closure in untreated (control) cells, and in cultures that were exposed to either CDDO-Me (500 nM), ENZ (20 μM), or CDDO-Me and ENZ combination. (**B**) Fold change in wound width is expressed as the mean ± SEM of two independent experiments, and significant differences between groups are shown as p-values (* $p < 0.05$; ** $p < 0.005$).

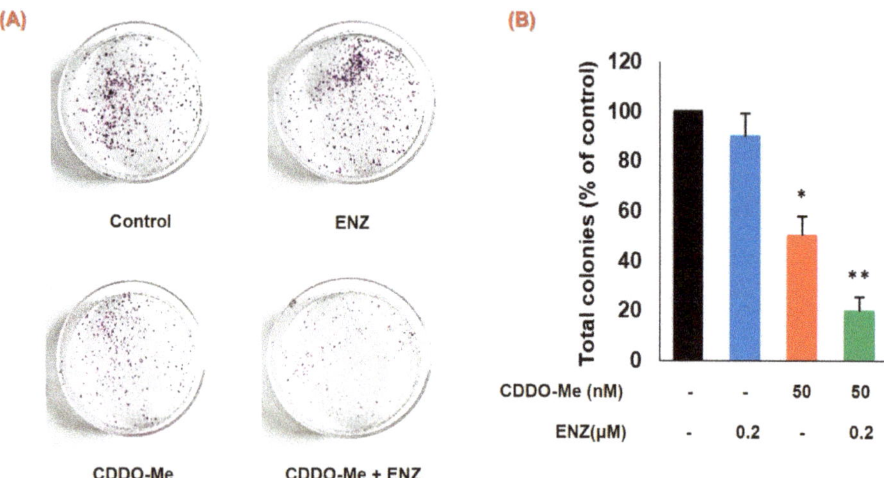

Figure 6. Long-term effects of CDDO-Me and ENZ on the clonogenic ability of PC cells. The 22Rv1 cells (500 cells/plate) were exposed to CDDO-Me (50 nM) alone or in combination with ENZ (0.2 μM) for two weeks. (**A**) A representative image of colony forming units (CFU) is shown. (**B**) Effect of drug exposure on percent change in total number of CFUs in 22Rv1 cells as compared to the control is shown. Data are expressed as the mean ± SEM of two independent experiments and significant differences between groups are shown as p-values (* $p < 0.05$; ** $p < 0.005$).

4. Discussion

Second messenger signaling via the androgen receptor is indispensable for the prostate epithelial cells, not only for the normal functioning and homeostasis of the prostate gland but also for the development of prostatic neoplasms and the progression of PC to CRPC [2]. Therefore, therapeutic strategies to suppress AR signaling have remained the mainstay of PC treatment [5]. Despite the initial efficacy of ADT, PC patients eventually become resistant to even high doses of antiandrogens, leading to the development of CRPC phenotype [2]. One of the most crucial factors implicated in the emergence of CRPC cells is the expression of constitutively active AR splice variants (AR-Vs); AR-V7 being predominantly documented in clinical samples [12]. Interestingly, AR-FL is still persistently expressed in the CRPC cells, and is known to dimerize with truncated AR splice variants [22,23]. Furthermore, the crosstalk of AR with other signaling pathways involved in tumorigenesis results in the synergistic aberrant expression of the target genes associated with mitogenesis, increased cell viability, clonogenicity, and migratory behavior, leading to aggressive and invasive CRPC recurrence [54,55]. Therefore, therapeutic agents which can suppress both AR-FL and AR-V7 at physiologically achievable concentrations are urgently needed. Recently, we had documented that sulforaphane (SFN), a phytochemical derived from broccoli sprouts, and other cruciferous vegetables, can suppress the expression of both AR-FL and AR-V7 [56,57]. This AR-suppressive effect of SFN, a known Nrf2 inducer, was also shown to enhance the anticancer efficacy of anti-androgens [56,57]. Since CDDO-Me is also a potent Nrf2 inducer, in this study, we investigated the effect of CDDO-Me on AR expression in CRPC cells. Our findings demonstrated that low-dose CDDO-Me suppresses both AR-FL and AR-V7 expression and augments the efficacy of the second-generation anti-androgen, ENZ. At nanomolar concentrations (100–500 nM), exposure to CDDO-Me suppressed the protein expression of AR-FL in the androgen dependent LNCaP cells and in the androgen independent C4-2B cells [Figure 1]. In addition, it also inhibited the protein expression of both AR-FL and AR-V7 in the highly aggressive CRPC cell line, 22Rv1 [Figure 1]. Since CDDO-Me is in late-stage clinical trials for chronic kidney disease (CKD), we envision that its utility as an adjunct to ADT will be very valuable in treating CRPC and should be clinically tested.

LNCaP is an androgen-dependent cell line that was first isolated from a human metastatic prostate adenocarcinoma in the lymph node and expresses mRNA/protein of both AR and PSA [58,59]. C4-2B is a bone metastatic CRPC subline derived from LNCaP which also expresses both AR and PSA although it also can grow in androgen deprived conditions [45,59]. C4-2B cells have been shown to grow both in intact, as well as castrated mice. 22Rv1 is a CRPC cell line expressing AR-FL and several AR splice variants, out of which AR-V7 is the most prominent [59]. It was isolated from the xenograft CWR22R derived from a patient with bone metastasis. In vitro studies using these three PC lines provide a model for both the early stages of PC and the progression of CRPC cells to AR-variant expressing aggressive and hormone-resistant phenotype. Thus, our observations on the potent suppression of both AR-FL and AR-V7 by the low-dose CDDO-Me implicate its potential to be used in both the early stage PC where the cancer cells are hormone dependent, as well as in the late stage of CRPC tumors where the cancer cells have selected for an androgen independent phenotype, but are still expressing AR-FL and AR-V7.

The CDDO-Me-mediated suppression of AR-FL and AR-V7 was found to be regulated both at the transcriptional, as well as the translational level [Figure 2]. Indeed, suppression of AR-FL and AR-V7 protein levels by combination of CDDO-Me and CHX was more significant as compared to the CHX treatment alone, suggesting that post translational regulation of AR may be a possible mechanism of action of CDDO-Me in these cells. We documented in our previously published study, that SFN can increase proteasomal degradation of AR [57]. Likewise, our unpublished observations suggest increased proteasomal activity in CDDO-Me exposed cells. CDDO-Me on binding to sulfhydryl (SH) groups of cysteine residues [60,61] on DNA binding domain (DBD) of AR-FL and AR-V7, may also increase the accumulation of misfolded AR proteins [62]. Moreover, since CDDO-Me has been shown to directly interact with HSP90 and degrade HSP90 client proteins [63], this may also be one of the mechanisms responsible for the AR suppressive effect of CDDO-Me in PC cells since AR is one of the client proteins of HSP90 [64]. Most interestingly, a significant decrease in mRNA levels of both AR-FL and AR-V7 was evident even at 3 h post exposure to CDDO-Me, suggesting that second messenger signaling by CDDO-Me may alter the transcriptional machinery at the AR promoter/enhancer regions, and thus rapidly downregulate AR gene expression. We had previously documented that overexpression of Nrf2 can suppress AR expression and function in LNCaP and C4-2B cells [65]. Exposure to the Nrf2 inducer, SFN, can also decrease AR gene expression [56]. Thus, transcriptional suppression of AR gene expression via Nrf2 inducers may be a novel direction in PC therapy. Ultimately, our observations on the dual effects of CDDO-Me on AR suppression may be highly beneficial in the long-term ablation of AR mediated protumorigenic effects in aggressive PC cells.

Several other Nrf2 inducing antioxidant phytochemicals such as SFN and curcumin have promising anticancer effects and are implicated to function by first inducing oxidative stress followed by increasing nuclear Nrf2 levels [66–69]. Similarly, CDDO-Me, being a much more potent Nrf2 inducing agent, has also been shown to rapidly induce ROS to cause cytotoxicity in various cancer cell lines in vitro [51]. In our studies, we similarly observed a biphasic effect of CDDO-Me on ROS production in PC cells [Figure 3A,B]. Exposure to CDDO-Me caused a transient induction of ROS within 2 h, whereas it suppressed ROS levels in the long-term (24 h), most likely by increasing Nrf2 levels [Figure 3C]. Nrf2 protein levels were found to be significantly increased at 24 h post exposure to CDDO-Me. Most importantly, the transient induction of ROS was found to be crucial for the potent AR suppressive effect of CDDO-Me on both AR-FL and AR-V7 expression. This was clearly evident from the rescue experiments, where pretreatment with NAC was able to nullify the AR suppressive effects of CDDO-Me [Figure 3D,E]. Post treatment with NAC, even within 6 h following CDDO-Me exposure, was not able to salvage the suppression of AR. Therefore, in addition to the role of Nrf2, a direct effect of oxidative stress in regulating both AR-FL and AR-V7 mRNA and protein in CRPC cells is possible. Oxidative stress has been associated with changes in the expression of several splicing factors, e.g., the hetero-nuclear ribonucleoproteins (hnRNPs) [70]. HnRNPs are important for promoting AR expression and production of variants in PC [71]. A putative molecular mechanism/s linked to the therapeutic effects of CDDO-Me is presented in Figure 7.

Importantly, functional assays addressing the tumorigenic ability of PC cells, i.e., viability, migration and clonogenicity, clearly demonstrated that the AR-suppressive effects of CDDO-Me can enhance the efficacy of the clinically approved anti-androgen ENZ. Exposure to nanomolar doses of CDDO-Me was able to potentiate the efficacy of ENZ in LNCaP, C4-2B, and the ENZ resistant 22Rv1 cells by significantly decreasing the cell viability of PC cells [Figure 4]. Androgen signaling enhances the migratory behavior of PC cells [2]. Higher invasive ability of cells, augmented apoptotic resistance, and epithelial mesenchymal transition (EMT) are the chief characteristics of migration, which are also critical elements of tumor metastasis [46,72]. In this respect, although antiandrogens have potent antiproliferative effects, they do not suppress PC cell migration significantly [73]. Our study shows that CDDO-Me treatment can decrease the migratory ability of PC cells and this effect is further augmented in the presence of ENZ [Figure 5]. Our in vitro finding thus advocates the benefit of using CDDO-Me in combination with ENZ to inhibit the metastatic behavior of PC cells. These findings may be of significant therapeutic value, especially in CRPC patients that overexpress AR-FL and/or AR-V7 and continue to utilize the AR signaling pathway.

Although we have not carried out in vivo studies in tumor-bearing animals, the potent antitumor effects seen with our drug combination were validated by measuring the clonogenic ability of PC cells in vitro [Figure 6]. Exposure to even a very low-dose of CDDO-Me (50 nM) showed significant suppression of CFUs, and co-exposure to CDDO-Me significantly enhanced the ability of ENZ to suppress the number of colonies. In vitro clonogenic assays simulate the seeding and proliferation of tumor initiating cells, i.e., cancer stem cells [74]. A number of previous studies have already been carried out in tumor xenografts in vivo using either clinically approved ENZ [75–79] or

CDDO-Me [26,37,40,41,44] alone; and therefore, studies using the combination of CDDO-Me and ENZ in tumor-bearing mice, would be the next feasible direction of our current findings. Taken together, our in vitro finding suggests a possible benefit of using safe concentrations of CDDO-Me alone to suppress the growth of micro-metastatic foci, and as an adjunct therapy to enhance the efficacy of antiandrogens. The long-term synergistic effects observed in our CFU assays, especially with very low doses of each of the agents, suggest the potential of this combination therapy.

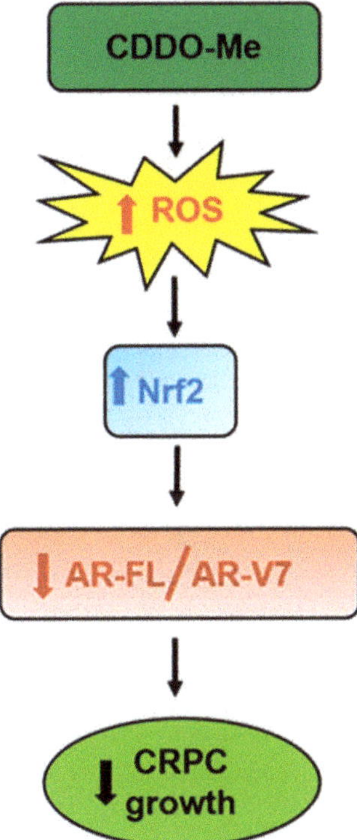

Figure 7. Mechanism of AR suppressive actions of CDDO-Me in PC cells. Exposure to Bardoxolone-methyl (CDDO-Me) causes a transient induction of reactive oxygen species (ROS) which in turn activates Nrf2 protein levels. Ultimately, this antioxidant transcription factor decreases oxidative stress and the aggressive properties of CRPC cells, e.g., growth, migration, clonogenic ability, etc. Both Nrf2 and ROS may be directly involved in suppressing AR expression (both at the gene and protein levels). Decreased AR signaling (via AR-FL and AR-V7) by CDDO-Me inhibits the growth of CRPC cells and can augment the anticancer efficacy of approved drugs like enzalutamide (ENZ).

5. Conlusions

Nanomolar (nM) concentrations of Bardoxolone-methyl (CDDO-Me) decreased both AR-FL and AR-V7 expression in PC cells. Therefore, CDDO-Me, which is in several late-stage clinical trials, may be used in combination with clinically approved anti-androgens such as ENZ in PC patients; both at the initial stage where tumors are overexpressing AR-FL only and at the later stage of CRPC where AR-V7 overexpressing tumors are frequently observed.

Author Contributions: Conceptualization, N.K. and D.M.; Methodology, N.K., P.K.C., and H.K.; Software, N.K.; Validation, N.K. and D.M.; Formal analysis, N.K., H.K. and D.M.; Investigation, N.K. and D.M.; Resources, D.M and S.C.S.; Data curation, N.K. and D.M.; Writing—N.K. and D.M.; Writing—review and editing, N.K., D.M. A.B.A.-M. and S.C.S.; Visualization, N.K., P.K.C., H.K., D.M., and S.C.S.; Supervision, D.M., A.B.A.-M. and S.C.S.;

Project administration, D.M. and S.C.S.; Funding acquisition, D.M. and S.C.S. All authors have read and agreed to the published version of the manuscript.

Funding: This study was supported by research funds of the Tulane Urology and Pharmacology departments to S.C.S. and D.M.

Conflicts of Interest: The authors declare no conflict of interest.

References

1. Yap, T.A.; Zivi, A.; Omlin, A.; de Bono, J.S. The changing therapeutic landscape of castration-resistant prostate cancer. *Nat. Rev. Clin. Oncol.* **2011**, *8*, 597–610. [CrossRef] [PubMed]

2. Hodgson, M.C.; Bowden, W.A.; Agoulnik, I.U. Androgen receptor footprint on the way to prostate cancer progression. *World, J. Urol.* **2012**, *30*, 279–285. [CrossRef] [PubMed]

3. Harris, W.P.; Mostaghel, E.A.; Nelson, P.S.; Montgomery, B. Androgen deprivation therapy: Progress in understanding mechanisms of resistance and optimizing androgen depletion. *Nat. Clin. Pract. Urol.* **2009**, *6*, 76–85. [CrossRef] [PubMed]

4. Kim, W.; Ryan, C.J. Androgen Receptor Directed Therapies in Castration-Resistant Metastatic Prostate Cancer. *Curr. Treat. Options Oncol.* **2012**, *13*, 189–200. [CrossRef]

5. Godbole, A.M.; Njar, V.C.O. New insights into the androgen-targeted therapies and epigenetic therapies in prostate cancer. *Prostate Cancer* **2011**, *2011*, 918707. [CrossRef]

6. Scher, H.I.; Buchanan, G.; Gerald, W.; Butler, L.M.; Tilley, W.D. Targeting the androgen receptor: Improving outcomes for castration-resistant prostate cancer. *Endocr. Relat. Cancer* **2004**, *11*, 459–476. [CrossRef]

7. Chang, K.H.; Ercole, C.E.; Sharifi, N. Androgen metabolism in prostate cancer: From molecular mechanisms to clinical consequences. *Br. J. Cancer* **2014**, *111*, 1249–1254. [CrossRef]

8. Lamont, K.R.; Tindall, D.J. Minireview: Alternative activation pathways for the androgen receptor in prostate cancer. *Mol. Endocrinol.* **2011**, *25*, 897–907. [CrossRef]

9. Brooke, G.; Bevan, C. The Role of Androgen Receptor Mutations in Prostate Cancer Progression. *Curr. Genom.* **2009**, *10*, 18–25. [CrossRef]

10. Armstrong, C.M.; Gao, A.C. Drug resistance in castration resistant prostate cancer: Resistance mechanisms and emerging treatment strategies. *Am. J. Clin. Exp. Urol.* **2015**, *3*, 64–76.

11. Guo, Z.; Yang, X.; Sun, F.; Jiang, R.; Linn, D.E.; Chen, H.; Chen, H.; Kong, X.; Melamed, J.; Tepper, C.G.; et al. A novel androgen receptor splice variant is up-regulated during prostate cancer progression and promotes androgen depletion-resistant growth. *Cancer Res.* **2009**, *69*, 2305–2313. [CrossRef] [PubMed]

12. Dehm, S.M.; Tindall, D.J. Alternatively spliced androgen receptor variants. *Endocr. Relat. Cancer* **2011**, *18*, R183–R196. [CrossRef] [PubMed]

13. Zhang, X.; Morrissey, C.; Sun, S.; Ketchandji, M.; Nelson, P.S.; True, L.D.; Vakar-Lopez, F.; Vessella, R.L.; Plymate, S.R. Androgen receptor variants occur frequently in castration resistant prostate cancer metastases. *PLoS ONE* **2011**, *6*. [CrossRef]

14. Sun, S.; Sprenger, C.C.T.; Vessella, R.L.; Haugk, K.; Soriano, K.; Mostaghel, E.A.; Page, S.T.; Coleman, I.M.; Nguyen, H.M.; Sun, H.; et al. Castration resistance in human prostate cancer is conferred by a frequently occurring androgen receptor splice variant. *J. Clin. Invest.* **2010**, *120*, 2715–2730. [CrossRef] [PubMed]

15. Hu, R.; Dunn, T.A.; Wei, S.; Isharwal, S.; Veltri, R.W.; Humphreys, E.; Han, M.; Partin, A.W.; Vessella, R.L.; Isaacs, W.B.; et al. Ligand-independent androgen receptor variants derived from splicing of cryptic exons signify hormone-refractory prostate cancer. *Cancer Res.* **2009**, *69*, 16–22. [CrossRef] [PubMed]

16. Dehm, S.M.; Schmidt, L.J.; Heemers, H.V.; Vessella, R.L.; Tindall, D.J. Splicing of a novel androgen receptor exon generates a constitutively active androgen receptor that mediates prostate cancer therapy resistance. *Cancer Res.* **2008**, *68*, 5469–5477. [CrossRef] [PubMed]

17. Del Re, M.; Biasco, E.; Crucitta, S.; Derosa, L.; Rofi, E.; Orlandini, C.; Miccoli, M.; Galli, L.; Falcone, A.; Jenster, G.W.; et al. The Detection of Androgen Receptor Splice Variant 7 in Plasma-derived Exosomal RNA Strongly Predicts Resistance to Hormonal Therapy in Metastatic Prostate Cancer Patients. *Eur. Urol.* **2017**, *71*, 680–687. [CrossRef]

18. Antonarakis, E.S.; Armstrong, A.J.; Dehm, S.M.; Luo, J. Androgen receptor variant-driven prostate cancer: Clinical implications and therapeutic targeting. *Prostate Cancer Prostatic Dis.* **2016**, *19*, 231–241. [CrossRef]

19. Lokhandwala, P.M.; Riel, S.L.; Haley, L.; Lu, C.; Chen, Y.; Silberstein, J.; Zhu, Y.; Zheng, G.; Lin, M.T.; Gocke, C.D.; et al. Analytical Validation of Androgen Receptor Splice Variant 7 Detection in a Clinical Laboratory Improvement Amendments (CLIA) Laboratory Setting. *J. Mol. Diagnostics* **2017**, *19*, 115–125. [CrossRef]

20. Antonarakis, E.S.; Lu, C.; Wang, H.; Luber, B.; Nakazawa, M.; Roeser, J.C.; Chen, Y.; Mohammad, T.A.; Chen, Y.; Fedor, H.L.; et al. AR-V7 and Resistance to Enzalutamide and Abiraterone in Prostate Cancer. *N. Engl. J. Med.* **2014**, *371*, 1028–1038. [CrossRef]

21. Sarwar, M.; Semenas, J.; Miftakhova, R.; Simoulis, A.; Robinson, B.; Görloff Wingren, A.; Mongan, N.P.; Heery, D.M.; Johnsson, H.; Abrahamsson, P.-A.; et al. Targeted suppression of AR-V7 using PIP5K1α inhibitor overcomes enzalutamide resistance in prostate cancer cells. *Oncotarget* **2016**, *7*, 63065–63081. [CrossRef] [PubMed]

22. Watson, P.A.; Chen, Y.F.; Balbas, M.D.; Wongvipat, J.; Socci, N.D.; Viale, A.; Kim, K.; Sawyers, C.L. Constitutively active androgen receptor splice variants expressed in castration-resistant prostate cancer require full-length androgen receptor. *Proc. Natl. Acad. Sci. USA* **2010**, *107*, 16759–16765. [CrossRef] [PubMed]

23. Xu, D.; Zhan, Y.; Qi, Y.; Cao, B.; Bai, S.; Xu, W.; Gambhir, S.S.; Lee, P.; Sartor, O.; Flemington, E.K.; et al. Androgen receptor splice variants dimerize to transactivate target genes. *Cancer Res.* **2015**, *75*, 3663–3671. [CrossRef] [PubMed]

24. Wang, Y.Y.; Yang, Y.X.; Zhe, H.; He, Z.X.; Zhou, S.F. Bardoxolone methyl (CDDO-Me) as a therapeutic agent: An update on its pharmacokinetic and pharmacodynamic properties. *Drug Des. Devel. Ther.* **2014**, *8*, 2075–2088. [CrossRef]

25. Wang, Y.Y.; Zhe, H.; Zhao, R. Preclinical evidences toward the use of triterpenoid CDDO-Me for solid cancer prevention and treatment. *Mol. Cancer* **2014**, *13*, 30. [CrossRef]

26. Deeb, D.; Gao, X.; Jiang, H.; Dulchavsky, S.A.; Gautam, S.C. Oleanane triterpenoid CDDO-Me inhibits growth and induces apoptosis in prostate cancer cells by independently targeting pro-survival Akt and mTOR. *Prostate* **2009**, *69*, 851–860. [CrossRef]

27. Kim, E.H.; Deng, C.; Sporn, M.B.; Royce, D.B.; Risingsong, R.; Williams, C.R.; Liby, K.T. CDDO-methyl ester delays breast cancer development in Brca1-mutated mice. *Cancer Prev. Res.* **2012**, *5*, 89–97. [CrossRef]

28. Gao, X.; Liu, Y.; Deeb, D.; Arbab, A.S.; Guo, A.M.; Dulchavsky, S.A.; Gautam, S.C. Synthetic oleanane triterpenoid, CDDO-Me, induces apoptosis in ovarian cancer cells by inhibiting prosurvival AKT/NF-κB/mTOR signaling. *Anticancer Res.* **2011**, *31*, 3673–3681.

29. Liby, K.; Royce, D.B.; Williams, C.R.; Risingsong, R.; Yore, M.M.; Honda, T.; Gribble, G.W.; Dmitrovsky, E.; Sporn, T.A.; Sporn, M.B. The synthetic triterpenoids CDDO-methyl ester and CDDO-ethyl amide prevent lung cancer induced by vinyl carbamate in A/J mice. *Cancer Res.* **2007**, *67*, 2414–2419. [CrossRef]

30. Konopleva, M.; Tsao, T.; Ruvolo, P.; Stiouf, I.; Estrov, Z.; Leysath, C.E.; Zhao, S.; Harris, D.; Chang, S.; Jackson, C.E.; et al. Novel triterpenoid CDDO-Me is a potent inducer of apoptosis and differentiation in acute myelogenous leukemia. *Blood* **2002**, *99*, 326–335. [CrossRef]

31. Liby, K.T.; Royce, D.B.; Risingsong, R.; Williams, C.R.; Maitra, A.; Hruban, R.H.; Sporn, M.B. Synthetic triterpenoids prolong survival in a transgenic mouse model of pancreatic cancer. *Cancer Prev. Res.* **2010**, *3*, 1427–1434. [CrossRef] [PubMed]

32. Ryu, K.; Susa, M.; Choy, E.; Yang, C.; Hornicek, F.J.; Mankin, H.J.; Duan, Z. Oleanane triterpenoid CDDO-Me induces apoptosis in multidrug resistant osteosarcoma cells through inhibition of Stat3 pathway. *BMC Cancer* **2010**, *10*. [CrossRef] [PubMed]

33. Dinkova-Kostova, A.T.; Liby, K.T.; Stephenson, K.K.; Holtzclaw, W.D.; Gao, X.; Suh, N.; Williams, C.; Risingsong, R.; Honda, T.; Gribble, G.W.; et al. Extremely potent triterpenoid inducers of the phase 2 response: Correlations of protection against oxidant and inflammatory stress. *Proc. Natl. Acad. Sci. USA* **2005**, *102*, 4584–4589. [CrossRef] [PubMed]

34. Yates, M.S.; Tauchi, M.; Katsuoka, F.; Flanders, K.C.; Liby, K.T.; Honda, T.; Gribble, G.W.; Johnson, D.A.; Johnson, J.A.; Burton, N.C.; et al. Pharmacodynamic characterization of chemopreventive triterpenoids as exceptionally potent inducers of Nrf2-regulated genes. *Mol. Cancer Ther.* **2007**, *6*, 154–162. [CrossRef] [PubMed]

35. Ahmad, R.; Raina, D.; Meyer, C.; Kharbanda, S.; Kufe, D. Triterpenoid CDDO-Me blocks the NF-κB pathway by direct inhibition of IKKβ on Cys-179. *J. Biol. Chem.* **2006**, *281*, 35764–35769. [CrossRef] [PubMed]

36. Ahmad, R.; Raina, D.; Meyer, C.; Kufe, D. Triterpenoid CDDO-methyl ester inhibits the Janus-activated kinase-1 (JAK1)→signal transducer and activator of transcription-3 (STAT3) pathway by direct inhibition of JAK1 and STAT3. *Cancer Res.* **2008**, *68*, 2920–2926. [CrossRef]

37. Hyer, M.L.; Shi, R.; Krajewska, M.; Meyer, C.; Lebedeva, I.V.; Fisher, P.B.; Reed, J.C. Apoptotic activity and mechanism of 2-cyano-3,12-dioxoolean-1,9-dien-28- oic-acid and related synthetic triterpenoids in prostate cancer. *Cancer Res.* **2008**, *68*, 2927–2933. [CrossRef]

38. Ikeda, T.; Sporn, M.; Honda, T.; Gribble, G.W.; Kufe, D. The novel triterpenoid CDDO and its derivatives induce apoptosis by disruption of intracellular redox balance. *Cancer Res.* **2003**, *63*, 5551–5558.

39. Chintharlapalli, S.; Papineni, S.; Konopleva, M.; Andreef, M.; Samudio, I.; Safe, S. 2-Cyano-3,12-dioxoolean-1,9-dien-28-oic acid and related compounds inhibit growth of colon cancer cells through peroxisome proliferator-activated receptor γ-dependent and -independent pathways. *Mol. Pharmacol.* **2005**, *68*, 119–128. [CrossRef]

40. Deeb, D.; Gao, X.; Liu, Y.; Jiang, D.; Divine, G.W.; Arbab, A.S.; Dulchavsky, S.A.; Gautam, S.C. Synthetic triterpenoid CDDO prevents the progression and metastasis of prostate cancer in TRAMP mice by inhibiting survival signaling. *Carcinogenesis* **2011**, *32*, 757–764. [CrossRef]

41. Gao, X.; Deeb, D.; Liu, Y.; Arbab, A.S.; Divine, G.W.; Dulchavsky, S.A.; Gautam, S.C. Prevention of prostate cancer with oleanane synthetic triterpenoid CDDO-Me in the TRAMP mouse model of prostate cancer. *Cancers* **2011**, *3*, 3353–3369. [CrossRef]

42. Hong, D.S.; Kurzrock, R.; Supko, J.G.; He, X.; Naing, A.; Wheler, J.; Lawrence, D.; Eder, J.P.; Meyer, C.J.; Ferguson, D.A.; et al. A phase I first-in-human trial of bardoxolone methyl in patients with advanced solid tumors and lymphomas. *Clin. Cancer Res.* **2012**, *18*, 3396–3406. [CrossRef] [PubMed]

43. Deeb, D.; Gao, X.; Dulchavsky, S.A.; Gautam, S.C. CDDO-me induces apoptosis and inhibits Akt, mTOR and NF-kappaB signaling proteins in prostate cancer cells. *Anticancer Res.* **2007**, *27*, 3035–3044. [PubMed]

44. Liu, Y.; Gao, X.; Deeb, D.; Arbab, A.S.; Gautam, S.C. Telomerase reverse transcriptase (TERT) is a therapeutic target of oleanane triterpenoid cddo-me in prostate cancer. *Molecules* **2012**, *17*, 14795–14809. [CrossRef] [PubMed]

45. Wu, H.-C.; Hsieh, J.-T.; Gleave, M.E.; Brown, N.M.; Pathak, S.; Chung, L.W.K. Derivation of androgen-independent human LNCaP prostatic cancer cell sublines: Role of bone stromal cells. *Int. J. Cancer* **1994**, *57*, 406–412. [CrossRef]

46. Uygur, B.; Wu, W.S. SLUG promotes prostate cancer cell migration and invasion via CXCR4/CXCL12 axis. *Mol. Cancer* **2011**, *10*, 139. [CrossRef]

47. Chou, T.C. Drug combination studies and their synergy quantification using the chou-talalay method. *Cancer Res.* **2010**, *70*, 440–446. [CrossRef]

48. Shiota, M.; Yokomizo, A.; Naito, S. Oxidative stress and androgen receptor signaling in the development and progression of castration-resistant prostate cancer. *Free Radic. Biol. Med.* **2011**, *51*, 1320–1328. [CrossRef]

49. Schultz, M.A.; Abdel-Mageed, A.B.; Mondal, D. The NrF1 and NrF2 balance in oxidative stress regulation and androgen signaling in prostate cancer cells. *Cancers* **2010**, *2*, 1354–1378. [CrossRef]

50. Jutooru, I.; Guthrie, A.S.; Chadalapaka, G.; Pathi, S.; Kim, K.; Burghardt, R.; Jin, U.-H.; Safe, S. Mechanism of Action of Phenethylisothiocyanate and Other Reactive Oxygen Species-Inducing Anticancer Agents. *Mol. Cell. Biol.* **2014**, *34*, 2382–2395. [CrossRef]

51. Jin, U.H.; Cheng, Y.; Zhou, B.; Safe, S. Bardoxolone methyl and a related triterpenoid downregulate cMyc expression in leukemia cells. *Mol. Pharmacol.* **2017**, *91*, 438–450. [CrossRef] [PubMed]

52. Matuszak, E.A.; Kyprianou, N. Androgen regulation of epithelial-mesenchymal transition in prostate tumorigenesis. *Expert Rev. Endocrinol. Metab.* **2011**, *6*, 469–482. [CrossRef] [PubMed]

53. Nomura, Y.; Tashiro, H.; Hisamatsu, K. In Vitro Clonogenic Growth and Metastatic Potential of Human Operable Breast Cancer. *Cancer Res.* **1989**, *49*, 5288–5293. [PubMed]

54. Khurana, N.; Sikka, S. Targeting Crosstalk between Nrf-2, NF-κB and Androgen Receptor Signaling in Prostate Cancer. *Cancers* **2018**, *10*, 352. [CrossRef] [PubMed]

55. Khurana, N.; Sikka, S.C. Interplay Between SOX9, Wnt/β-Catenin and Androgen Receptor Signaling in Castration-Resistant Prostate Cancer. *Int. J. Mol. Sci.* **2019**, *20*, 2066. [CrossRef] [PubMed]

56. Khurana, N.; Talwar, S.; Chandra, P.K.; Sharma, P.; Abdel-Mageed, A.B.; Mondal, D.; Sikka, S.C. Sulforaphane increases the efficacy of anti-androgens by rapidly decreasing androgen receptor levels in prostate cancer cells. *Int. J. Oncol.* **2016**, *49*, 1609–1619. [CrossRef]

57. Khurana, N.; Kim, H.; Chandra, P.K.; Talwar, S.; Sharma, P.; Abdel-Mageed, A.B.; Sikka, S.C.; Mondal, D. Multimodal actions of the phytochemical sulforaphane suppress both AR and AR-V7 in 22Rv1 cells: Advocating a potent pharmaceutical combination against castration-resistant prostate cancer. *Oncol. Rep.* **2017**, *38*, 2774–2786. [CrossRef]

58. Horoszewicz, J.S.; Leong, S.S.; Chu, T.M.; Wajsman, Z.L.; Friedman, M.; Papsidero, L.; Kim, U.; Chai, L.S.; Kakati, S.; Arya, S.K.; et al. The LNCaP cell line–a new model for studies on human prostatic carcinoma. *Prog. Clin. Biol. Res.* **1980**, *37*, 115–132.

59. Cunningham, D.; You, Z. In vitro and in vivo model systems used in prostate cancer research. *J. Biol. Methods* **2015**, *2*, 17. [CrossRef]

60. Sporn, M.B.; Liby, K.T.; Yore, M.M.; Fu, L.; Lopchuk, J.M.; Gribble, G.W. New synthetic triterpenoids: Potent agents for prevention and treatment of tissue injury caused by inflammatory and oxidative stress. *J. Nat. Prod.* **2011**, *74*, 537–545. [CrossRef]

61. Yore, M.M.; Kettenbach, A.N.; Sporn, M.B.; Gerber, S.A.; Liby, K.T. Proteomic analysis shows synthetic oleanane triterpenoid binds to mTOR. *PLoS ONE* **2011**, *6*. [CrossRef] [PubMed]

62. Jeong, S.A.; Kim, I.Y.; Lee, A.R.; Yoon, M.J.; Cho, H.; Lee, J.S.; Choi, K.S. Ca 2+ influx-mediated dilation of the endoplasmic reticulum and c-FLIP L downregulation trigger CDDO-Me-induced apoptosis in breast cancer cells. *Oncotarget* **2015**, *6*, 21173–21192. [CrossRef] [PubMed]

63. Qin, D.-J.; Tang, C.-X.; Yang, L.; Lei, H.; Wei, W.; Wang, Y.-Y.; Ma, C.-M.; Gao, F.-H.; Xu, H.-Z.; Wu, Y.-L. Hsp90 Is a Novel Target Molecule of CDDO-Me in Inhibiting Proliferation of Ovarian Cancer Cells. *PLoS ONE* **2015**, *10*, e0132337. [CrossRef] [PubMed]

64. Tan, M.E.; Li, J.; Xu, H.E.; Melcher, K.; Yong, E.L. Androgen receptor: Structure, role in prostate cancer and drug discovery. *Acta Pharmacol. Sin.* **2015**, *36*, 3–23. [CrossRef] [PubMed]

65. Schultz, M.A.; Hagan, S.S.; Datta, A.; Zhang, Y.; Freeman, M.L.; Sikka, S.C.; Abdel-Mageed, A.B.; Mondal, D. Nrf1 and Nrf2 Transcription Factors Regulate Androgen Receptor Transactivation in Prostate Cancer Cells. *PLoS ONE* **2014**, *9*, e87204. [CrossRef]

66. Park, H.S.; Han, M.H.; Kim, G.-Y.; Moon, S.-K.; Kim, W.-J.; Hwang, H.J.; Park, K.Y.; Choi, Y.H. Sulforaphane induces reactive oxygen species-mediated mitotic arrest and subsequent apoptosis in human bladder cancer 5637 cells. *Food Chem. Toxicol.* **2014**, *64*, 157–165. [CrossRef]

67. Kocyigit, A.; Guler, E.M. Curcumin induce DNA damage and apoptosis through generation of reactive oxygen species and reducing mitochondrial membrane potential in melanoma cancer cells. *Cell. Mol. Biol. (Noisy-le-grand).* **2017**, *63*, 97–105. [CrossRef]

68. Gersey, Z.C.; Rodriguez, G.A.; Barbarite, E.; Sanchez, A.; Walters, W.M.; Ohaeto, K.C.; Komotar, R.J.; Graham, R.M. Curcumin decreases malignant characteristics of glioblastoma stem cells via induction of reactive oxygen species. *BMC Cancer* **2017**, *17*, 99. [CrossRef] [PubMed]

69. Choi, W.Y.; Choi, B.T.; Lee, W.H.; Choi, Y.H. Sulforaphane generates reactive oxygen species leading to mitochondrial perturbation for apoptosis in human leukemia U937 cells. *Biomed. Pharmacother.* **2008**, *62*, 637–644. [CrossRef]

70. Helness, A.; Gaudreau, M.-C.; Grapton, D.; Vadnais, C.; Fraszczak, J.; Wilhelm, B.; Robert, F.; Heyd, F.; Moroy, T. Loss of heterogeneous nuclear ribonucleoprotein L (HNRNP L) leads to mitochondrial dysfunction, DNA damage response and caspase-dependent cell death in hematopoietic stem cells. *Exp. Hematol.* **2016**, *44*, S78–S79. [CrossRef]

71. Takayama, K.I. Splicing factors have an essential role in prostate cancer progression and androgen receptor signaling. *Biomolecules* **2019**, *9*, 131. [CrossRef] [PubMed]

72. Huo, C.; Kao, Y.H.; Chuu, C.P. Androgen receptor inhibits epithelial-mesenchymal transition, migration, and invasion of PC-3 prostate cancer cells. *Cancer Lett.* **2015**, *369*, 103–111. [CrossRef] [PubMed]

73. Demir, A.; Cecen, K.; Karadag, M.A.; Kocaaslan, R.; Turkeri, L. The course of metastatic prostate cancer under treatment. *Springerplus* **2014**, *3*, 725. [CrossRef] [PubMed]

74. Aapro, M.S.; Eliason, J.F.; Krauer, F.; Alberto, P. Colony formation in vitro as a prognostic indicator for primary breast cancer. *J. Clin. Oncol.* **1987**, *5*, 890–896. [CrossRef]

75. Sekhar, K.R.; Wang, J.; Freeman, M.L.; Kirschner, A.N. Radiosensitization by enzalutamide for human prostate cancer is mediated through the DNA damage repair pathway. *PLoS ONE* **2019**, *14*. [CrossRef]

76. Ardiani, A.; Farsaci, B.; Rogers, C.J.; Protter, A.; Guo, Z.; King, T.H.; Apelian, D.; Hodge, J.W. Combination therapy with a second-generation androgen receptor antagonist and a metastasis vaccine improves survival in a spontaneous prostate cancer model. *Clin. Cancer Res.* **2013**, *19*, 6205–6218. [CrossRef]

77. Luk, I.S.U.; Shrestha, R.; Xue, H.; Wang, Y.; Zhang, F.; Lin, D.; Haegert, A.; Wu, R.; Dong, X.; Collins, C.C.; et al. BIRC6 Targeting as Potential Therapy for Advanced, Enzalutamide-Resistant Prostate Cancer. *Clin. Cancer Res.* **2017**, *23*, 1542–1551. [CrossRef]

78. Cheng, J.; Moore, S.; Gomez-Galeno, J.; Lee, D.-H.; Okolotowicz, K.J.; Cashman, J.R. A Novel Small Molecule Inhibits Tumor Growth and Synergizes Effects of Enzalutamide on Prostate Cancer. *J. Pharmacol. Exp. Ther.* **2019**, *371*, 703–712. [CrossRef]

79. Toren, P.; Kim, S.; Johnson, F.; Zoubeidi, A. Combined AKT and MEK Pathway Blockade in Pre-Clinical Models of Enzalutamide-Resistant Prostate Cancer. *PLoS ONE* **2016**, *11*, e0152861. [CrossRef]

Review

Redox-Mediated Mechanism of Chemoresistance in Cancer Cells

Eun-Kyung Kim [1], MinGyeong Jang [1], Min-Jeong Song [1], Dongwoo Kim [2], Yosup Kim [2] and Ho Hee Jang [1,2,*]

1   Department of Biochemistry, College of Medicine, Gachon University, Incheon 21999, Korea;
    ekkim@gachon.ac.kr (E.-K.K.); apnea4001@naver.com (M.J.); neptune6nrg@hanmail.net (M.-J.S.)
2   Department of Health Sciences and Technology, GAIHST, Gachon University, Incheon 21999, Korea;
    cd0575@naver.com (D.K.); youandkys@naver.com (Y.K.)
*   Correspondence: hhjang@gachon.ac.kr; Tel.: +82-32-899-6317

Received: 18 September 2019; Accepted: 8 October 2019; Published: 10 October 2019

Abstract: Cellular reactive oxygen species (ROS) status is stabilized by a balance of ROS generation and elimination called redox homeostasis. ROS is increased by activation of endoplasmic reticulum stress, nicotinamide adenine dinucleotide phosphate (NADPH) oxidase family members and adenosine triphosphate (ATP) synthesis of mitochondria. Increased ROS is detoxified by superoxide dismutase, catalase, and peroxiredoxins. ROS has a role as a secondary messenger in signal transduction. Cancer cells induce fluctuations of redox homeostasis by variation of ROS regulated machinery, leading to increased tumorigenesis and chemoresistance. Redox-mediated mechanisms of chemoresistance include endoplasmic reticulum stress-mediated autophagy, increased cell cycle progression, and increased conversion to metastasis or cancer stem-like cells. This review discusses changes of the redox state in tumorigenesis and redox-mediated mechanisms involved in tolerance to chemotherapeutic drugs in cancer.

Keywords: reactive oxygen species; antioxidant proteins; chemoresistance; oxaliplatin; 5-Fluorouracil

1. Introduction

Reactive oxygen species (ROS), as second messengers, function in various cellular signal pathways in normal cells and cancer cells [1]. Redox homeostasis is regulated by a balanced status between ROS production and scavenging (Figure 1) [1,2]. Signal cascades induced by stimuli can lead to ROS generation from ligand-receptor interactions [2–4]. Molecules that can directly penetrate the cell membrane, such as lipophilic growth hormones (steroid hormones and thyroid hormones) and chemical drugs, can activate mitochondrial-mediated ROS generation [5–7]. Although various stimuli can induce changes in ROS and affect the physiological response in cells, the antioxidant proteins stabilize ROS levels to maintain redox homeostasis [8]. Superoxide dismutase (SOD), catalase, peroxiredoxin (Prx), and nuclear factor erythroid 2-related factor 2 (Nrf2) are antioxidant modules [9]. Local ROS level, as a second messenger, amplifies only the specific region where receptor activation transduces a linear signal response. [3,10]. This process is regulated locally by ROS inducers and antioxidant modules to overcome the possibility that the alternative ROS can affect whole cells [3].

Many studies have shown that redox imbalances can induce signaling pathways that promote cancer progression, senescence, differentiation, and apoptosis [8]. Cancer cells show enhanced glycolysis-mediated metabolisms to overcome over-utilized ATP or alter cellular signal pathways [11]. Thus, many cancer cells upregulate antioxidants as protection against their high levels of ROS. Chemotherapeutic agents can induce increased ROS levels, and most cancer cells treated with chemotherapy suffer from ROS-mediated apoptosis [12]. Some cancer cells evolve mechanisms to escape ROS-mediated apoptosis and acquire tolerance to anti-cancer drugs [13]. The ROS system has a dual function that can either induce apoptosis or allow cells to adapt to various environments. ROS regulation has thus been a critical target for the development of anticancer drugs [14]. In this review, we discuss the change of redox balance by the generation or removal of ROS in tumorigenesis and redox-mediated mechanisms of the chemoresistance in chemotherapy.

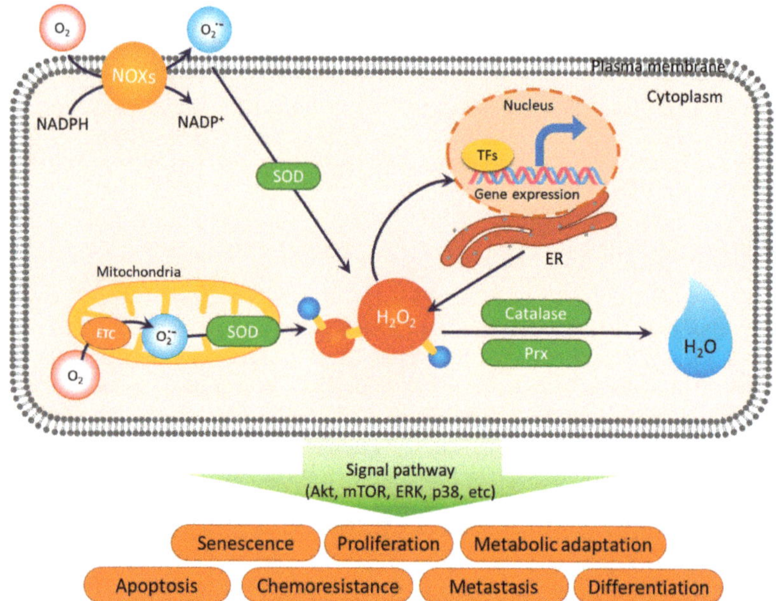

Figure 1. Redox homeostasis between generation and elimination of reactive oxygen species (ROS). ROS production is regulated by the nicotinamide adenine dinucleotide phosphate (NADPH) oxidases (NOXs) in membranes, the electron transport chain (ETC) of the adenosine triphosphate (ATP) synthesis process in mitochondria, and the protein synthesis process in endoplasmic reticulum (ER) during O$_2$ consumption. Alternative levels of ROS induce DNA damage or transcription factors (TFs)-mediated gene expression in the nucleus. The superoxide anion (O$_2\bullet^-$) produced intracellularly is neutralized to hydrogen peroxide (H$_2$O$_2$) by the superoxide dismutase (SOD) family. H$_2$O$_2$ are detoxified to H$_2$O by catalase and peroxiredoxin (Prx). ROS regulate cellular processes such as proliferation, apoptosis, chemoresistance, and differentiation through a variety of signaling pathways.

2. Redox Homeostasis in Tumorigenesis

2.1. ROS Generation

Intracellular redox functions as an oncogenic factor for the activation of signal transduction in tumorigenesis [9]. ROS consists of both free radical and non-radical groups. The free radical group includes superoxide anion (O$_2\bullet^-$), peroxyl radical (RO$_2\bullet$), hydroxyl radical ($\bullet$OH), and hydroperoxyl radical (HO$_2\bullet$). Hydrogen peroxide (H$_2$O$_2$) and single oxygen (^{1}O$_2$) are classified as non-radical ROS. Production of intracellular ROS is generated by ATP synthesis in mitochondria, protein synthesis in the endoplasmic reticulum (ER), and activation of (NADPH) oxidase NOX family members [5].

2.1.1. ATP Synthesis in Mitochondria

Mitochondria generate intracellular ROS during the electron transport chain (ETC) of the ATP synthesis process [15]. The homeostasis of ROS in mitochondria is maintained by antioxidant proteins. Upon electron leakage of the ETC, the abnormal ROS status of mitochondria can activate apoptosis in carcinoma cells [15,16].

Cancer cells show increased metabolism for their elevated proliferation and migration. Cancer cells have significantly increased the ATP production as well as the ROS [15–18]. Chemoresistant cancer cells require the active pump of the ATP-driven multidrug efflux, such as ATP-binding cassette (ABC) transporters [19]. The role of these transporters is to pump out intracellular toxic chemical drugs into the extracellular region by ATP hydrolysis [20]. ABC transporters include multidrug resistance-associated protein 1 (MRP1/ABCC1), breast cancer resistance protein/ABC subfamily G member 2 (BCRP/ABCG2), ABC subfamily B member 5 (ABCB5), and multidrug resistance protein 1/ABC subfamily B member 1 (MDR1/ABCB1) [19–21]. Enhanced ROS level is generated by the ETC, but the antioxidant machinery is also induced to adapt to the higher ROS level. Thus, regulation of the ETC in mitochondria may be a good approach to overcome chemoresistance via the blockage of routes that generate ATP or the dysregulation of ROS production that induces apoptosis.

2.1.2. Endoplasmic Reticulum (ER)

The ER is a dynamical cellular organelle that plays a role the protein folding system, which regulates almost all of the membrane proteins and secretory proteins for post-translational modification [22,23]. The intracellular H_2O_2 in the process of protein synthesis in ER is generated by the formation of disulfide bridges during the induction of functional three-dimensional structures via protein disulfide isomerase (PDI) and other oxidoreductases [24,25]. Thus, the intracellular ROS status of ER maintains the relatively high level [26,27]. The ER-stress induced response is involved in the survival, metastasis, and angiogenesis in cancer cells under rough microenvironmental situations [28,29].

2.1.3. NADPH Oxidases (NOXs)

The NOX family members consist of NOX1–5 and dual oxidase (DUOX) 1 and 2 [30]. H_2O_2 and $O_2\bullet-$ produced by NOXs function as secondary messengers to transduce signals in response to various growth-related factors and chemical drugs [31–33]. NOX-induced ROS production provokes the acquisition of chemoresistance and contributes to cancer progression [34,35].

NOX1 is mainly located at the plasma membrane and endosome. NOX2 and NOX5 are located in the ER and plasma membrane [36,37]. NOX3 is localized mostly at the plasma membrane as well as mitochondria. NOX4 is also localized at the plasma membrane, the ER, the inner membrane of mitochondria and nucleus [38]. DUOX1 is located in the plasma membrane and ER, while DUOX2 is localized to the plasma membrane, ER, and cytosol as well as mitochondria and nucleus [30].

NOX1 and NOX5 regulate the drug efficacy of chemotherapy in prostate cancer [39]. NOX2 expression is related to invasion and progression in gastric cancer and acute myeloid leukemia (AML) [40,41]. NOX4 has a function of regulation in drug resistance [34]. Overexpression of DUOX and NOX4 has been detected in human thyroid tumors [35]. Due to the limits of expression in the inner ear epithelial cells and cochlea, the precise role of NOX3 in cancer is unknown [30,42]. In pancreatic ductal carcinoma, the elevated ROS level by activated NOXs induces tolerance against chemotherapy and radiation therapy [43–45]. The target of NOXs is a druggable strategy to treat cancer by drugs that inhibit NOXs, and cancer cells can be treated by inducing redox state-mediated triggers of apoptosis [46].

2.2. ROS Elimination

Redox homeostasis is regulated by the antioxidant enzymes. Cancer cells maintain sustained overexpression of antioxidant proteins to detoxify the ROS byproducts of over-activated cellular metabolisms.

2.2.1. SODs

SOD is an enzyme that catalyzes the partitioning of two superoxide anions into hydrogen peroxide and molecular oxygen by the metalloenzymatic reaction [47]. SOD dependent-neutralization is important as the cell's first barricade to ROS in the antioxidant systems. SOD has specific metal cofactors for the enzymatic activity such as SOD1 with copper (Cu) and zinc (Zn), SOD2 with manganese (Mn), and SOD3 with copper (Cu) and zinc (Zn) [48]. The SOD family is consisted of SOD1 (Cu/ZnSOD), SOD2 (MnSOD), and SOD3 (Cu/ZnSOD) [47,48]. The expression of SOD1, the most abundant SOD protein in cytoplasm, is increased in mammary carcinomas and lung carcinomas [48,49]. SOD2 was identified as downregulated in tumors in early studies, and thus SOD2 was initially considered a tumor suppressor [50]. However, recent studies have shown that SOD2 exhibits tumor-type dependent function [51,52]. SOD2 levels are higher in late-stage tumors as well as in invasive and metastatic cancers [48,50,53]. SOD2 also functions in the regulation of mitochondrial integrity and function [51,53]. Thus, SOD2 plays an important role in tumor progression. SOD3, or extracellular superoxide dismutase EcSOD, is localized in the extracellular matrix and binds to the glycocalyx in cell surfaces [54]. SOD3 functions neutralize from $O_2\bullet^-$ by the membrane-bound NOXs to H_2O_2 [54–56]. The role of SOD3 in cancer is less known.

2.2.2. Catalase

Catalase is a 62 kDa enzyme and consists of four identical subunits, including an N-terminal region for catalase reaction, a beta-barrel region for three-dimensional structure, a connection region for binding heme groups, and an alpha-helix region for NADPH binding [57,58]. The major function of catalase is to metabolize high concentrations of H_2O_2 for the protection of ROS-induced damage in cells. The reaction mechanism of catalase occurs in two-steps using heme groups. The first reaction is a process in which the heme cofactor reacts with a single molecule of H_2O_2 to produce an oxidative heme group (an oxoferryl porphyrin cation radical, which reduces the return to the previous step). In the second reaction, the oxoferryl porphyrin cation radical of catalase rapidly reacts with the second molecule of H_2O_2 to produce oxygen byproducts and water [58].

Another role of catalase is in the regulation of the integrin pathway during proliferation or migration [58]. Overexpression of catalase has been detected in various carcinomas, such as chronic myeloid leukemia, gastric cancer, and skin cancer [59–61]. Anticancer drugs also increased catalase levels in oral cancer cells, bladder cancer cells, pancreatic cancer cells, and gastric cancer cells [62–65]. Catalase expression is controlled by various mechanisms. At the transcriptional level, the expression of catalase is regulated by the activity of transcription factors on the catalase promoters, mRNA stability, and epigenetic chromatin structure [57,58]. At the protein level, the expression of catalase is affected by post-translational modification such as ubiquitination and phosphorylation [57,58].

2.2.3. Prxs

Prx is a thiol-specific peroxidase protein without other cofactors for detoxification of H_2O_2 to H_2O. The reaction mechanism by which Prx decomposes H_2O_2 into H_2O occurs through a cycle, where peroxidatic Cys (C_P-SH) of Prx reacts with H_2O_2 to oxidize to sulfenic acid (C_P-SOH) and then back to a reduced peroxidatic Cys (C_P-SH) state with the presence of reducing equivalents [66]. Prxs maintain cellular ROS homeostasis through this catalytic cycle [66,67]. The Prx family includes Prx1 to Prx6. Prx1 and Prx2 are located in the cytosol and nucleus, and Prx3 is localized in mitochondria. Prx4 is localized to the ER, the cytosol, and secretion. Prx5 is located in the cytosol, nucleus, mitochondria, and peroxisomes [66,68]. Prx6 is located in the nucleus, cytosol, extracellular space, and lysosome [69]. Most Prx family members are overexpressed in various carcinomas and may serve as biomarkers for cancer diagnosis [70]. Prx1 functions as an oncogenic factor in tumorigenesis. Prx1 leads to reduced DNA damage and apoptosis by detoxifying ROS. Prx1 also regulates cell signaling including NF-κB, JNK, Akt, p38 activity, VEGF, and ERK pathways [66]. Therefore, overexpression of Prx1 causes aberrant cell signaling that is beneficial for cancer cells. Prx2 function is paradoxical; Prx2 not only induces activation of the ERK pathway for promotion of metastasis but also stabilizes E-cadherin for suppression of metastasis [71,72]. Prx3 is an oncogenic factor and induces carcinogenesis via tolerance to ROS [66]. Prx4, Prx5, and Prx6 promote metastasis via clearance of increased ROS level in cancers [68–70].

2.2.4. Nrf2

Nrf2 is a transcription factor containing the basic-region leucine zipper domain. Nrf2 is maintained at low levels by Keap1-mediated ubiquitin-dependent proteasomal degradation in the normal condition of cells [73]. Under oxidative stress or exposure to different stressors, Nrf2 is released form Keap1 due to the modification of the Keap1 Cys residue, which prevents ubiquitin-dependent proteasomal degradation of Nrf2 [73]. Nrf2 plays a role in protecting cells from oxidative stress-mediated damages through expression of target genes involved in detoxification. Nrf2-dependent gene families include antioxidant genes (SOD, CAT, Prx, GR, and TR) and genes involved in drug metabolism/transport (MRP1 and BCRP/ABCG2) (Figure 2) [74–76]. MRP belong to a family of membrane-anchored transporters and pump out a wide range of compounds, including peptides, lipids, organic anions, and drugs through ATP hydrolysis [76]. Carcinogenesis or chemoresistance in various cancers such as breast cancer, leukemia, neuroblastoma, and lung cancer increases the expression of Nrf2 or induces hyper-activation [77–80]. Therefore, a combination of Nrf2 inhibitors with anticancer drugs may derive therapeutic effects in patients [76].

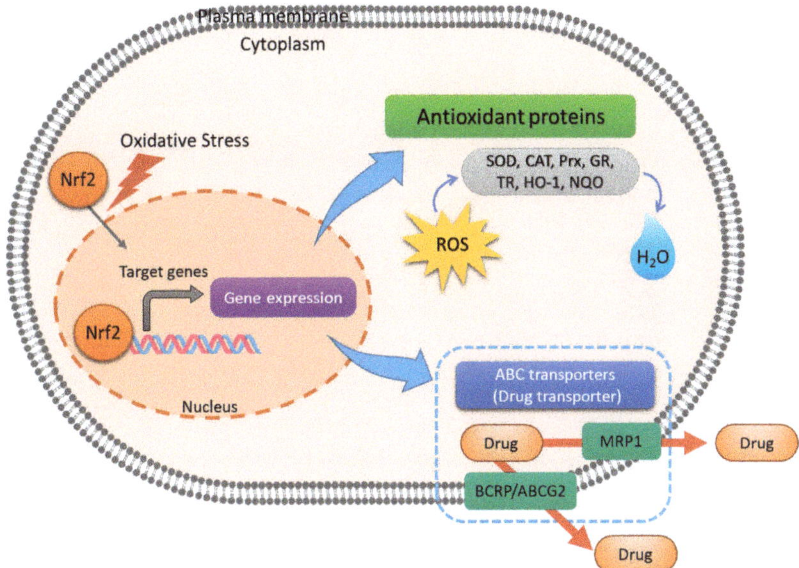

Figure 2. Nuclear factor erythroid 2-related factor 2 (Nrf2) regulates redox-homeostasis and chemoresistance in cells. Nrf2 induces antioxidant proteins such as superoxide dismutase (SOD), catalase (CAT), peroxiredoxin (Prx), glutathione reductase (GR), thioredoxin reductase (TR), heme oxygenase-1 (HO-1), and NAD(P)H quinone oxidoreductase 1 (NQO). Multidrug resistance protein 1 (MRP1) and breast cancer resistance protein/ATP-binding cassette subfamily G member 2 (BCRP/ABCG2) are related with drug transport and are upregulated by activation of Nrf2.

2.3. Redox Homeostasis of Chemoresistance

Chemoresistance arises after long-time exposures to anticancer drugs [81,82]. The difference between temporary treatment and continuous treatment leads to different levels of ROS homeostasis in cancer cells. Cancer cells regulate ROS levels to acquire chemoresistance [13].

2.3.1. Oxaliplatin Resistance

Oxaliplatin is a platinum-based drug that is widely used in lung cancer, breast cancer, pancreatic cancer, and gastric cancer [6]. The cytotoxicity mechanism of oxaliplatin involves its binding genomic DNA, which induces apoptosis in cancer cells, and the generation of ribosome biogenesis stress [6,83].

Resistance to oxaliplatin decreases the production of ROS levels [83–85]. However, alterations of NAPDH oxidase, ER stress, and ETC in mitochondria in oxaliplatin-resistant cells have not been deeply investigated. Oxaliplatin can also form mitochondrial DNA adducts and affect protein synthesis in mitochondria, resulting in mitochondrial abnormalities [86]. Treatment of oxaliplatin induces dysfunction in the mitochondrial respiratory chain and permeability [86,87]. High concentrations of oxaliplatin enhance ROS levels in mitochondria [86,87].

Oxaliplatin-resistant cell lines show altered expression of antioxidant proteins. Upregulated SOD1 and SOD2 detoxify drug-mediated radical species in oxaliplatin-resistant colon cancer cells [88]. SOD3 is also highly expressed in the mouse model of oxaliplatin-induced liver injury [89]. Oxaliplatin-resistant colon cancer cells have increased Nrf2 expression and exhibit chemotherapeutic effects via inhibition of Nrf2 signaling [90,91].

2.3.2. 5-Fluorouracil (5-FU) Resistance

5-FU is an uracil analog drug that is widely used in various cancers. 5-FU induces DNA and RNA damage in the nucleus and mitochondria from byproducts of cellular 5-FU metabolism [92].

5-FU resistance causes a high level of intracellular ROS in colon cancer cells [93–95]. DUOX2 expression is altered in response to 5-FU resistance. The expression of DUOX2 is enhanced in 5-FU chemoresistant colon cancer cell lines. The upregulation of DUOX2 induces high levels of ROS and invasion ability [96]. ER-stress related factors, including PERK, GRP78, and ATF6, are upregulated in 5-FU tolerant colon cancer cells [97].

5-FU incorporation into mitochondrial DNA induces destabilization of mitochondrial DNA and protein synthesis. 5-FU can also induce ROS-mediated damage in mitochondria [98,99]. 5-FU resistance leads to a down-regulation of ATP synthesis via lower expression of ATP synthase subunits or reduced activity of ATP synthase [100,101]. Drug resistance to 5-FU induces a high level of SOD1 and Prx1 in the adaption of increased intracellular ROS conditions [102,103]. In several cancer cells, the expression or intracellular location of Nrf2 is associated with 5-FU resistance [95,103–107]. For example, Nrf2 is overexpressed in 5-FU resistant gastric cancer and nuclear localization, and the expression of Nrf2 is increased in 5-FU resistant colorectal cancer cells [95,105–107].

3. Redox-Mediated Mechanism of Chemoresistance

Multidrug resistance (MDR) is regulated by the upregulation of antioxidant proteins (hemooxygenase-1, superoxide dismutase, catalase). These antioxidant factors detoxify the altered intracellular ROS levels. Variation of the ROS level is required for chemoresistance or for the upregulation of drug efflux against chemotherapy [13,102]. ROS-mediated mechanisms for acquired chemoresistance involve ER stress and autophagy, cell cycle perturbation for overcoming cell cycle arrest, and reprogramming for promoting epithelial to mesenchymal transition or conversion to cancer stem-like cells (Figure 3) [31,104,108–112].

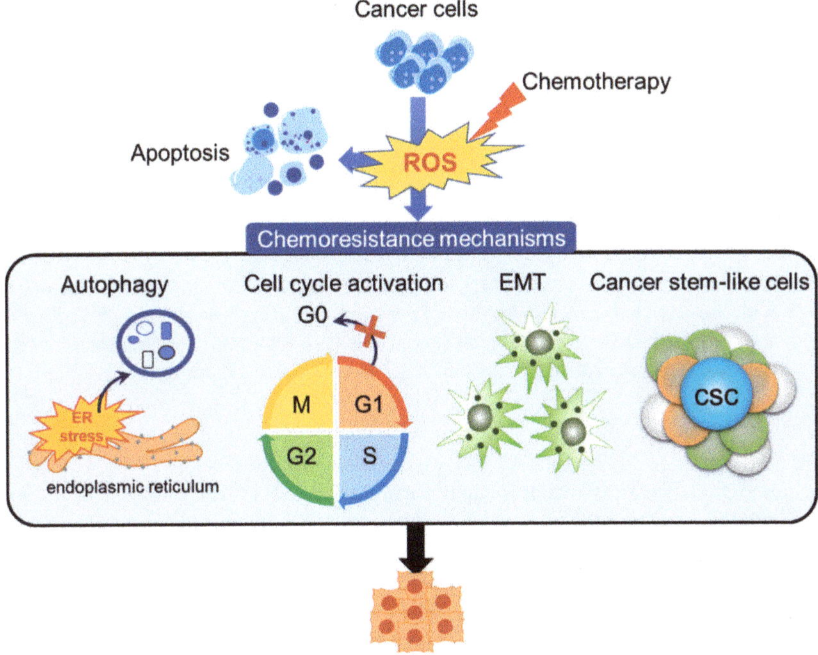

Figure 3. Redox homeostasis between generation and elimination of ROS. Cancer cells increase ROS-mediated apoptosis when exposed to chemotherapy treatments. Some cancer cells adapt to fluctuating ROS states through chemoresistance mechanisms. Activation of autophagy by ER stress gets rid of damaged organelles and protein aggregation. Cell cycle activation by the ignored entrance of the G0 phase increases cell proliferation in cancers. Epithelial-mesenchymal transition (EMT) enhances migration to other organs for the escapement of damaging environments. Cancer stem-like cells (CSCs) increase the expression of drug metabolic enzymes/transporters for cell survival from drug-mediated apoptosis. The specific microenvironment, called the niche, of CSCs are protected from chemotherapy. These mechanisms lead to the birth of chemoresistant cancer cells.

3.1. ER Stress-Mediated Autophagy

ROS play an important role in switching from ER stress-mediated apoptosis to autophagy in drug-resistant carcinomas [110]. Acquired resistance to anti-cancer drugs in cancer cells also results in tolerance to ER stress-mediated cell death [113]. The ER stress response also serves as a survival signal for chemoresistance in cancers [114,115]. The ER stress response is controlled by inositol-requiring enzyme-1 (IRE1), protein kinase R-like endoplasmic reticulum kinase (PERK), and activating transcription factor 6 (ATF6) in normal cells [116,117]. The loss of tumor suppressor factors or activation of oncogenes often induces activation of ER stress response-related factors to generate tumorigenesis in response to chemotherapy [97,118]. PERK is associated with the upregulation of the MDR-related protein MRP1 in chemoresistant colon adenocarcinoma cell lines [119].

Autophagy is a highly controlled degradation system of damaged organelles and protein aggregation. Autophagy is activated by starvation, organelle damage, and ER stress, and various cancers show dysfunction in autophagy [120,121]. ROS is another stimulus that can activate autophagy. H_2O_2 accumulation can result in oxidization of ATG4, a factor involved in the autophagic process [122–124]. Oxidation of ATG4 induces initiation of autophagy [125,126].

3.1.1. Oxaliplatin-Resistance

Regulation of autophagy is one of the mechanisms of oxaliplatin resistance. The tolerance of colon cancer cell lines to oxaliplatin involves the down-regulation of Bcl2-modifying factor (BMF), ATG7, and Beclin-1 [127]. BMF induces necrosis, apoptosis, and autophagy [128–130]. Chemoresistance can be acquired by activated autophagy depending on the cell type. Oxaliplatin-resistant hepatocellular carcinoma shows autophagy activation via ROS generation and induced modulation of autophagosomes, LC3-II accumulation, and LC3 redistribution [86,131].

3.1.2. 5-FU-Resistance

Chemoresistance to 5-FU is regulated by autophagy systems, although autophagy is increased or decreased in various cancers. Autophagy-related proteins, including Beclin-1, ATG5, and LC3-II, are downregulated in 5-FU-resistant human colon cancer cells [94]. However, 5-FU-resistant breast cancer cell lines show overexpression of ATG5, Beclin-1, LC3-II, and increased autophagy [132].

3.2. Overcoming Cell Cycle Arrest

Aberrant levels of ROS result in increased cell cycle progression to bypass arrest and acquire chemoresistance against cancer chemotherapies [108,112]. The cell cycle is regulated by positive regulators, cyclin and cyclin-dependent kinases (CDKs), and negative factors, including cyclin-dependent kinase inhibitors [108,133]. One of the hallmarks of the early stage of cancer is abnormal cell cycle progression resulting from the dysregulation of cell cycle-related factors [82,108,134].

The cell cycle progression is responsive to changes in the redox state of metabolism [112]. The redox signaling pathways alters cell cycle progression that converges as a regulator of CDK [135]. Altered ROS levels promote increased cell cycle via phosphorylation of cell cycle regulatory factors or upregulation of the cyclin family [136–139]. Cyclin D1 is overexpressed in various human carcinomas [140–142]. Overproduction of ROS leads to metastasis via regulation of cyclin D1, which functions in the invasion and metastatic properties of tumors [143,144].

3.2.1. Oxaliplatin-Resistance

Acquired oxaliplatin resistance affects cell cycle progression. The oxaliplatin-resistant LoVo cell line has overcome oxaliplatin-mediated G2 phase arrest [145]. Overexpression of the cell cycle 5regulators cyclin D1 and B1 are reported in the oxaliplatin-resistant HT-29 cell line [146].

3.2.2. 5-FU-Resistance

5-FU affects cell cycle perturbation by its incorporation in DNA and interfering with DNA synthesis [100]. Cancer cells with treatment of 5-FU have sufficient time during the cell cycle to correct the mis-incorporated fluoronucleotides and prolong DNA synthesis during the cell cycle [100,147]. However, cell lines with acquired chemoresistance show high expressions of cell cycle-related activators that result in resistance to cell cycle arrest. Breast cancer cell lines with 5-FU resistance show upregulation of cell cycle-related proteins and enhanced cellular proliferation [148,149]. 5-FU resistance results in modified expression of G1 phase cyclins in oral cancer cell lines and aberrant cell cycle regulation [100,150].

3.3. Epithelial-Mesenchymal Transition (EMT) and Cancer Stem-Like Cells (CSC)

The EMT process allows metastatic tumor cells to migrate to organs. Altered intracellular ROS levels may lead to the promotion of EMT in cancer cell lines that are resistant to anti-cancer drugs [81,108]. Several proteins function in the development of chemoresistance in metastatic advanced carcinomas [151,152]. Chemoresisance in cancer cells results in switching from chemotherapy-mediated apoptosis to EMT properties. Moreover, EMT-related signaling pathways, such as sonic hedgehog

(SHH), Notch, TGF-β, and Wnt, overlap with renewal and maintenance of CSC [153,154]. Metastatic characterized cancer cells show phenotypes of both EMT and CSC [155,156].

Some chemoresistant cancer cell lines have characteristics of CSCs, including decreased proliferation, a greater proportion of G0/G1 cells, and increased ability of sphere-forming capacity [157,158]. The ABC transporter is related with drug resistance and cancer stem like-cells [159].

3.3.1. Oxaliplatin-Resistance

Resistant to oxaliplatin results in enhanced migration and invasion abilities in colon cancer cells [160,161]. CCN2 and ID-1 are upregulated in oxaliplatin-resistant tumor cells [162,163]. CCN2 regulates cell proliferation, chemotaxis, and migration, while ID-1 is involved in blocking cell differentiation [162–164]. The Cx32 tumor suppressor protein is associated with positive expression of E-cadherin and negative expression of vimentin [165,166]. Cx32 level is decreased in hepatocellular carcinoma resistant to oxaliplatin [165]. Ataxin-2-line (ATXN2L) promotes migration and invasion and is elevated in oxaliplatin-resistant gastric cancer [167,168]. The cancer stem-like cell markers Oct4 and Sox2 are increased in oxaliplatin-resistant colon cancer cell lines [169].

3.3.2. 5-FU-Resistance

Acquired 5-FU resistance results in altered EMT-related morphological phenotypes, such as reduced cellular adhesion, down-regulation of E-cadherin, up-regulation of N-cadherin and twist, the enlarged formation of pseudopodia and spindle-shaped morphology [170,171]. 5-FU-resistant carcinoma cell lines show high expression of vimentin, ZEB1, ZEB2, slug, snail, twist, and N-cadherin [170,172–175]. Low level of E-cadherin has been reported during 5-FU resistance of various cancer cell lines [172,174]. Increased mesenchymal factors enhanced migration. TGF-beta-mediated EMT and cancer stem-like cell capacities are reported in 5-FU-resistant pancreatic cancer cell lines [176]. CD44 is a cell surface marker of cancer stem-like cells. CD44 variant 9 is high in 5-FU-resistant gastric cancer cell lines [177,178]. Some colon cancer cells with 5-FU-resistance show features of cancer stem-like cells [179].

4. Conclusions

ROS are involved in physiological signal cascades in normal and cancer cells. Most of drug or growth factors induce downstream cascades that result in short-lived ROS generation. Antioxidant proteins, as ROS scavengers, play a role in the detoxification of ROS and can regulate the intensity of ROS-mediated signal transduction. Thus, cancer cells regulate the redox homeostasis to survival. ROS-mediated chemoresistance is regulated by the control of ER stress-mediated autophagy, overactivation of cell proliferation, and promotion of EMT and cancer stem-like cells.

The antioxidant system includes diverse proteins such as SOD, catalase, Prx, glutathione peroxidase, and thiol peroxidase, among others. Nrf2 has increased the expression and activity in oxaliplatin or 5-FU resistant cancer cells. Several antioxidant proteins are up-regulated in chemoresistance such as SOD1, SOD2, and SOD3 in oxaliplatin-resistant cancer cells or SOD1 and Prx1 in 5-FU chemoresistance. However, studies of variation in function or expression of other antioxidant proteins in chemoresistance are limited. Investigation of the regulation of antioxidant proteins is required for overcoming chemoresistance by regulation of the redox state, and better understanding of this process may provide new targets for the development of anti-cancer drugs.

The mechanism for acquired chemoresistance may be paradoxical. Regulation of autophagy in chemoresistance results in different responses depending on the cell type. Although chemoresistant cancer cell lines show upregulated proliferation, some chemoresistant cells become cancer stem-like cells, which are characterized by low proliferation. Thus, whether the mechanism of chemoresistance is cell type-specific should be examined in future studies. Due to the characteristic of each cell, this phenomenon (that uses different mechanisms of chemoresistance acquisition in cancer cells) is induced by signal transduction proteins such as Akt, mTOR, ERK, p38, SHH, and Wnt, depending on the activity or expression level of kinases. Most drugs change ROS status in cancers. However, ROS-mediated mechanism can occur by different pathways. Therefore, investigation of chemoresistance can reveal some of kinases with hyperactivation or hypoactivation. These results provide clues to the development of drugs in chemoresistant-related therapies.

In conclusion, clarifying the mechanisms underlying the regulation of redox-mediated chemoresistance may provide targets for drug development for overcoming chemoresistance in preclinical and clinical settings.

Author Contributions: Conceptualization, E.-K.K. and H.H.J.; data curation, M.J.; writing—original draft preparation, E.-K.K., M.-J.S., and H.H.J.; writing—review and editing, E.-K.K., D.K., Y.K., and H.H.J.; visualization, M.J.; supervision, H.H.J.; funding acquisition, M.J., and E.-K.K.

Funding: This work was supported by the National Research Foundation of Korea (NRF) grant funded by the Korea government (NRF-2017R1D1A1B03035550) to E.-K.K., and by the National Research Foundation of Korea (NRF) and the Center for Women In Science, Engineering and Technology (WISET) Grant funded by the Ministry of Science, ICT & Future Planning of Korea (MSIP) under the Program for Returners into R&D to M.J.

Conflicts of Interest: The authors declare no conflict of interest.

References

1. Ray, P.D.; Huang, B.W.; Tsuji, Y. Reactive oxygen species (ROS) homeostasis and redox regulation in cellular signaling. *Cell. Signal.* **2012**, *24*, 981–990. [CrossRef] [PubMed]
2. Lu, Q.B.; Wan, M.Y.; Wang, P.Y.; Zhang, C.X.; Xu, D.Y.; Liao, X.; Sun, H.J. Chicoric acid prevents PDGF-BB-induced VSMC dedifferentiation, proliferation and migration by suppressing ROS/NFkappaB/mTOR/P70S6K signaling cascade. *Redox Biol.* **2018**, *14*, 656–668. [CrossRef] [PubMed]
3. Nordzieke, D.E.; Medrano-Fernandez, I. The Plasma Membrane: A Platform for Intra- and Intercellular Redox Signaling. *Antioxidants* **2018**, *7*, 168. [CrossRef] [PubMed]
4. Choi, M.H.; Lee, I.K.; Kim, G.W.; Kim, B.U.; Han, Y.H.; Yu, D.Y.; Park, H.S.; Kim, K.Y.; Lee, J.S.; Choi, C.; et al. Regulation of PDGF signalling and vascular remodelling by peroxiredoxin II. *Nature* **2005**, *435*, 347–353. [CrossRef] [PubMed]
5. Dharmaraja, A.T. Role of Reactive Oxygen Species (ROS) in Therapeutics and Drug Resistance in Cancer and Bacteria. *J. Med. Chem.* **2017**, *60*, 3221–3240. [CrossRef] [PubMed]
6. Martinez-Balibrea, E.; Martinez-Cardus, A.; Gines, A.; Ruiz de Porras, V.; Moutinho, C.; Layos, L.; Manzano, J.L.; Bugés, C.; Bystrup, S.; Esteller, M.; et al. Tumor-Related Molecular Mechanisms of Oxaliplatin Resistance. *Mol. Cancer Ther.* **2015**, *14*, 1767–1776. [CrossRef]
7. Nobel, S.; Abrahmsen, L.; Oppermann, U. Metabolic conversion as a pre-receptor control mechanism for lipophilic hormones. *Eur. J. Biochem.* **2001**, *268*, 4113–4125. [CrossRef]
8. Moloney, J.N.; Cotter, T.G. ROS signalling in the biology of cancer. *Semin. Cell Dev. Biol.* **2018**, *80*, 50–64. [CrossRef]
9. Yang, H.; Villani, R.M.; Wang, H.; Simpson, M.J.; Roberts, M.S.; Tang, M.; Liang, X. The role of cellular reactive oxygen species in cancer chemotherapy. *J. Exp. Clin. Cancer Res.* **2018**, *37*, 266. [CrossRef]
10. Sauer, H.; Wartenberg, M.; Hescheler, J. Reactive oxygen species as intracellular messengers during cell growth and differentiation. *Cell. Physiol. Biochem.* **2001**, *11*, 173–186. [CrossRef]
11. Gorrini, C.; Harris, I.S.; Mak, T.W. Modulation of oxidative stress as an anticancer strategy. *Nat. Rev. Drug Discov.* **2013**, *12*, 931–947. [CrossRef]
12. Galadari, S.; Rahman, A.; Pallichankandy, S.; Thayyullathil, F. Reactive oxygen species and cancer paradox: To promote or to suppress? *Free Radic. Biol. Med.* **2017**, *104*, 144–164. [CrossRef]
13. Cui, Q.; Wang, J.Q.; Assaraf, Y.G.; Ren, L.; Gupta, P.; Wei, L.; Ashby, C.R., Jr.; Yang, D.-H.; Chen, Z.-S. Modulating ROS to overcome multidrug resistance in cancer. *Drug Resist. Updat.* **2018**, *41*, 1–25. [CrossRef]
14. Zou, Z.; Chang, H.; Li, H.; Wang, S. Induction of reactive oxygen species: An emerging approach for cancer therapy. *Apoptosis* **2017**, *22*, 1321–1335. [CrossRef]
15. Porporato, P.E.; Filigheddu, N.; Pedro, J.M.B.; Kroemer, G.; Galluzzi, L. Mitochondrial metabolism and cancer. *Cell Res.* **2018**, *28*, 265–280. [CrossRef]
16. Porporato, P.E.; Payen, V.L.; Baselet, B.; Sonveaux, P. Metabolic changes associated with tumor metastasis, part 2: Mitochondria, lipid and amino acid metabolism. *Cell. Mol. Life Sci.* **2016**, *73*, 1349–1363. [CrossRef]

17. Oliva, C.R.; Markert, T.; Ross, L.J.; White, E.L.; Rasmussen, L.; Zhang, W.; Everts, M.; Moellering, D.R.; Bailey, S.M.; Suto, M.J.; et al. Identification of Small Molecule Inhibitors of Human Cytochrome c Oxidase That Target Chemoresistant Glioma Cells. *J. Biol. Chem.* **2016**, *291*, 24188–24199. [CrossRef]

18. Cheng, C.W.; Kuo, C.Y.; Fan, C.C.; Fang, W.C.; Jiang, S.S.; Lo, Y.K.; Kao, M.C.; Lee, A.Y.L. Overexpression of Lon contributes to survival and aggressive phenotype of cancer cells through mitochondrial complex I-mediated generation of reactive oxygen species. *Cell Death Dis.* **2013**, *4*, e681. [CrossRef]

19. Prieto-Vila, M.; Takahashi, R.U.; Usuba, W.; Kohama, I.; Ochiya, T. Drug Resistance Driven by Cancer Stem Cells and Their Niche. *Int. J. Mol. Sci.* **2017**, *18*, 2574. [CrossRef]

20. Lobo, N.A.; Shimono, Y.; Qian, D.; Clarke, M.F. The biology of cancer stem cells. *Annu. Rev. Cell Dev. Biol.* **2007**, *23*, 675–699. [CrossRef]

21. Vasiliou, V.; Vasiliou, K.; Nebert, D.W. Human ATP-binding cassette (ABC) transporter family. *Hum. Genom.* **2009**, *3*, 281–290. [CrossRef]

22. Bahar, E.; Kim, J.Y.; Yoon, H. Chemotherapy Resistance Explained through Endoplasmic Reticulum Stress-Dependent Signaling. *Cancers* **2019**, *11*, 338. [CrossRef]

23. Zhang, Z.; Zhang, L.; Zhou, L.; Lei, Y.; Zhang, Y.; Huang, C. Redox signaling and unfolded protein response coordinate cell fate decisions under ER stress. *Redox Biol.* **2018**. [CrossRef] [PubMed]

24. Qin, M.; Wang, W.; Thirumalai, D. Protein folding guides disulfide bond formation. *Proc. Natl. Acad. Sci. USA* **2015**, *112*, 11241–11246. [CrossRef] [PubMed]

25. Malhotra, J.D.; Kaufman, R.J. Endoplasmic reticulum stress and oxidative stress: A vicious cycle or a double-edged sword? *Antioxid. Redox Signal.* **2007**, *9*, 2277–2293. [CrossRef] [PubMed]

26. Araki, K.; Iemura, S.; Kamiya, Y.; Ron, D.; Kato, K.; Natsume, T.; Nagata, K. Ero1-alpha and PDIs constitute a hierarchical electron transfer network of endoplasmic reticulum oxidoreductases. *J. Cell Biol.* **2013**, *202*, 861–874. [CrossRef] [PubMed]

27. Enyedi, B.; Varnai, P.; Geiszt, M. Redox state of the endoplasmic reticulum is controlled by Ero1L-alpha and intraluminal calcium. *Antioxid. Redox Signal.* **2010**, *13*, 721–729. [CrossRef] [PubMed]

28. Cubillos-Ruiz, J.R.; Bettigole, S.E.; Glimcher, L.H. Tumorigenic and Immunosuppressive Effects of Endoplasmic Reticulum Stress in Cancer. *Cell* **2017**, *168*, 692–706. [CrossRef]

29. Oakes, S.A. Endoplasmic reticulum proteostasis: A key checkpoint in cancer. *Am. J. Physiol. Cell Physiol.* **2017**, *312*, C93–C102. [CrossRef]

30. Brandes, R.P.; Weissmann, N.; Schroder, K. Nox family NADPH oxidases: Molecular mechanisms of activation. *Free Radic. Biol. Med.* **2014**, *76*, 208–226. [CrossRef]

31. Weng, M.S.; Chang, J.H.; Hung, W.Y.; Yang, Y.C.; Chien, M.H. The interplay of reactive oxygen species and the epidermal growth factor receptor in tumor progression and drug resistance. *J. Exp. Clin. Cancer Res.* **2018**, *37*, 61. [CrossRef] [PubMed]

32. Miyata, Y.; Matsuo, T.; Sagara, Y.; Ohba, K.; Ohyama, K.; Sakai, H. A Mini-Review of Reactive Oxygen Species in Urological Cancer: Correlation with NADPH Oxidases, Angiogenesis, and Apoptosis. *Int. J. Mol. Sci.* **2017**, *18*, 2214. [CrossRef] [PubMed]

33. Vara, D.; Watt, J.M.; Fortunato, T.M.; Mellor, H.; Burgess, M.; Wicks, K.; Mace, K.; Reeksting, S.; Lubben, A.; Wheeler-Jones, C.P.D.; et al. Direct Activation of NADPH Oxidase 2 by 2-Deoxyribose-1-Phosphate Triggers Nuclear Factor Kappa B-Dependent Angiogenesis. *Antioxid. Redox Signal.* **2018**, *28*, 110–130. [CrossRef] [PubMed]

34. Shanmugasundaram, K.; Nayak, B.K.; Friedrichs, W.E.; Kaushik, D.; Rodriguez, R.; Block, K. NOX4 functions as a mitochondrial energetic sensor coupling cancer metabolic reprogramming to drug resistance. *Nat. Commun.* **2017**, *8*, 997. [CrossRef] [PubMed]

35. Ameziane-El-Hassani, R.; Schlumberger, M.; Dupuy, C. NADPH oxidases: New actors in thyroid cancer? *Nat. Rev. Endocrinol.* **2016**, *12*, 485–494. [CrossRef] [PubMed]

36. Laurindo, F.R.; Araujo, T.L.; Abrahao, T.B. Nox NADPH oxidases and the endoplasmic reticulum. *Antioxid. Redox Signal.* **2014**, *20*, 2755–2775. [CrossRef]

37. Fulton, D.J. Nox5 and the regulation of cellular function. *Antioxid. Redox Signal.* **2009**, *11*, 2443–2452. [CrossRef] [PubMed]

38. Mason, C.V.; Hines, M.C. Alpha, beta, and gamma crystallins in the ocular lens of rabbits: Preparation and partial characterization. *Investig. Ophthalmol.* **1966**, *5*, 601–609.

39. Holl, M.; Koziel, R.; Schafer, G.; Pircher, H.; Pauck, A.; Hermann, M.; Klocker, H.; Jansen-Dürr, P.; Sampson, N. ROS signaling by NADPH oxidase 5 modulates the proliferation and survival of prostate carcinoma cells. *Mol. Carcinog.* **2016**, *55*, 27–39. [CrossRef]

40. Marlein, C.R.; Zaitseva, L.; Piddock, R.E.; Robinson, S.D.; Edwards, D.R.; Shafat, M.S.; Zhou, Z.; Lawes, M.; Bowles, K.M.; Rushworth, S.A. NADPH oxidase-2 derived superoxide drives mitochondrial transfer from bone marrow stromal cells to leukemic blasts. *Blood* **2017**, *130*, 1649–1660. [CrossRef]

41. Wang, P.; Shi, Q.; Deng, W.H.; Yu, J.; Zuo, T.; Mei, F.C.; Wang, W.X. Relationship between expression of NADPH oxidase 2 and invasion and prognosis of human gastric cancer. *World J. Gastroenterol.* **2015**, *21*, 6271–6279. [CrossRef] [PubMed]

42. Skonieczna, M.; Hejmo, T.; Poterala-Hejmo, A.; Cieslar-Pobuda, A.; Buldak, R.J. NADPH Oxidases: Insights into Selected Functions and Mechanisms of Action in Cancer and Stem Cells. *Oxid Med. Cell. Longev.* **2017**, *2017*, 9420539. [CrossRef] [PubMed]

43. Grasso, C.; Jansen, G.; Giovannetti, E. Drug resistance in pancreatic cancer: Impact of altered energy metabolism. *Crit. Rev. Oncol. Hematol.* **2017**, *114*, 139–152. [CrossRef] [PubMed]

44. Wang, P.; Sun, Y.C.; Lu, W.H.; Huang, P.; Hu, Y. Selective killing of K-ras-transformed pancreatic cancer cells by targeting NAD(P)H oxidase. *Chin. J. Cancer* **2015**, *34*, 166–176. [CrossRef] [PubMed]

45. Garrido-Laguna, I.; Hidalgo, M. Pancreatic cancer: From state-of-the-art treatments to promising novel therapies. *Nat. Rev. Clin. Oncol.* **2015**, *12*, 319–334. [CrossRef] [PubMed]

46. Weyemi, U.; Redon, C.E.; Parekh, P.R.; Dupuy, C.; Bonner, W.M. NADPH Oxidases NOXs and DUOXs as putative targets for cancer therapy. *Anti-Cancer Agents Med. Chem.* **2013**, *13*, 502–514.

47. Che, M.; Wang, R.; Li, X.; Wang, H.Y.; Zheng, X.F.S. Expanding roles of superoxide dismutases in cell regulation and cancer. *Drug Discov. Today* **2016**, *21*, 143–149. [CrossRef]

48. Dhar, S.K.; St Clair, D.K. Manganese superoxide dismutase regulation and cancer. *Free Radic. Biol. Med.* **2012**, *52*, 2209–2222. [CrossRef]

49. Papa, L.; Hahn, M.; Marsh, E.L.; Evans, B.S.; Germain, D. SOD2 to SOD1 switch in breast cancer. *J. Biol. Chem.* **2014**, *289*, 5412–5416. [CrossRef]

50. Glasauer, A.; Sena, L.A.; Diebold, L.P.; Mazar, A.P.; Chandel, N.S. Targeting SOD1 reduces experimental non-small-cell lung cancer. *J. Clin. Investig.* **2014**, *124*, 117–128. [CrossRef]

51. Oberley, L.W.; Buettner, G.R. Role of superoxide dismutase in cancer: A review. *Cancer Res.* **1979**, *39*, 1141–1149. [PubMed]

52. Hempel, N.; Carrico, P.M.; Melendez, J.A. Manganese superoxide dismutase (Sod2) and redox-control of signaling events that drive metastasis. *Anti-Cancer Agents Med. Chem.* **2011**, *11*, 191–201. [CrossRef]

53. Hempel, N.; Melendez, J.A. Intracellular redox status controls membrane localization of pro- and anti-migratory signaling molecules. *Redox Biol.* **2014**, *2*, 245–250. [CrossRef] [PubMed]

54. Jung, I.; Kim, T.Y.; Kim-Ha, J. Identification of Drosophila SOD3 and its protective role against phototoxic damage to cells. *FEBS Lett.* **2011**, *585*, 1973–1978. [CrossRef] [PubMed]

55. Fisher, A.B. Redox signaling across cell membranes. *Antioxid. Redox Signal.* **2009**, *11*, 1349–1356. [CrossRef] [PubMed]

56. Fukai, T.; Folz, R.J.; Landmesser, U.; Harrison, D.G. Extracellular superoxide dismutase and cardiovascular disease. *Cardiovasc. Res.* **2002**, *55*, 239–249. [CrossRef]

57. Glorieux, C.; Calderon, P.B. Catalase, a remarkable enzyme: Targeting the oldest antioxidant enzyme to find a new cancer treatment approach. *Biol. Chem.* **2017**, *398*, 1095–1108. [CrossRef] [PubMed]

58. Glorieux, C.; Zamocky, M.; Sandoval, J.M.; Verrax, J.; Calderon, P.B. Regulation of catalase expression in healthy and cancerous cells. *Free Radic. Biol. Med.* **2015**, *87*, 84–97. [CrossRef] [PubMed]

59. Rainis, T.; Maor, I.; Lanir, A.; Shnizer, S.; Lavy, A. Enhanced oxidative stress and leucocyte activation in neoplastic tissues of the colon. *Dig. Dis. Sci.* **2007**, *52*, 526–530. [CrossRef]

60. Hwang, T.S.; Choi, H.K.; Han, H.S. Differential expression of manganese superoxide dismutase, copper/zinc superoxide dismutase, and catalase in gastric adenocarcinoma and normal gastric mucosa. *Eur. J. Surg. Oncol.* **2007**, *33*, 474–479. [CrossRef]

61. Sander, C.S.; Hamm, F.; Elsner, P.; Thiele, J.J. Oxidative stress in malignant melanoma and non-melanoma skin cancer. *Br. J. Dermatol.* **2003**, *148*, 913–922. [CrossRef] [PubMed]

62. Yen, H.C.; Li, S.H.; Majima, H.J.; Huang, Y.H.; Chen, C.P.; Liu, C.C.; Tu, Y.C.; Chen, C.W. Up-regulation of antioxidant enzymes and coenzyme Q(10) in a human oral cancer cell line with acquired bleomycin resistance. *Free Radic. Res.* **2011**, *45*, 707–716. [CrossRef] [PubMed]

63. Kuramitsu, Y.; Taba, K.; Ryozawa, S.; Yoshida, K.; Zhang, X.; Tanaka, T.; Maehara, S.I.; Maehara, Y.; Sakaida, I.; Nakamura, K. Identification of up- and down-regulated proteins in gemcitabine-resistant pancreatic cancer cells using two-dimensional gel electrophoresis and mass spectrometry. *Anti-Cancer Res.* **2010**, *30*, 3367–3372.

64. Xu, H.; Choi, S.M.; An, C.S.; Min, Y.D.; Kim, K.C.; Kim, K.J.; Choi, C.H. Concentration-dependent collateral sensitivity of cisplatin-resistant gastric cancer cell sublines. *Biochem. Biophys. Res. Commun.* **2005**, *328*, 618–622. [CrossRef] [PubMed]

65. Xu, B.H.; Gupta, V.; Singh, S.V. Characterization of a human bladder cancer cell line selected for resistance to mitomycin C. *Int. J. Cancer* **1994**, *58*, 686–692. [CrossRef] [PubMed]

66. Kim, Y.; Jang, H.H. Role of Cytosolic 2-Cys Prx1 and Prx2 in Redox Signaling. *Antioxidants* **2019**, *8*, 169. [CrossRef] [PubMed]

67. Sharapov, M.G.; Novoselov, V.I. Catalytic and Signaling Role of Peroxiredoxins in Carcinogenesis. *Biochemistry* **2019**, *84*, 79–100. [CrossRef]

68. Knoops, B.; Goemaere, J.; Van der Eecken, V.; Declercq, J.P. Peroxiredoxin 5: Structure, mechanism, and function of the mammalian atypical 2-Cys peroxiredoxin. *Antioxid. Redox Signal.* **2011**, *15*, 817–829. [CrossRef]

69. Fisher, A.B. Peroxiredoxin 6: A bifunctional enzyme with glutathione peroxidase and phospholipase A(2) activities. *Antioxid. Redox Signal.* **2011**, *15*, 831–844. [CrossRef]

70. Nicolussi, A.; D'Inzeo, S.; Capalbo, C.; Giannini, G.; Coppa, A. The role of peroxiredoxins in cancer. *Mol. Clin. Oncol.* **2017**, *6*, 139–153. [CrossRef]

71. Park, Y.H.; Kim, S.U.; Kwon, T.H.; Kim, J.M.; Song, I.S.; Shin, H.J.; Lee, B.K.; Bang, D.H.; Lee, S.J.; Lee, D.S. Peroxiredoxin II promotes hepatic tumorigenesis through cooperation with Ras/Forkhead box M1 signaling pathway. *Oncogene* **2016**, *35*, 3503–3513. [CrossRef] [PubMed]

72. Lee, D.J.; Kang, D.H.; Choi, M.; Choi, Y.J.; Lee, J.Y.; Park, J.H.; Park, Y.J.; Lee, K.W.; Kang, S.W. Peroxiredoxin-2 represses melanoma metastasis by increasing E-Cadherin/beta-Catenin complexes in adherens junctions. *Cancer Res.* **2013**, *73*, 4744–4757. [CrossRef] [PubMed]

73. Giudice, A.; Arra, C.; Turco, M.C. Review of molecular mechanisms involved in the activation of the Nrf2-ARE signaling pathway by chemopreventive agents. *Methods Mol. Biol.* **2010**, *647*, 37–74. [PubMed]

74. Giudice, A.; Barbieri, A.; Bimonte, S.; Cascella, M.; Cuomo, A.; Crispo, A.; D'Arena, G.; Galdiero, M.; Della Pepa, M.E.; Botti, G.; et al. Dissecting the prevention of estrogen-dependent breast carcinogenesis through Nrf2-dependent and independent mechanisms. *Onco Targets Ther.* **2019**, *12*, 4937–4953. [CrossRef] [PubMed]

75. Grewal, G.K.; Kukal, S.; Kanojia, N.; Saso, L.; Kukreti, S.; Kukreti, R. Effect of Oxidative Stress on ABC Transporters: Contribution to Epilepsy Pharmacoresistance. *Molecules* **2017**, *22*, 365. [CrossRef] [PubMed]

76. Bai, X.; Chen, Y.; Hou, X.; Huang, M.; Jin, J. Emerging role of NRF2 in chemoresistance by regulating drug-metabolizing enzymes and efflux transporters. *Drug Metab. Rev.* **2016**, *48*, 541–567. [CrossRef] [PubMed]

77. Brufsky, A.M. Current Approaches and Emerging Directions in HER2-resistant Breast Cancer. *Breast Cancer* **2014**, *8*, 109–118. [CrossRef]

78. Ganan-Gomez, I.; Wei, Y.; Yang, H.; Boyano-Adanez, M.C.; Garcia-Manero, G. Oncogenic functions of the transcription factor Nrf2. *Free Radic. Biol. Med.* **2013**, *65*, 750–764. [CrossRef]

79. Rushworth, S.A.; Macewan, D.J. The role of nrf2 and cytoprotection in regulating chemotherapy resistance of human leukemia cells. *Cancers* **2011**, *3*, 1605–1621. [CrossRef]

80. Hayes, J.D.; McMahon, M. NRF2 and KEAP1 mutations: Permanent activation of an adaptive response in cancer. *Trends Biochem. Sci.* **2009**, *34*, 176–188. [CrossRef]

81. Marin, J.J.G.; Cives-Losada, C.; Asensio, M.; Lozano, E.; Briz, O.; Macias, R.I.R. Mechanisms of Anticancer Drug Resistance in Hepatoblastoma. *Cancers* **2019**, *11*, 407. [CrossRef] [PubMed]

82. Housman, G.; Byler, S.; Heerboth, S.; Lapinska, K.; Longacre, M.; Snyder, N.; Sarkar, S. Drug resistance in cancer: An overview. *Cancers* **2014**, *6*, 1769–1792. [CrossRef] [PubMed]

83. Bruno, P.M.; Liu, Y.; Park, G.Y.; Murai, J.; Koch, C.E.; Eisen, T.J.; Pritchard, J.R.; Pommier, Y.; Lippard, S.J.; Hemann, M.T. A subset of platinum-containing chemotherapeutic agents kills cells by inducing ribosome biogenesis stress. *Nat. Med.* **2017**, *23*, 461–471. [CrossRef] [PubMed]

84. Zhang, Y.; Huang, L.; Shi, H.; Chen, H.; Tao, J.; Shen, R.; Wang, T. Ursolic acid enhances the therapeutic effects of oxaliplatin in colorectal cancer by inhibition of drug resistance. *Cancer Sci.* **2018**, *109*, 94–102. [CrossRef] [PubMed]

85. Shi, Y.; Tang, B.; Yu, P.W.; Tang, B.; Hao, Y.X.; Lei, X.; Luo, H.X.; Zeng, D.Z. Autophagy protects against oxaliplatin-induced cell death via ER stress and ROS in Caco-2 cells. *PLoS ONE* **2012**, *7*, e51076. [CrossRef] [PubMed]

86. Ding, Z.B.; Hui, B.; Shi, Y.H.; Zhou, J.; Peng, Y.F.; Gu, C.Y.; Yang, H.; Shi, G.M.; Ke, A.W.; Wang, X.Y. Autophagy activation in hepatocellular carcinoma contributes to the tolerance of oxaliplatin via reactive oxygen species modulation. *Clin. Cancer Res.* **2011**, *17*, 6229–6238. [CrossRef]

87. Canta, A.; Pozzi, E.; Carozzi, V.A. Mitochondrial Dysfunction in Chemotherapy-Induced Peripheral Neuropathy (CIPN). *Toxics* **2015**, *3*, 198–223. [CrossRef]

88. Kelley, M.R.; Jiang, Y.; Guo, C.; Reed, A.; Meng, H.; Vasko, M.R. Role of the DNA base excision repair protein, APE1 in cisplatin, oxaliplatin, or carboplatin induced sensory neuropathy. *PLoS ONE* **2014**, *9*, e106485. [CrossRef]

89. Plasencia, C.; Martinez-Balibrea, E.; Martinez-Cardus, A.; Quinn, D.I.; Abad, A.; Neamati, N. Expression analysis of genes involved in oxaliplatin response and development of oxaliplatin-resistant HT29 colon cancer cells. *Int. J. Oncol.* **2006**, *29*, 225–235. [CrossRef]

90. Pirpour Tazehkand, A.; Akbarzadeh, M.; Velaie, K.; Sadeghi, M.R.; Samadi, N. The role of Her2-Nrf2 axis in induction of oxaliplatin resistance in colon cancer cells. *Biomed. Pharmacother.* **2018**, *103*, 755–766. [CrossRef]

91. Chian, S.; Li, Y.Y.; Wang, X.J.; Tang, X.W. Luteolin sensitizes two oxaliplatin-resistant colorectal cancer cell lines to chemotherapeutic drugs via inhibition of the Nrf2 pathway. *Asian Pac. J. Cancer Prev.* **2014**, *15*, 2911–2916. [CrossRef] [PubMed]

92. Robinson, S.M.; Mann, J.; Vasilaki, A.; Mathers, J.; Burt, A.D.; Oakley, F.; White, S.A.; Mann, D.A. Pathogenesis of FOLFOX induced sinusoidal obstruction syndrome in a murine chemotherapy model. *J. Hepatol.* **2013**, *59*, 318–326. [CrossRef] [PubMed]

93. Longley, D.B.; Harkin, D.P.; Johnston, P.G. 5-fluorouracil: Mechanisms of action and clinical strategies. *Nat. Rev. Cancer* **2003**, *3*, 330–338. [CrossRef] [PubMed]

94. Yao, C.W.; Kang, K.A.; Piao, M.J.; Ryu, Y.S.; Fernando, P.; Oh, M.C.; Park, J.E.; Shilinkova, K.; Na, S.Y.; Jeong, S.U.; et al. Reduced Autophagy in 5-Fluorouracil Resistant Colon Cancer Cells. *Biomol. Ther. (Seoul)* **2017**, *25*, 315–320. [CrossRef] [PubMed]

95. Kang, K.A.; Piao, M.J.; Kim, K.C.; Kang, H.K.; Chang, W.Y.; Park, I.C.; Keum, Y.S.; Surh, Y.J.; Hyun, J.W. Epigenetic modification of Nrf2 in 5-fluorouracil-resistant colon cancer cells: Involvement of TET-dependent DNA demethylation. *Cell Death Dis.* **2014**, *5*, e1183. [CrossRef] [PubMed]

96. Kang, K.A.; Ryu, Y.S.; Piao, M.J.; Shilnikova, K.; Kang, H.K.; Yi, J.M.; Boulanger, M.; Paolillo, R.; Bossis, G.; Yoon, S.Y.; et al. DUOX2-mediated production of reactive oxygen species induces epithelial mesenchymal transition in 5-fluorouracil resistant human colon cancer cells. *Redox Biol.* **2018**, *17*, 224–235. [CrossRef]

97. Kim, J.K.; Kang, K.A.; Piao, M.J.; Ryu, Y.S.; Han, X.; Fernando, P.M.; Oh, M.C.; Park, J.E.; Shilnilova, K.; Boo, S.Y.; et al. Endoplasmic reticulum stress induces 5-fluorouracil resistance in human colon cancer cells. *Environ. Toxicol. Pharmacol.* **2016**, *44*, 128–133. [CrossRef]

98. Chan, J.Y.; Phoo, M.S.; Clement, M.V.; Pervaiz, S.; Lee, S.C. Resveratrol displays converse dose-related effects on 5-fluorouracil-evoked colon cancer cell apoptosis: The roles of caspase-6 and p53. *Cancer Biol. Ther.* **2008**, *7*, 1305–1312. [CrossRef]

99. Hwang, P.M.; Bunz, F.; Yu, J.; Rago, C.; Chan, T.A.; Murphy, M.P.; Kelso, G.F.; Smith, R.A.; Kinzler, K.W.; Vogelstein, B. Ferredoxin reductase affects p53-dependent, 5-fluorouracil-induced apoptosis in colorectal cancer cells. *Nat. Med.* **2001**, *7*, 1111–1117. [CrossRef]

100. Zhang, N.; Yin, Y.; Xu, S.J.; Chen, W.S. 5-Fluorouracil: Mechanisms of resistance and reversal strategies. *Molecules* **2008**, *13*, 1551–1569. [CrossRef]

101. Shin, Y.K.; Yoo, B.C.; Chang, H.J.; Jeon, E.; Hong, S.H.; Jung, M.S.; Lim, S.J.; Park, J.G. Down-regulation of mitochondrial F1F0-ATP synthase in human colon cancer cells with induced 5-fluorouracil resistance. *Cancer Res.* **2005**, *65*, 3162–3170. [CrossRef] [PubMed]

102. Hwang, I.T.; Chung, Y.M.; Kim, J.J.; Chung, J.S.; Kim, B.S.; Kim, H.J.; Kim, J.S.; Yoo, Y.D. Drug resistance to 5-FU linked to reactive oxygen species modulator 1. *Biochem. Biophys. Res. Commun.* **2007**, *359*, 304–310. [CrossRef] [PubMed]

103. Shi, L.; Wu, L.; Chen, Z.; Yang, J.; Chen, X.; Yu, F.; Zheng, F.; Lin, X. MiR-141 Activates Nrf2-Dependent Antioxidant Pathway via Down-Regulating the Expression of Keap1 Conferring the Resistance of Hepatocellular Carcinoma Cells to 5-Fluorouracil. *Cell. Physiol. Biochem.* **2015**, *35*, 2333–2348. [CrossRef] [PubMed]

104. Del Vecchio, C.A.; Feng, Y.; Sokol, E.S.; Tillman, E.J.; Sanduja, S.; Reinhardt, F.; Gupta, P.B. De-differentiation confers multidrug resistance via noncanonical PERK-Nrf2 signaling. *PLoS Biol.* **2014**, *12*, e1001945. [CrossRef] [PubMed]

105. Hu, X.F.; Yao, J.; Gao, S.G.; Wang, X.S.; Peng, X.Q.; Yang, Y.T.; Feng, X.S. Nrf2 overexpression predicts prognosis and 5-FU resistance in gastric cancer. *Asian Pac. J. Cancer Prev.* **2013**, *14*, 5231–5235. [CrossRef] [PubMed]

106. Kang, K.A.; Piao, M.J.; Ryu, Y.S.; Kang, H.K.; Chang, W.Y.; Keum, Y.S.; Hyun, J.W. Interaction of DNA demethylase and histone methyltransferase upregulates Nrf2 in 5-fluorouracil-resistant colon cancer cells. *Oncotarget* **2016**, *7*, 40594–40620. [CrossRef]

107. Akhdar, H.; Loyer, P.; Rauch, C.; Corlu, A.; Guillouzo, A.; Morel, F. Involvement of Nrf2 activation in resistance to 5-fluorouracil in human colon cancer HT-29 cells. *Eur. J. Cancer* **2009**, *45*, 2219–2227. [CrossRef]

108. Alimbetov, D.; Askarova, S.; Umbayev, B.; Davis, T.; Kipling, D. Pharmacological Targeting of Cell Cycle, Apoptotic and Cell Adhesion Signaling Pathways Implicated in Chemoresistance of Cancer Cells. *Int. J. Mol. Sci.* **2018**, *19*, 1690. [CrossRef]

109. Jiang, J.; Wang, K.; Chen, Y.; Chen, H.; Nice, E.C.; Huang, C. Redox regulation in tumor cell epithelial-mesenchymal transition: Molecular basis and therapeutic strategy. *Signal Transduct. Target. Ther.* **2017**, *2*, 17036. [CrossRef]

110. Shen, Y.; Yang, J.; Zhao, J.; Xiao, C.; Xu, C.; Xiang, Y. The switch from ER stress-induced apoptosis to autophagy via ROS-mediated JNK/p62 signals: A survival mechanism in methotrexate-resistant choriocarcinoma cells. *Exp. Cell Res.* **2015**, *334*, 207–218. [CrossRef]

111. Dayem, A.A.; Choi, H.Y.; Kim, J.H.; Cho, S.G. Role of oxidative stress in stem, cancer, and cancer stem cells. *Cancers* **2010**, *2*, 859–884. [CrossRef] [PubMed]

112. Burhans, W.C.; Heintz, N.H. The cell cycle is a redox cycle: Linking phase-specific targets to cell fate. *Free Radic. Biol. Med.* **2009**, *47*, 1282–1293. [CrossRef] [PubMed]

113. Salaroglio, I.C.; Panada, E.; Moiso, E.; Buondonno, I.; Provero, P.; Rubinstein, M.; Kopecka, J.; Riganti, C. PERK induces resistance to cell death elicited by endoplasmic reticulum stress and chemotherapy. *Mol. Cancer* **2017**, *16*, 91. [CrossRef] [PubMed]

114. Madden, E.; Logue, S.E.; Healy, S.J.; Manie, S.; Samali, A. The role of the unfolded protein response in cancer progression: From oncogenesis to chemoresistance. *Biol. Cell* **2019**, *111*, 1–17. [CrossRef] [PubMed]

115. Huang, Z.; Zhou, L.; Chen, Z.; Nice, E.C.; Huang, C. Stress management by autophagy: Implications for chemoresistance. *Int. J. Cancer* **2016**, *139*, 23–32. [CrossRef] [PubMed]

116. Wang, M.; Kaufman, R.J. The impact of the endoplasmic reticulum protein-folding environment on cancer development. *Nat. Rev. Cancer* **2014**, *14*, 581–597. [CrossRef] [PubMed]

117. Dufey, E.; Sepulveda, D.; Rojas-Rivera, D.; Hetz, C. Cellular mechanisms of endoplasmic reticulum stress signaling in health and disease. 1. An overview. *Am. J. Physiol. Cell Physiol.* **2014**, *307*, C582–C594. [CrossRef]

118. Song, S.; Tan, J.; Miao, Y.; Zhang, Q. Crosstalk of ER stress-mediated autophagy and ER-phagy: Involvement of UPR and the core autophagy machinery. *J. Cell. Physiol.* **2018**, *233*, 3867–3874. [CrossRef]

119. Healy, S.J.; Gorman, A.M.; Mousavi-Shafaei, P.; Gupta, S.; Samali, A. Targeting the endoplasmic reticulum-stress response as an anticancer strategy. *Eur. J. Pharmacol.* **2009**, *625*, 234–246. [CrossRef]

120. He, L.; He, T.; Farrar, S.; Ji, L.; Liu, T.; Ma, X. Antioxidants Maintain Cellular Redox Homeostasis by Elimination of Reactive Oxygen Species. *Cell Physiol. Biochem.* **2017**, *44*, 532–553. [CrossRef]

121. Ma, X.; Zhang, S.; He, L.; Rong, Y.; Brier, L.W.; Sun, Q.; Liu, R.; Fan, W.; Chen, S.; Yue, S.; et al. MTORC1-mediated NRBF2 phosphorylation functions as a switch for the class III PtdIns3K and autophagy. *Autophagy* **2017**, *13*, 592–607. [CrossRef] [PubMed]

122. Woo, J.; Park, E.; Dinesh-Kumar, S.P. Differential processing of Arabidopsis ubiquitin-like Atg8 autophagy proteins by Atg4 cysteine proteases. *Proc. Natl. Acad. Sci. USA* **2014**, *111*, 863–868. [CrossRef] [PubMed]

123. Azad, M.B.; Chen, Y.; Gibson, S.B. Regulation of autophagy by reactive oxygen species (ROS): Implications for cancer progression and treatment. *Antioxid. Redox Signal.* **2009**, *11*, 777–790. [CrossRef] [PubMed]

124. Scherz-Shouval, R.; Shvets, E.; Fass, E.; Shorer, H.; Gil, L.; Elazar, Z. Reactive oxygen species are essential for autophagy and specifically regulate the activity of Atg4. *EMBO J.* **2007**, *26*, 1749–1760. [CrossRef] [PubMed]

125. Zhang, X.; Cheng, X.; Yu, L.; Yang, J.; Calvo, R.; Patnaik, S.; Hu, X.; Gao, Q.; Yang, M.; Lawas, M.; et al. MCOLN1 is a ROS sensor in lysosomes that regulates autophagy. *Nat. Commun.* **2016**, *7*, 12109. [CrossRef] [PubMed]

126. Scherz-Shouval, R.; Shvets, E.; Elazar, Z. Oxidation as a post-translational modification that regulates autophagy. *Autophagy* **2007**, *3*, 371–373. [CrossRef] [PubMed]

127. Gines, A.; Bystrup, S.; Ruiz de Porras, V.; Guardia, C.; Musulen, E.; Martinez-Cardus, A.; Manzano, J.L.; Layos, L.; Abad, A.; Martinez-Balibrea, E. PKM2 Subcellular Localization Is Involved in Oxaliplatin Resistance Acquisition in HT29 Human Colorectal Cancer Cell Lines. *PLoS ONE* **2015**, *10*, e0123830. [CrossRef]

128. Delgado, M.; Tesfaigzi, Y. Is BMF central for anoikis and autophagy? *Autophagy* **2014**, *10*, 168–169. [CrossRef]

129. Tischner, D.; Manzl, C.; Soratroi, C.; Villunger, A.; Krumschnabel, G. Necrosis-like death can engage multiple pro-apoptotic Bcl-2 protein family members. *Apoptosis* **2012**, *17*, 1197–1209. [CrossRef]

130. Grespi, F.; Soratroi, C.; Krumschnabel, G.; Sohm, B.; Ploner, C.; Geley, S.; Hengst, L.; Hacker, G.; Villunnger, A. BH3-only protein Bmf mediates apoptosis upon inhibition of CAP-dependent protein synthesis. *Cell Death Differ.* **2010**, *17*, 1672–1683. [CrossRef]

131. Du, H.; Yang, W.; Chen, L.; Shi, M.; Seewoo, V.; Wang, J.; Lin, A.; Liu, Z.; Qui, W. Role of autophagy in resistance to oxaliplatin in hepatocellular carcinoma cells. *Oncol. Rep.* **2012**, *27*, 143–150. [PubMed]

132. Das, C.K.; Linder, B.; Bonn, F.; Rothweiler, F.; Dikic, I.; Michaelis, M.; Cinatl, J.; Mandal, M.; Kogel, D. BAG3 Overexpression and Cytoprotective Autophagy Mediate Apoptosis Resistance in Chemoresistant Breast Cancer Cells. *Neoplasia* **2018**, *20*, 263–279. [CrossRef] [PubMed]

133. Sherr, C.J.; Roberts, J.M. CDK inhibitors: Positive and negative regulators of G1-phase progression. *Genes Dev.* **1999**, *13*, 1501–1512. [CrossRef] [PubMed]

134. Campbell, P.J.; Yachida, S.; Mudie, L.J.; Stephens, P.J.; Pleasance, E.D.; Stebbings, L.A.; Morsberger, L.A.; Latimer, C.; McLaren, S.; Lin, M.L.; et al. The patterns and dynamics of genomic instability in metastatic pancreatic cancer. *Nature* **2010**, *467*, 1109–1113. [CrossRef]

135. Menon, S.G.; Goswami, P.C. A redox cycle within the cell cycle: Ring in the old with the new. *Oncogene* **2007**, *26*, 1101–1109. [CrossRef] [PubMed]

136. Maziere, C.; Trecherel, E.; Ausseil, J.; Louandre, C.; Maziere, J.C. Oxidized low density lipoprotein induces cyclin A synthesis. Involvement of ERK, JNK and NFkappaB. *Atherosclerosis* **2011**, *218*, 308–313. [CrossRef]

137. Tickner, J.; Fan, L.M.; Du, J.; Meijles, D.; Li, J.M. Nox2-derived ROS in PPARgamma signaling and cell-cycle progression of lung alveolar epithelial cells. *Free Radic. Biol. Med.* **2011**, *51*, 763–772. [CrossRef]

138. Havens, C.G.; Ho, A.; Yoshioka, N.; Dowdy, S.F. Regulation of late G1/S phase transition and APC Cdh1 by reactive oxygen species. *Mol. Cell. Biol.* **2006**, *26*, 4701–4711. [CrossRef]

139. Deshpande, S.S.; Irani, K. Oxidant signalling in carcinogenesis: A commentary. *Hum. Exp. Toxicol.* **2002**, *21*, 63–64. [CrossRef]

140. Stamatakos, M.; Palla, V.; Karaiskos, I.; Xiromeritis, K.; Alexiou, I.; Pateras, I.; Kontzoglou, K. Cell cyclins: Triggering elements of cancer or not? *World J. Surg. Oncol.* **2010**, *8*, 111. [CrossRef]

141. Fu, M.; Wang, C.; Li, Z.; Sakamaki, T.; Pestell, R.G. Minireview: Cyclin D1: Normal and abnormal functions. *Endocrinology* **2004**, *145*, 5439–5447. [CrossRef] [PubMed]

142. Imanishi, Y.; Hosokawa, Y.; Yoshimoto, K.; Schipani, E.; Mallya, S.; Papanikolaou, A.; Kifor, O.; Tokura, T.; Sablosky, M.; Ledgard, F.; et al. Primary hyperparathyroidism caused by parathyroid-targeted overexpression of cyclin D1 in transgenic mice. *J. Clin. Investig.* **2001**, *107*, 1093–1102. [CrossRef] [PubMed]

143. Yasui, M.; Yamamoto, H.; Ngan, C.Y.; Damdinsuren, B.; Sugita, Y.; Fukunaga, H.; Gu, J.; Maeda, M.; Takemasa, I.; Ikeda, M.; et al. Antisense to cyclin D1 inhibits vascular endothelial growth factor-stimulated growth of vascular endothelial cells: Implication of tumor vascularization. *Clin. Cancer Res.* **2006**, *12*, 4720–4729. [CrossRef] [PubMed]

144. Sakamaki, T.; Casimiro, M.C.; Ju, X.; Quong, A.A.; Katiyar, S.; Liu, M.; Jiao, X.; Li, A.; Zhang, X.; Lu, Y.; et al. Cyclin D1 determines mitochondrial function in vivo. *Mol. Cell. Biol.* **2006**, *26*, 5449–5469. [CrossRef] [PubMed]

145. Hsu, H.H.; Chen, M.C.; Baskaran, R.; Lin, Y.M.; Day, C.H.; Lin, Y.J.; Tu, C.C.; Vijaya Padma, V.; Kuo, W.W.; Huang, C.Y. Oxaliplatin resistance in colorectal cancer cells is mediated via activation of ABCG2 to alleviate ER stress induced apoptosis. *J. Cell Physiol.* **2018**, *233*, 5458–5467. [CrossRef]

146. El Khoury, F.; Corcos, L.; Durand, S.; Simon, B.; Le Jossic-Corcos, C. Acquisition of anticancer drug resistance is partially associated with cancer stemness in human colon cancer cells. *Int. J. Oncol.* **2016**, *49*, 2558–2568. [CrossRef] [PubMed]

147. Guo, X.; Goessl, E.; Jin, G.; Collie-Duguid, E.S.; Cassidy, J.; Wang, W.; O'Brien, W. Cell cycle perturbation and acquired 5-fluorouracil chemoresistance. *Anti-Cancer Res.* **2008**, *28*, 9–14.

148. Wang, W.B.; Yang, Y.; Zhao, Y.P.; Zhang, T.P.; Liao, Q.; Shu, H. Recent studies of 5-fluorouracil resistance in pancreatic cancer. *World J. Gastroenterol.* **2014**, *20*, 15682–15690. [CrossRef]

149. Wang, W.; Cassidy, J.; O'Brien, V.; Ryan, K.M.; Collie-Duguid, E. Mechanistic and predictive profiling of 5-Fluorouracil resistance in human cancer cells. *Cancer Res.* **2004**, *64*, 8167–8176. [CrossRef]

150. Li, M.H.; Ito, D.; Sanada, M.; Odani, T.; Hatori, M.; Iwase, M.; Nagumo, M. Effect of 5-fluorouracil on G1 phase cell cycle regulation in oral cancer cell lines. *Oral Oncol.* **2004**, *40*, 63–70. [CrossRef]

151. Phi, L.T.H.; Sari, I.N.; Yang, Y.G.; Lee, S.H.; Jun, N.; Kim, K.S.; Lee, Y.K.; Kwon, H.Y. Cancer Stem Cells (CSCs) in Drug Resistance and their Therapeutic Implications in Cancer Treatment. *Stem Cells Int.* **2018**, *2018*, 5416923. [CrossRef] [PubMed]

152. Du, B.; Shim, J.S. Targeting Epithelial-Mesenchymal Transition (EMT) to Overcome Drug Resistance in Cancer. *Molecules* **2016**, *21*, 965. [CrossRef] [PubMed]

153. Singh, A.; Settleman, J. EMT, cancer stem cells and drug resistance: An emerging axis of evil in the war on cancer. *Oncogene* **2010**, *29*, 4741–4751. [CrossRef] [PubMed]

154. Hermann, P.C.; Huber, S.L.; Herrler, T.; Aicher, A.; Ellwart, J.W.; Guba, M.; Bruns, C.J.; Heeschen, C. Distinct populations of cancer stem cells determine tumor growth and metastatic activity in human pancreatic cancer. *Cell Stem Cell* **2007**, *1*, 313–323. [CrossRef] [PubMed]

155. Wielenga, V.J.; Smits, R.; Korinek, V.; Smit, L.; Kielman, M.; Fodde, R.; Clevers, H.; Pals, S.T. Expression of CD44 in Apc and Tcf mutant mice implies regulation by the WNT pathway. *Am. J. Pathol.* **1999**, *154*, 515–523. [CrossRef]

156. Shibue, T.; Weinberg, R.A. EMT, CSCs, and drug resistance: The mechanistic link and clinical implications. *Nat. Rev. Clin. Oncol.* **2017**, *14*, 611–629. [CrossRef] [PubMed]

157. Kojima, H.; Okumura, T.; Yamaguchi, T.; Miwa, T.; Shimada, Y.; Nagata, T. Enhanced cancer stem cell properties of a mitotically quiescent subpopulation of p75NTR-positive cells in esophageal squamous cell carcinoma. *Int. J. Oncol.* **2017**, *51*, 49–62. [CrossRef] [PubMed]

158. Wang, J.; Liu, X.; Jiang, Z.; Li, L.; Cui, Z.; Gao, Y.; Kong, D.; Liu, X. A novel method to limit breast cancer stem cells in states of quiescence, proliferation or differentiation: Use of gel stress in combination with stem cell growth factors. *Oncol. Lett.* **2016**, *12*, 1355–1360. [CrossRef]

159. Begicevic, R.R.; Falasca, M. ABC Transporters in Cancer Stem Cells: Beyond Chemoresistance. *Int. J. Mol. Sci.* **2017**, *18*, 2362. [CrossRef]

160. Jiao, L.; Li, D.D.; Yang, C.L.; Peng, R.Q.; Guo, Y.Q.; Zhang, X.S.; Zhu, X.F. Reactive oxygen species mediate oxaliplatin-induced epithelial-mesenchymal transition and invasive potential in colon cancer. *Tumour Biol.* **2016**, *37*, 8413–8423. [CrossRef]

161. Yang, A.D.; Fan, F.; Camp, E.R.; van Buren, G.; Liu, W.; Somcio, R.; Gray, M.J.; Cheng, H.; Hoff, P.M.; Ellis, L.M. Chronic oxaliplatin resistance induces epithelial-to-mesenchymal transition in colorectal cancer cell lines. *Clin. Cancer Res.* **2006**, *12*, 4147–4153. [CrossRef] [PubMed]

162. Liao, X.; Bu, Y.; Jiang, S.; Chang, F.; Jia, F.; Xiao, X.; Song, G.; Zhang, M.; Ning, P.; Jia, Q. CCN2-MAPK-Id-1 loop feedback amplification is involved in maintaining stemness in oxaliplatin-resistant hepatocellular carcinoma. *Hepatol. Int.* **2019**, *13*, 440–453. [CrossRef]

163. Yin, X.; Tang, B.; Li, J.H.; Wang, Y.; Zhang, L.; Xie, X.Y.; Zhang, B.H.; Qiu, S.J.; Wu, W.Z.; Ren, Z.G. ID1 promotes hepatocellular carcinoma proliferation and confers chemoresistance to oxaliplatin by activating pentose phosphate pathway. *J. Exp. Clin. Cancer Res.* **2017**, *36*, 166. [CrossRef] [PubMed]

164. Dhar, A.; Ray, A. The CCN family proteins in carcinogenesis. *Exp. Oncol.* **2010**, *32*, 2–9. [PubMed]

165. Yang, Y.; Yao, J.H.; Du, Q.Y.; Zhou, Y.C.; Yao, T.J.; Wu, Q.; Liu, J.; Ou, Y.R. Connexin 32 downregulation is critical for chemoresistance in oxaliplatin-resistant HCC cells associated with EMT. *Cancer Manag. Res.* **2019**, *11*, 5133–5146. [CrossRef] [PubMed]

166. Fujimoto, E.; Sato, H.; Shirai, S.; Nagashima, Y.; Fukumoto, K.; Hagiwara, H.; Negishi, E.; Ueno, K.; Omori, Y.; Yamasaki, H.; et al. Connexin32 as a tumor suppressor gene in a metastatic renal cell carcinoma cell line. *Oncogene* **2005**, *24*, 3684–3690. [CrossRef] [PubMed]

167. Lin, L.; Li, X.; Pan, C.; Lin, W.; Shao, R.; Liu, Y.; Zhang, J.; Luo, Y.; Qian, K.; Shi, M.; et al. ATXN2L upregulated by epidermal growth factor promotes gastric cancer cell invasiveness and oxaliplatin resistance. *Cell Death Dis.* **2019**, *10*, 173. [CrossRef] [PubMed]

168. Kaehler, C.; Isensee, J.; Nonhoff, U.; Terrey, M.; Hucho, T.; Lehrach, H.; Krobitsch, S. Ataxin-2-like is a regulator of stress granules and processing bodies. *PLoS ONE* **2012**, *7*, e50134. [CrossRef]

169. Yang, D.; Wang, H.; Zhang, J.; Li, C.; Lu, Z.; Liu, J.; Lin, C.; Li, G.; Qian, H. In vitro characterization of stem cell-like properties of drug-resistant colon cancer subline. *Oncol. Res.* **2013**, *21*, 51–57. [CrossRef]

170. Kim, A.Y.; Kwak, J.H.; Je, N.K.; Lee, Y.H.; Jung, Y.S. Epithelial-mesenchymal Transition is Associated with Acquired Resistance to 5-Fluorocuracil in HT-29 Colon Cancer Cells. *Toxicol. Res.* **2015**, *31*, 151–156. [CrossRef]

171. Harada, K.; Ferdous, T.; Ueyama, Y. Establishment of 5-fluorouracil-resistant oral squamous cell carcinoma cell lines with epithelial to mesenchymal transition changes. *Int. J. Oncol.* **2014**, *44*, 1302–1308. [CrossRef] [PubMed]

172. Sun, L.; Ke, J.; He, Z.; Chen, Z.; Huang, Q.; Ai, W.; Wang, G.; Wei, Y.; Zou, X.; Zhang, S.; et al. HES1 Promotes Colorectal Cancer Cell Resistance To 5-Fu by Inducing Of EMT and ABC Transporter Proteins. *J. Cancer* **2017**, *8*, 2802–2808. [CrossRef] [PubMed]

173. Zhang, C.; Ma, Q.; Shi, Y.; Li, X.; Wang, M.; Wang, J.; Ge, J.; Chen, Z.; Wang, Z.; Jiang, H. A novel 5-fluorouracil-resistant human esophageal squamous cell carcinoma cell line Eca-109/5-FU with significant drug resistance-related characteristics. *Oncol. Rep.* **2017**, *37*, 2942–2954. [CrossRef]

174. Vishnoi, K.; Mahata, S.; Tyagi, A.; Pandey, A.; Verma, G.; Jadli, M.; Singh, T.; Singh, S.M.; Bharti, A.C. Human papillomavirus oncoproteins differentially modulate epithelial-mesenchymal transition in 5-FU-resistant cervical cancer cells. *Tumour Biol.* **2016**, *37*, 13137–13154. [CrossRef] [PubMed]

175. Zhang, W.; Feng, M.; Zheng, G.; Chen, Y.; Wang, X.; Pen, B.; Yin, J.; Yu, Y.; He, Z. Chemoresistance to 5-fluorouracil induces epithelial-mesenchymal transition via up-regulation of Snail in MCF7 human breast cancer cells. *Biochem. Biophys. Res. Commun.* **2012**, *417*, 679–685. [CrossRef] [PubMed]

176. Izumiya, M.; Kabashima, A.; Higuchi, H.; Igarashi, T.; Sakai, G.; Iizuka, H.; Nakamura, S.; Adachi, M.; Hamamoto, Y.; Funakoshi, S.; et al. Chemoresistance is associated with cancer stem cell-like properties and epithelial-to-mesenchymal transition in pancreatic cancer cells. *Anti-Cancer Res.* **2012**, *32*, 3847–3853.

177. Miyoshi, S.; Tsugawa, H.; Matsuzaki, J.; Hirata, K.; Mori, H.; Saya, H.; Kanai, T.; Suzuki, H. Inhibiting xCT Improves 5-Fluorouracil Resistance of Gastric Cancer Induced by CD44 Variant 9 Expression. *Anti-Cancer Res.* **2018**, *38*, 6163–6170. [CrossRef] [PubMed]

178. Terzioglu, G.; Turksoy, O.; Bayrak, O.F. Identification of An mtDNA Setpoint Associated with Highest Levels of CD44 Positivity and Chemoresistance in HGC-27 and MKN-45 Gastric Cancer Cell Lines. *Cell J.* **2018**, *20*, 312–317. [PubMed]

179. Denise, C.; Paoli, P.; Calvani, M.; Taddei, M.L.; Giannoni, E.; Kopetz, S.; Kazmi, S.M.; Pia, M.M.; Pettazzoni, P.; Sacco, E.; et al. 5-fluorouracil resistant colon cancer cells are addicted to OXPHOS to survive and enhance stem-like traits. *Oncotarget* **2015**, *6*, 41706–41721. [CrossRef] [PubMed]

 *antioxidants*

Article

Chronic Oxidative Stress Promotes Molecular Changes Associated with Epithelial Mesenchymal Transition, NRF2, and Breast Cancer Stem Cell Phenotype

Ana Čipak Gašparović [1,*,†], Lidija Milković [1,†], Nadia Dandachi [2], Stefanie Stanzer [2], Iskra Pezdirc [3], Josip Vrančić [4,5], Sanda Šitić [6], Christoph Suppan [2] and Marija Balic [2,*]

[1] Division of Molecular Medicine, Ruđer Bošković Institute, HR-10000 Zagreb, Croatia; lidija.milkovic@irb.hr

[2] Department of Internal Medicine, Division of Oncology, Medical University, Graz 8036, Austria; nadia.dandachi@medunigraz.at (N.D.); stefanie.stanzer@gmail.com (S.S.); christoph.suppan@klinikum-graz.at (C.S.)

[3] Outhospital Emergency Medicine Department of Krapina Zagorje County, HR-49000 Krapina, Croatia; iskra1511@gmail.com

[4] Institute of Cancer Sciences, University of Glasgow, Glasgow G12 8QQ, UK; josip.vrancic@glasgow.ac.uk

[5] Cancer Research UK Beatson Institute, Glasgow G61 1BD, UK

[6] Sestre milosrdnice University Hospital Centre, University Hospital for Tumors, HR-10000 Zagreb, Croatia, sanda.sitic@yahoo.com

* Correspondence: acipak@irb.hr (A.Č.G.); Marija.Balic@klinikum-graz.at (M.B.)

† Both authors contributed equally.

Received: 4 November 2019; Accepted: 10 December 2019; Published: 11 December 2019

Abstract: Oxidative stress plays a role in carcinogenesis, but it also contributes to the modulation of tumor cells and microenvironment caused by chemotherapeutics. One of the consequences of oxidative stress is lipid peroxidation, which can, through reactive aldehydes such as 4-hydroxy-2-nonenal (HNE), affect cell signaling pathways. On the other hand, cancer stem cells (CSC) are now recognized as a major factor of malignancy by causing metastasis, relapse, and therapy resistance. Here, we evaluated whether oxidative stress and HNE modulation of the microenvironment can influence CSC growth, modifications of the epithelial to mesenchymal transition (EMT) markers, the antioxidant system, and the frequency of breast cancer stem cells (BCSC). Our results showed that oxidative changes in the microenvironment of BCSC and particularly chronic oxidative stress caused changes in the proliferation and growth of breast cancer cells. In addition, changes associated with EMT, increase in glutathione (GSH) and Nuclear factor erythroid 2-related factor 2 (NRF2) were observed in breast cancer cells grown on HNE pretreated collagen and under chronic oxidative stress. Our results suggest that chronic oxidative stress can be a bidirectional modulator of BCSC fate. Low levels of HNE can increase differentiation markers in BCSC, while higher levels increased GSH and NRF2 as well as certain EMT markers, thereby increasing therapy resistance.

Keywords: breast cancer stem cells; 4-hydroxy-2-nonenal; extracellular matrix; NRF2

---

1. Introduction

Tumor cell heterogeneity has been a known fact for a long time, but evidence increasingly suggests that heterogeneity of tumors may be associated with a subpopulation of tumor-initiating cells, also called cancer stem cells (CSCs), as a subpopulation driving tumorigenesis and cancer progression [1]. These cells represent only a small proportion of tumor mass, but seem to have the capability of dissemination and may, for still unknown reasons, reactivate from the quiescent state and cause recurrence of the disease [2,3]. The fate of CSC seems to be highly dependent on their niche and state, either activating or quiescent, which may be determined by their microenvironment. This concept of the tumor being dependent on its microenvironment has been postulated early by Stephen Paget [4], and by clinical trials demonstrating that therapeutic interventions with bisphosphonates positively impact the clinical outcome of breast cancer patients, and confirmed the importance of these interactions [5,6]. Today, a wide array of evidence suggests that the network of interactions between the tumor, the microenvironment with the stroma, the extracellular matrix and the inflammatory cells bidirectionally modulate their tumorigenicity [7,8]. Despite recent advances, interactions between CSC and the microenvironment are difficult to study due to a lack of optimal methods for the isolation of CSC and efficient functional assays, as well as due to a variety of proteins, enzymes, and growth/inhibition factors forming the extracellular matrix (ECM) of the tumor and the CSC niche. In vitro sphere formation assays have been shown to be suitable surrogate models to study CSC biology [9,10].

Numerous factors govern cell growth to generate CSC, and epithelial to mesenchymal transition (EMT) is the process that strongly supports and/or generates the CSC phenotype [11]. EMT is a process normally occurring in embryological development, but if awakened latter in the adult organism, it becomes pathological and generates mesenchymal cells with the ability to migrate [11]. This process is reversible, but in the means of cancer, it is highly undesirable, and EMT and back to mesenchymal to epithelial transition (MET) is the process that causes metastasis [12]. EMT is accompanied by changes in many signaling pathways, which result in differential expression of EMT transcription factors such as snail family transcriptional repressor 1 (SNAIL), snail family transcriptional repressor 2 (SLUG), twist family bHLH transcription factor 1 (TWIST1) [13], but also Nanog homeobox (NANOG), POU class 5 homeobox 1 (OCT4), and SRY-box transcription factor 2 (SOX2) [14]. Studies suggest that these transcription factors, especially TWIST1, can translocate to the nucleus upon increased stiffness of ECM, represented by collagen I [15], which indicates that ECM has a role in this process.

Oxidative stress, a state of increased reactive oxygen species (ROS) production, affects all cell systems. It also represents an important factor contributing to the modulation of tumor cell and microenvironment reactions to chemotherapeutics. Increased ROS may lead to numerous consequences, such as genetic instability, one of the major characteristics of cancer, and the modification of lipids by peroxidation [16]. Lipid peroxidation (LPO) with its end-products—reactive aldehydes—have been increasingly recognized as a biomarker of different diseases, particularly cancer, where mitochondrial HNE plays an important role [17]. In addition, these reactive aldehydes, especially 4-hydroxy-2-nonenal (HNE), are involved in different signaling pathways influencing the cells' fate (e.g., differentiation, proliferation, or apoptosis) [18,19]. One of the signaling pathways affected by HNE is NRF2/KEAP1(Nuclear factor erythroid 2-related factor 2/Kelch-like ECH-associated protein 1) [20]. NRF2 is an antioxidant transcription factor that is bound to KEAP1 in its inactive state. HNE binds to KEAP1 cysteines and thereby releases its inhibition of NRF2. The release of NRF2 causes its translocation to the nucleus and activation of antioxidant genes' transcription and consequently enabling cells to survive oxidative challenge [20].

The present study aimed to elucidate if oxidative stress and HNE-modified collagen I, as a representative protein of ECM, in combination with HNE-induced chronic stress influence BCSC. The changes in the frequency of BCSC, antioxidative defense system, and transcriptional and protein expression of EMT markers were evaluated. These changes indicated that different surface modifications and chronic stress may bidirectionally modulate BCSC, supporting either differentiation or stress adaptation.

2. Materials and Methods

2.1. Cell Line and Medium

SUM159 cells (Asterand, Royston, Hertfordshire, UK), estrogen receptor, progesterone receptor, and Her2negative cell line, with the potential of generating stem-like subpopulation were cultured as mammospheres, according to previous publications [1,2]. Briefly, cells were cultured in Mammary Epithelial Basal Medium (MEBM; Lonza, Basel, Switzerland) supplemented with 10 ng/mL basic fibroblast growth factor (bFGF), 20 ng/mL epidermal growth factor (EGF, both from Peprotech, Rocky Hill, Hartford County, CT, USA), 5000 U/mL heparin (Sigma Aldrich, St Louis, MO, USA) and 20 μL/mL B27 supplement (Gibco/Invitrogen, Waltham, MA, USA) at 37 °C in a 5% CO_2 humidified atmosphere. Mammospheres larger than 40 μm were collected with 40 μm nylon cell strainers (Corning Incorporated-Life sciences, Durham, N.C., USA) and used for experiments.

2.2. Collagen Coating

To test cell growth characteristics on an extracellular matrix (ECM), collagen I was used as an ECM representative protein. Collagen I (Sigma Aldrich, St Louis, MO, USA) was dissolved in acetic acid (50 mM, Kemika, Zagreb, Croatia), diluted in redistilled sterile water in a final concentration of 2 mg/mL and used in the native state or modified by 1 or 10 μM HNE (Enzo Life Sciences, Lausen, Switzerland). Depending on the type of analysis, different formats of cell culture dishes were used with the same coating conditions: Native or modified collagen to its final concentration of 5 μg/cm^2. Thus, coated cell culture dishes were left to dry in a laminar flow cabinet overnight at room temperature (RT) and subsequently sterilized under UV light for 20 min. Dot-blot analysis with HNE-histidine monoclonal antibody was applied to confirm the binding of HNE to collagen I had occurred (Supplementary Figure S1). After confirmation that HNE did bind to histidine residues of collagen, we proceeded with evaluating the influence of collagen on measured parameters. Cells were also seeded on uncoated surfaces, further referred to as polystyrene (PS).

2.3. Cell Seeding and HNE Treatment

Mammospheres were dissociated to a single cell suspension by TrypLE (Gibco/Invitrogen Paisley, UK), and 10,000 cells/100 μL were plated in pre-coated or uncoated cell culture dishes and left to adhere for 3 h. Regardless of the cell culture dish format used, the experimental stoichiometry was maintained in all analyses. The formats of the cell culture dishes were as follows: 96-well microplates (cell viability and proliferation; TPP, Techno Plastic Products AG, Trasadingen, Switzerland); 6-well microplates (qRT-PCR, Western blot; Falcon, BD Biosciences, Franklin Lakes, NJ, USA); 8-well glass chamber slide (immunocytochemical analyses of hormone receptors; Nalgen Nunc Int, Naperville, IL, USA). Cells were then treated with different concentrations of HNE once, for a single treatment, or every 2nd day for 10 days, for multiple treatments. For cell viability and cell proliferation assays, these HNE concentrations varied from physiological (1 to 10 μM) to supraphysiological and pathological (25 to 100 μM). Controls of each growth surface were cultured without HNE. Analyses were performed after 48 h for single HNE treatments and 10 days for multiple HNE treatments as described for each analysis below. After the analysis of cell proliferation and cell viability, 10 μM HNE was selected for further analyses of putative breast cancer stem cell phenotypes, EMT marker expression, and immunocytochemical analyses of hormone receptors and antioxidative defense system. Untreated cells of each coating condition served as controls. All the mentioned analyses are described in more detail below.

2.4. Cell Viability—MTT Assay

The cell viability was determined by an MTT-based assay, EZ4U, following the manufacturer's recommendations (Biomedica, Vienna, Austria). Briefly after the treatment, 48 and 10 days after the seeding, cells were incubated with the MTT dye for an hour, and the absorbance was measured on a plate reader at 450 nm with a reference wavelength at 620 nm (Easy-Reader 400 FW; SLT Lab Instruments, GmbH, Salzburg, Austria).

2.5. Cell Proliferation—^{3}H-thymidine Incorporation Assay (^{3}HT)

The assay was based on the incorporation of radioactively labeled thymidine to the replicating DNA. The assay was performed as described previously [3]. Briefly, cells were treated as described in the previous chapter. ^{3}H-thymidine (1 µCi/well) was added to each well 24 h or 9 days after HNE treatment(s) and left for 24 h to allow thymidine incorporation into the DNA. Cells were then harvested, and the rate of ^{3}H-thymidine incorporation was measured on a Wallac 1904 DSA liquid scintillation counter (Perkin Elmer, Waltham, MA, USA).

2.6. Flow Cytometry Analyses of Putative Breast Cancer Stem Cell Phenotypes

For analyses of putative breast cancer stem cell markers, cells were treated as described above. After 10 days, cells were collected from culture dishes with accutase (PAA Laboratories GmbH, Pasching, Austria). Cells were then incubated for 5 min at 37 °C and rinsed twice with phosphate-buffered saline (PBS). Cells forming mammospheres during the experiments were singularized with TrypLE and finally resuspended in MEBM with supplementation for further analyses.

For the Aldefluor assay, cells were washed, counted, and finally resuspended in Aldefluor buffer [21]. To measure aldehyde dehydrogenase (ALDH) activity, the Aldefluor assay (STEMCELL Technologies, Grenoble, France) was performed according to the manufacturer's instructions and as previously published [21,22]. Briefly, 2 sets of samples with the Aldefluor substrate BODIPY-aminoacetaldehyde (BAAA) were prepared: (a) control: With diethylaminobenzaldehyde (DEAB, the specific inhibitor of ALDH) and (b) sample: Without DEAB. Controls were used for establishing the background fluorescence of these cells and defining the ALDH-positive region on the Fluorescence Channel 1 (FL1*) vs. the SSC dot plot. The absence of DEAB in the sample group converted BAAA to its fluorescent product, BODIPY-aminoacetate (BAA), defining the ALDH-positive population.

For analyses of CD44 and CD24 expression, cells were incubated with horse serum dilute 1:20 in 6% bovine serum albumin (BSA)/PBS for 30 min. Aliquots of antibodies anti-CD44 Allophycocyanin and anti-CD24 Fluorescein isothiocyanate (BD Bioscience, Schwechat, Austria) at a dilution of 1:40 in a final concentration of 0.08 µg/mL and 5 µg/mL, respectively, were then added and the samples were incubated at 4 °C for 30 min. The cells were washed and stored at 4 °C in the dark until the acquisition on the flow cytometer was performed. The protocol was performed as previously published [2,21]. Cells without staining and isotype controls, all from BD Bioscience, were integrated as controls in all experiments.

All samples were assayed on an LSRII flow cytometer (BD Bioscience), and the data were analyzed with the DIVA software version 8.0.1 (BD Bioscience Concorde Business Park 1/E/1/7, Schwechat, Austria).

2.7. Immunocytochemical Analyses of Hormone Receptors

For immunocytochemical analyses, cells were treated as described above. After 10 days, cells were fixed in ice-cold methanol for 20 min, dried, and stored until the staining. Cells were subjected to the antigen retrieval using Tris-EDTA solution, pH 9.0, by heating at 85 °C for 10 min to enable correct epitope folding. The monoclonal mouse anti-human estrogen receptor α (M7047, clone 1D5, DAKO, Glostrup, Denmark) and monoclonal mouse anti-human progesterone receptor (M3569, clone PgR636, DAKO, Glostrup, Denmark), both diluted to 1:50 in 1% BSA/PBS, were used. The secondary antibody EnVision (DAKO, Glostrup, Denmark), was used as recommended by the manufacturer. Finally, the reaction was visualized by DAB (3,3-diaminobenzidine tetrahydrochloride in organic solvent). Nuclei were counterstained by hematoxylin. The positive reaction was evaluated and scored by a trained pathologist (S.Š.) in a blinded manner.

2.8. Real-Time Quantitative PCR (qRT-PCR) Analyses of EMT Markers

After the cell treatment for 10 days, total RNA was extracted using a TRIzol Reagent (Invitrogen, Carlsbad, CA, USA) in accordance with the recommendation provided by the manufacturer. Nanodrop was used to quantify and asses the assay for purity (ThermoScientific, Waltham, MA, USA). The reverse transcription of one microgram of total RNA was performed using the QuantiTect Reverse Transcription Kit (Qiagen, Hilden, Germany) following the instructions of the manufacturer. LightCycler 480 (Roche) was used to perform qRT-PCR. Reactions were performed in 20 µL of total volume, consisting of 1× SYBR Green I Master Mix (Roche), 20 nanograms of cDNA as well as 25 µM of each primer (final concentration). All qRT-PCR reactions were conducted in duplicate, and afterward, the values of the quantification cycle were averaged. The comparative Ct method was utilized in the calculation of gene expression. Beta-2-microglobulin (B2M) and lactate dehydrogenase A (LDHA) were used as reference genes with the following primer sequences: B2M forward 5′TGCTGTCTCCATGTTTGATGTATCT 3′, B2M reverse 5′ TCTCTGCTCCCCACCTCTAAGT 3′ (NM_004048.3), LDHA forward 5′ TGTAGCAGATTTGGCAGAGAG 3′, LDHA reverse 5′ CATCATCCTTTATTCCGTAAAGAC 3′ (NM_005566.4). Primer sequences for fibronectin (FN), vimentin (VIM), N-cadherin (N CAD), SNAIL, SLUG, and TWIST were previously published [23].

2.9. ROS and Antioxidant Measurements

For ROS and antioxidant measurements, cells were treated as described above. On the 10th day of experiments, cells were incubated with 2′,7′–dichlorofluorescin diacetate (DCFDA) to allow the dye to overload the cells. Excess DCFDA was removed after 60 min when the cells were either incubated with medium alone (control) or with 10 µM HNE. ROS were measured with a Cary Eclipse Fluorescence Spectrophotometer (Varian Australia Pty Ltd., Mulgrave, Victoria, Australia) at λ_{ex} 500 nm and λ_{em} 529 nm.

For antioxidant measurements, cells were detached from the surface by TrypLE, and pelleted by centrifugation at the end of the 10-day treatment. Mammospheres were pelleted and dry pellets of all the experimental groups were stored till analyses. Prior to analyses, cells were lysed by 4 freeze/thaw cycles and afterward were centrifuged to remove cellular debris. Protein levels were then determined according to Bradford [24]. The catalase activity was measured according to the method by Goth with some modifications [25,26]. The activity of catalase was expressed as units per milligram of proteins in cell lysate (U mg^{-1}).

For the total GSH content, samples were diluted to 0.03 mg/mL and assayed using a modification of the Tietze method [26,27]. Concentrations of total GSH were expressed as µM of GSH per milligram of total protein (nmol mg^{-1}).

2.10. Western Blot

In order to perform Western blot analyses, cells were treated for 10 days, as described above. After 10 days, the cells were lysed in RIPA buffer (20 mM Tris-HCl (pH 7.5), 150 mM NaCl, 1% Triton X, 0.5% sodium deoxycholate, 0.1% sodium dodecyl sulfate (SDS)) containing protease inhibitors (Roche Diagnostics GmbH, Mannheim, Germany). The protein concentration of the thus obtained supernatant was quantified according to the Bradford method [24] by measuring absorbance at 595 nm using the microplate reader Multiskan EX (Thermo Electron Corporation, Shanghai, China) and interpolating from the standard curve. Protein samples were mixed with Laemmli buffer, boiled for 5 min at 95 °C and 40 µg of total proteins were resolved on the Tris-glycine SDS-PAGE gels (9% or 10%) and transferred to nitrocellulose membranes (Roti®-NC, Carl Roth, Karlsruhe, Germany). Membranes were stained with Ponceau S solution (Sigma Aldrich, St. Louis, MI, USA) for evaluation of transfer efficacy and scanned. Following blocking with 5% nonfat milk (Cell Signaling Technology (CST), Danvers, MA, USA) in Tris-buffered saline (TBS; 50 mM Tris-Cl, 150 mM NaCl, pH 7.6) containing 0.1% Tween-20 for 1 h, membranes were incubated with primary antibodies overnight at +4 °C. The primary antibodies used were: Rabbit monoclonal antibodies for NRF2 (CST:#12721), SLUG (CST:#9585), SNAIL (CST:#3879), NANOG (CST:#4903), OCT4 (CST:#2840), GAPDH (CST:#5174); rabbit polyclonal antibody for TWIST (Santa Cruz Biotechnology, sc-15393); mouse monoclonal antibody for Vimentin (Dako, M0725, Glostrup, Denmark). After incubation with horseradish peroxidase-conjugated secondary species-specific antibodies, immunoreactive bands were visualized using the SuperSignal™ West Pico PLUS Chemiluminescent Substrate (Thermo Scientific, Rockford, IL, USA) and Alliance 4.7 (UVITEC, Cambridge, UK). The analysis software Image Studio Lite (LI-COR, Lincoln, NE, USA) was used for quantification of levels of protein expression. Normalization was made with total proteins (Ponceau S staining) and with GAPDH as a loading control. Results are expressed as relative expression according to non-treated mammospheres (PS 0).

2.11. Statistical Analysis

All experiments were performed in at least 2 independent experiments with technical quadruplicates. For both single and multiple HNE treatments, inhibitory concentrations of 50% (IC_{50}) were calculated using non-linear regression curve fitting log (inhibitor) vs. response and variable slope with a least square (ordinary) fit, using GraphPadPrism 5 software (GraphPad Software, San Diego, CA, USA). Statistical analyses were performed using two-way ANOVA with Tukey's post hoc test. Values of $p < 0.05$ were considered significant.

3. Results

3.1. Effects of Single and Multiple Treatments of HNE on SUM159 Cells Growth

We have investigated the effects of single and multiple treatments of HNE as well as the influence of ECM represented by collagen type I, on the SUM159 growth. SUM159 cells grown in mammosphere-inducing conditions formed spheres on PS, in contrast to the adherent spread-like pattern observed on collagen-coated surfaces (Figure 1).

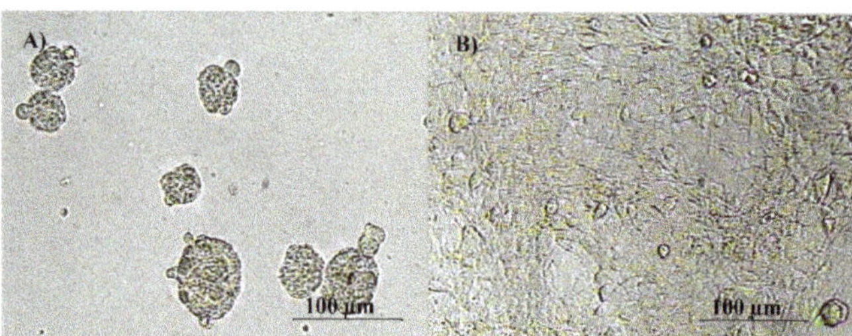

Figure 1. SUM159 cell growth morphology on different growth surfaces. (**A**) SUM159 cells in sphere inducing medium on low attaching growth surface (polystyrene (PS)) and (**B**) SUM159 cells growth in sphere inducing medium on the collagen I coated surface.

The MTT assay showed that SUM159 cell growth in mammosphere inducing conditions on PS had significantly lower viability regardless of HNE concentration used in comparison to coated surfaces and regardless of the time spent in the culture (3 and 10 days) ($p < 0.05$; Figure 2A,B). There was no difference in viability between cells grown on native or HNE-treated collagen when cells were treated with a range of HNE concentrations. The difference was observed in the concentrations causing inhibition, while 100 μM HNE showed inhibition between 50% to 60% after a single treatment, the viability was diminished at 50 μM HNE.

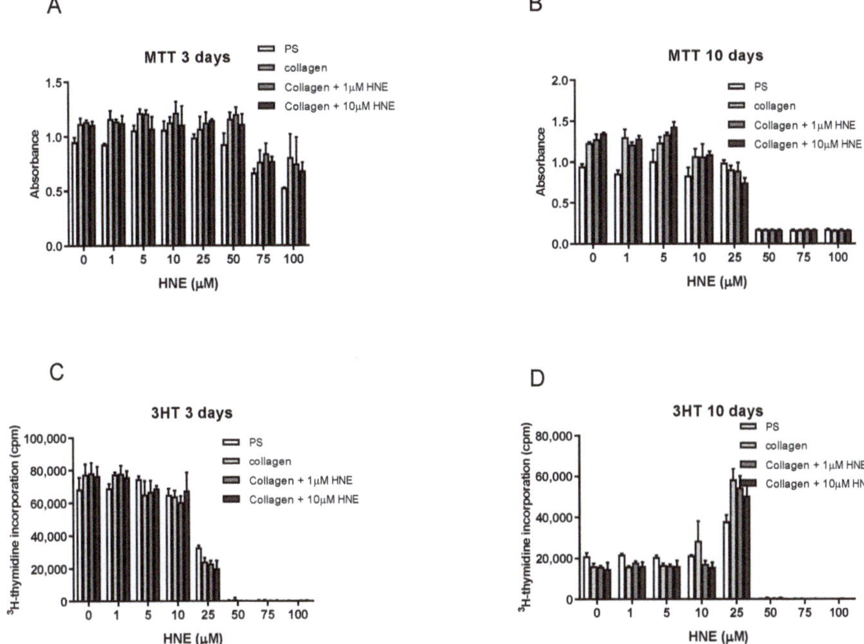

Figure 2. Effects of 4-hydroxy-2-nonenal (HNE) on SUM159 cell growth. SUM159 were exposed to single (**A,C**) and multiple HNE treatments (**B,D**). Their viability was evaluated by MTT (**A,B**), and their proliferation was evaluated by ^{3}H-thymidine incorporation assay (**C,D**).

Next, the proliferation of SUM159 cells with the ^{3}HT incorporation assay was assessed (Figure 2C,D). While the viability assay distinguished growth on PS and collagen, native, and HNE treated, the proliferation assay did not show any difference in proliferation rates on these surfaces. Inhibition of cell proliferation occurred at similar concentrations of HNE for all growth surfaces (IC$_{50}$ valued presented in Table 1). Multiple HNE treatment did not show differences in proliferation rate on different surfaces. Total growth inhibition was observed at 50 μM HNE and above. Interestingly, 25 μM HNE, which was IC$_{50}$ for single HNE treatment, was stimulating for multiple HNE treatments regardless of the growth surface, reaching more than 200% of the control value. Based on these results, 10 μM HNE was selected, as it did not alter the growth of mammospheres in either single or multiple treatments but did promote cell growth on native and HNE-modified collagen-coated surfaces.

In summary, the basic difference between different growth surfaces was observed by 50% inhibitory concentration (IC$_{50}$) measured by MTT and ^{3}H-thymidine assay (Table 1). In single HNE treatment, the IC$_{50}$ could not be determined in MTT assay as there was no total inhibitory concentration applied. In the ^{3}HT assay, a slight protective effect was observed for native collagen, and collagen treated with 1 μM HNE (24.05 μM for PS, and 25.54 and 24.83, respectively), while a slight decrease was observed on collagen with 10 μM HNE (23.60 μM). Multiple HNE treatments assayed by MTT sensitized the cells and decreased the IC$_{50}$ to 44.42 μM HNE on PS and to 28.78 μM, 27.74 μM and 26.48 μM for collagen-coated surfaces, native and treated with 1 μM and 10 μM HNE, respectively, while ^{3}HT assay showed no difference.

Table 1. Concentrations of HNE being inhibitory for 50% of the treated cells (IC_{50}).

Growth Surface	MTT IC_{50} (µM HNE)	^{3}HT IC_{50} (µM HNE)
Single HNE treatment		
PS	n.a.	24.05
Collagen I	n.a.	25.54
Collagen I + 1 µM HNE	n.a.	24.83
Collagen I + 10 µM HNE	n.a.	23.60
Multiple HNE treatments		
PS	44.42	44.59
Collagen I	28.78	45.04
Collagen I + 1 µM HNE	27.74	44.61
Collagen I + 10 µM HNE	26.48	44.64

n.a.—not applicable, concentrations used in the MTT assay did not cause total inhibition. PS—polystyrene.

3.2. Flow Cytometry Analyses for Putative Breast Cancer Stem Cell Phenotypes

In order to study possible changes in putative cancer stem cell markers due to HNE-pretreated collagen and due to multiple HNE treatments, the expression of CD44, CD24, and ALDH was assessed. The percentage of CD44$^+$CD24$^{-/low}$ (results not shown) was concordant with our previous results [21]. There were no significant changes in this phenotype during the treatment. On the other hand, the expression of ALDH-positive cells was different in regard to different growth surfaces and treatment conditions. As presented in Figure 3, untreated cells grown as a mammosphere culture showed the highest proportion of ALDH+ cells (10.5%). Growth on collagen decreased ALDH+ cells to 2.7%, and treatment of collagen with 1µM and 10 µM HNE additionally decreased ALDH+ cells to 0.2% and 0.1%, respectively. Next, HNE treatment was performed in order to assess ALDH activity under stress conditions. There was a decrease observed in the ALDH activity (2.9%) in mammospheres treated with HNE every second day for 10 days. When cells were grown on native collagen, there were small differences between the untreated and HNE-treated cells (2.7% vs. 2.4%). However, in the cells grown on HNE-pretreated collagen and treated with HNE every second day, an increase in ALDH activity was observed compared to the untreated cells on the same growth surface (untreated 0.2% vs. treated 0.9%), with even more pronounced difference of collagen pretreated with 10 µM HNE (0.1% vs. 3.8%). Therefore, our results indicated that HNE-modified collagen, in combination with chronic HNE treatment, caused concentration-dependent responses in ALDH positivity.

3.3. Expression of Hormone Receptors

As HNE caused concentration-dependent ALDH level changes, we wanted to asses if HNE collagen could induce differentiation. Therefore, we have determined estrogen and progesterone markers (ER and PR) by immunocytochemistry (Figure 4). The ER and PR positivity were validated by an experienced pathologist (S.Š.) by blindfold analysis. Mammospheres were completely negative for ER, while there was some insignificant positivity for PR, regardless of HNE treatment. On the other hand, cells grown on all collagen coatings had ER positivity. The highest ER positivity was observed on native collagen and pretreatment with HNE decreased the number of ER-positive cells in a concentration-dependent manner (40% for 1 µM HNE and 11.4% for 10 µM HNE, respectively). Moreover, treatment with 10 µM HNE did not change already-observed patterns with the exception of collagen pretreated with 1 µM HNE, where the treatment additionally increased the percentage of positive cells (40% vs. 80.2%). PR positivity was similar to ER, very low on PS. Growth on collagen increased PR positivity, with the highest levels on collagen pretreated with 1 µM HNE, and the lowest for 10 µM HNE. HNE treatment showed different trends, which were surface-specific: Increased PR positivity on PS and collagen pretreated with 1 µM HNE, decreased on native collagen, while it did not affect PR positivity on collagen pretreated with 10 µM HNE. These results are in line with ALDH results.

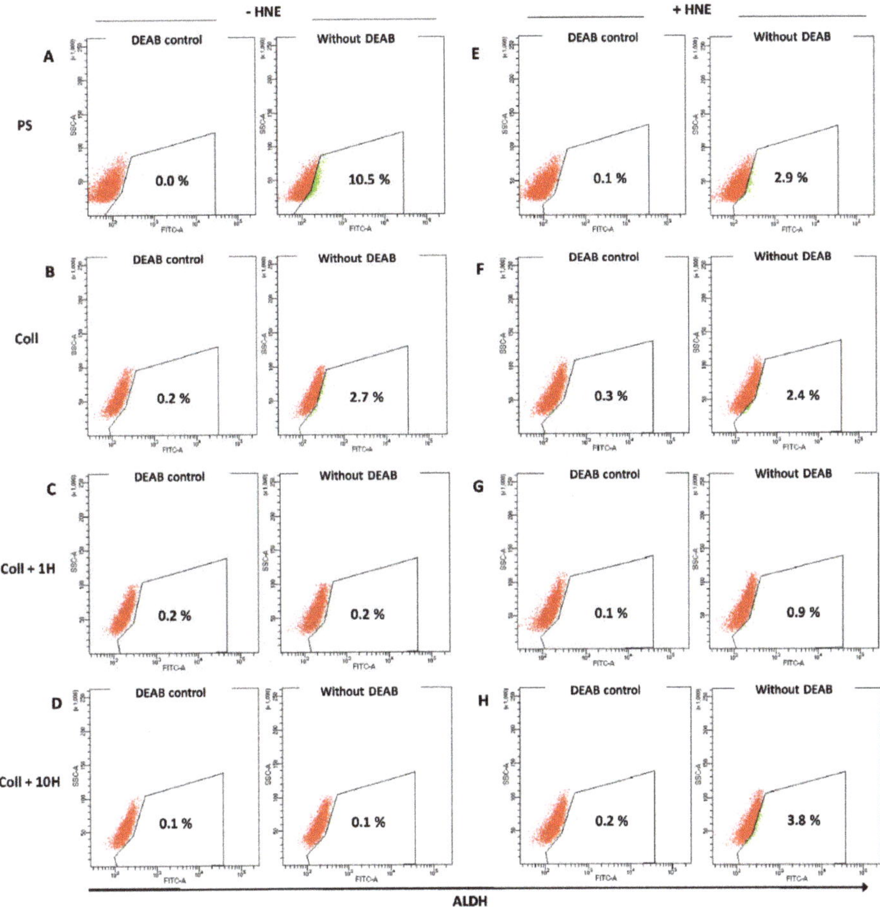

Figure 3. Effects of multiple HNE treatments on the expression of stem cell marker aldehyde dehydrogenase (ALDH) in SUM159 cells. SUM159 cells were cultured for 10 days on different growth surfaces: Polystyrene, PS (**A**), native collagen (**B**), collagen pretreated with 1 µM HNE (**C**), and on collagen pretreated with 10 µM HNE (**D**). Chronic stress was stimulated by the addition of 10 µM HNE every 2 days for 10 days in total on different growth surfaces: Polystyrene, PS (**E**), native collagen (**F**), collagen pretreated with 1 µM HNE (**G**) and on collagen pretreated with 10 µM HNE (**H**). For each panel, both the control and test samples are presented. Control is performed with the addition of diethylaminobenzaldehyde (DEAB), which is an inhibitor of ALDH.

3.4. Antioxidants and ROS

Further, as cells can adapt to the low level of stress, we have examined parts of the antioxidant defense system, particularly the levels of GSH and the activity of catalase (Figure 5). Catalase activity was the highest in mammospheres, and HNE treatment significantly reduced its activity. In cells grown on collagen, native HNE-pretreated ones had significantly lower catalase activity than in mammospheres ($p < 0.001$). HNE treatment reduced the catalase activity on native collagen while increasing the activity on collagen pretreated with 1 μM HNE. Treatment with HNE decreased catalase activity in mammospheres and cells grown on native collagen.

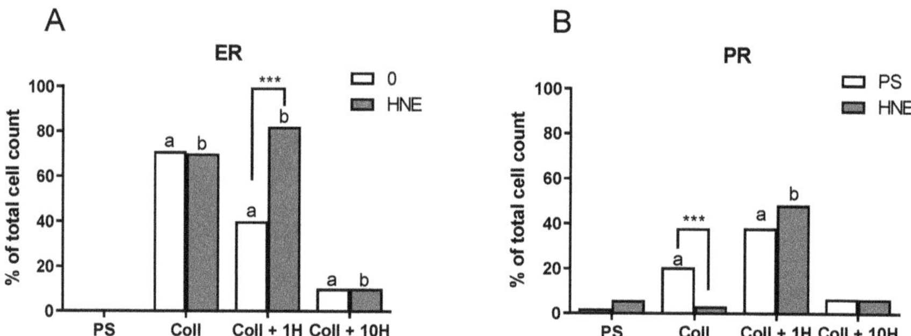

Figure 4. The presence of estrogen (ER) and progesterone (PR) receptors on SUM159 after multiple HNE treatments. After 10 days of treatment with 10 μM HNE every two days, positivity for ER (**A**) and PR (**B**) was evaluated on 1000 cells by the experienced pathologist (S.Š.). All results are expressed as percentages on a 1000 cell count, a—significantly different compared to the control on PS, at least $p < 0.05$, specified in the text; b—significantly different compared to HNE-treated PS at least $p < 0.05$, specified in the text; *** $p < 0.001$ control vs. HNE-treatment on the same growth surface.

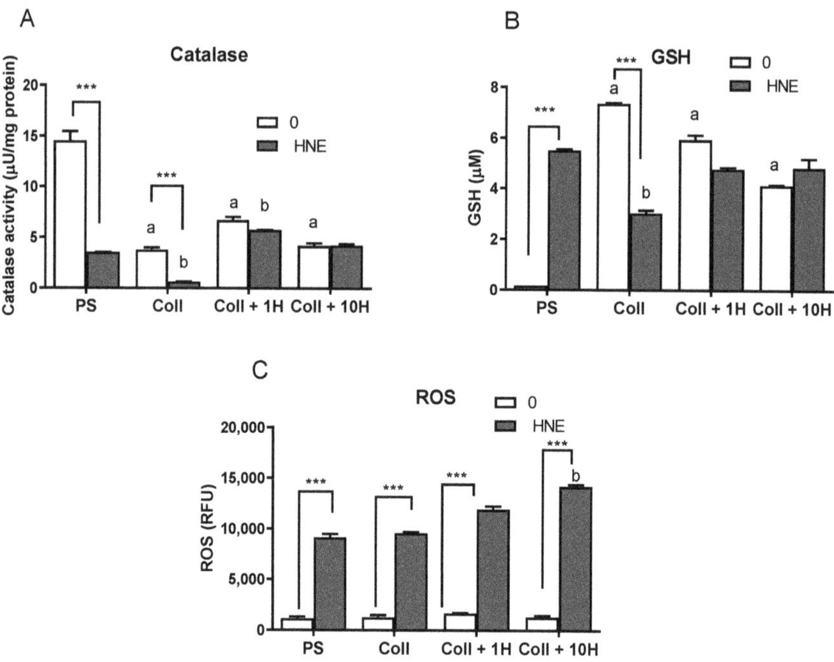

Figure 5. Effects of multiple HNE treatments on the catalase activity (**A**), Glutathione (GSH) levels (**B**), and (**C**) reactive oxygen species (ROS) in SUM159 cells grown on different surfaces. a—significantly different compared to the control on PS, at least $p < 0.05$, specified in the text; b—significantly different compared to HNE-treated PS at least $p < 0.05$, specified in the text; *** $p < 0.001$ control vs. HNE-treatment on the same growth surface.

Total GSH levels followed completely different patterns in comparison to catalase, which was expected as HNE is metabolized through the GSH system by binding to GSH [28]. Interestingly, the total GSH levels were the lowest in control mammospheres on PS. Growth on collagen, native, or HNE pretreated, increased GSH levels significantly ($p < 0.001$). HNE treatment decreased total GSH levels on native collagen, and this level was decreased when compared to mammospheres treated with HNE. Interestingly, the two tested antioxidants did not show similar patterns. As the main HNE scavenger GSH was increased with HNE treatment, but also with growth on collagen, native or HNE pretreated, indicating the need for this part of the antioxidant system.

Although both catalase activity and GSH levels varied on different growth surfaces, ROS levels were not changed on different surfaces. HNE addition to cultures significantly increased levels of ROS on all surfaces, with a more pronounced concentration-dependent increase on HNE-pretreated collagen.

3.5. EMT Markers

Changes in the expression of the selected EMT markers were assessed by qPCR, and the results are presented in Figure 6. Among the tested markers, fibronectin, and SLUG did not show any significant changes. The expression of N CAD was significantly increased only in HNE-treated SUM159 cells grown on native collagen ($p = 0.0096$) and collagen pretreated with 10 µM HNE ($p = 0.0185$) when compared to PS. Opposite patterns were observed for vimentin depending on the growth surface conditions and HNE treatment. Repeated HNE-treatment significantly decreased vimentin in mammospheres ($p = 0.0277$). When comparing different growth surfaces/conditions to PS, there was a slight decline within non-treated cells in vimentin levels with increasing HNE concentration, while in HNE-treated cells, surface pretreatments increased vimentin levels especially in cells grown on collagen pretreated with 10 µM HNE ($p = 0.0416$). Similar patterns were observed for NANOG, SNAIL, and TWIST. While multiple HNE-treatments significantly decreased levels of NANOG ($p = 0.0108$), SNAIL ($p = 0.033$), and TWIST ($p = 0.0004$) in mammospheres and NANOG ($p = 0.0219$) and TWIST ($p < 0.0001$) in cells grown on collagen, a slight increase can be observed for all three proteins in cells grown on collagen pretreated with 10 µM HNE. In addition, HNE pretreatment of collagen adversely affected the levels of NANOG, SNAIL, and TWIST in non-treated and HNE-treated cells when growth surfaces were compared to PS. Thus, in cells grown on collagen pretreated with 1 µM HNE, NANOG ($p = 0.0433$) and TWIST ($p = 0.0027$) significantly decreased in non-treated SUM159 cells. The growth on collagen pretreated with 10 µM HNE additionally decreased the levels of NANOG ($p = 0.006$), SNAIL ($p = 0.0313$), and TWIST ($p = 0.0002$) while in HNE-treated cells, the same growth surface increased the levels of these proteins, especially NANOG ($p = 0.0435$) and SNAIL ($p = 0.0063$). Similarly, multiple HNE treatment decreased the expression of OCT4 in cells grown on PS ($p = 0.011$) and on native collagen ($p = 0.047$). Depending on the different growth surface conditions to PS, the levels of OCT4, while revealing similar patterns to NANOG, SNAIL, and TWIST in non-treated cells, differed when cells were exposed to multiple HNE treatments.

As expected, EMT markers were the highest in mammospheres, and HNE treatment either caused no changes or caused a high decrease. Further, collagen and its pretreatment with HNE changed EMT markers, but combinations of HNE treatments and collagen pretreated with 10 µM HNE increased some of the markers to the levels found in mammospheres.

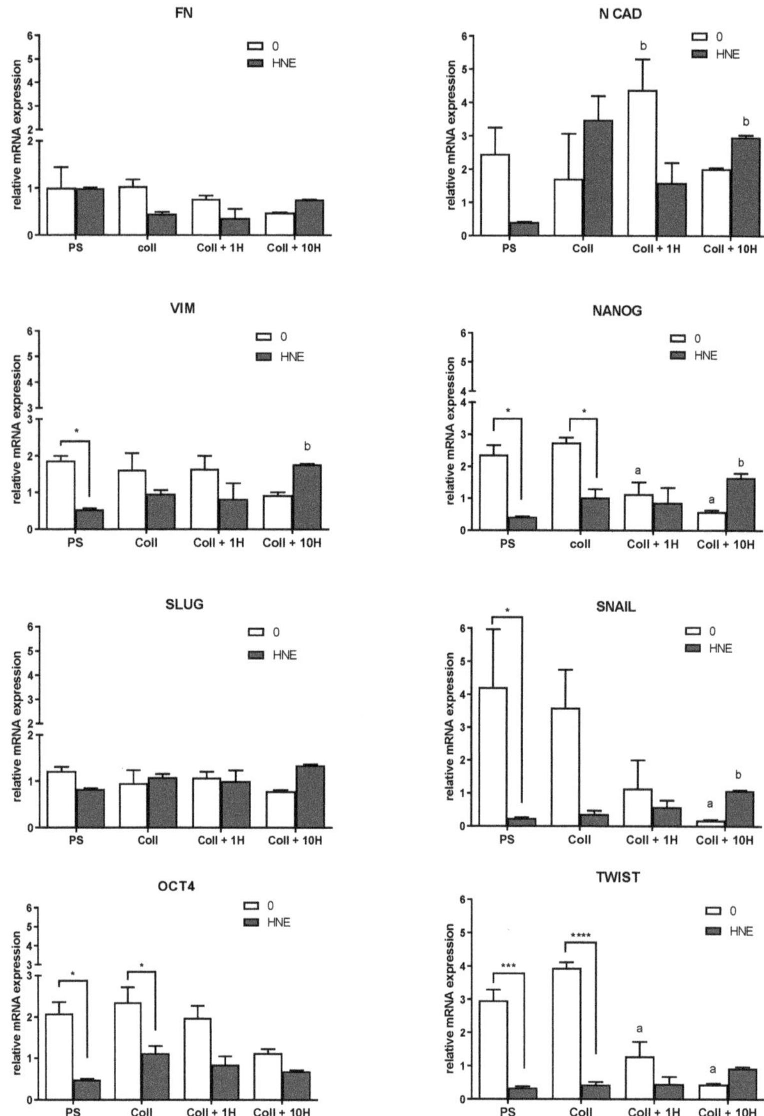

Figure 6. Expression of epithelial to mesenchymal transition (EMT) genes, including N-cadherin (N CAD), vimentin (VIM), fibronectin (FN), NANOG, OCT4, SLUG, SNAIL, and TWIST. The relative mRNA expression was analyzed by qRT-PCR. Bars represent mean +/–SEM of two biological replicates. a—significantly different compared to the control on PS, at least $p < 0.05$, specified in the text; b—significantly different compared to HNE-treated PS at least $p < 0.05$, specified in the text; * $p < 0.05$, *** $p < 0.001$, **** $p < 0.0001$ both control vs. HNE-treatment on the same growth surface.

3.6. Western Blot

In order to assess if multiple treatments with HNE caused an increase in antioxidant transcription factor NRF2 levels and to validate mRNA analysis of EMT markers, we performed Western blot analyses of these proteins (Figure 7). For NRF2, it was shown that HNE did not affect its levels when

SUM159 cells were grown as mammospheres on PS. In contrast to PS, on collagen, native, or HNE treated, SUM159 cells significantly increased NRF2 levels regardless of HNE treatment ($p < 0.02$ and $p < 0.0001$, for non-treated and HNE-treated, respectively). Multiple HNE treatments additionally increased NRF2 levels in SUM159 cells grown on native collagen and collagen pretreated with 10 μM HNE ($p < 0.0001$ and $p = 0.0083$). In the case of EMT markers, their reaction patterns differed. SLUG and SNAIL did not show any differences regardless of growth surface and HNE treatment. Vimentin was significantly increased in non-treated SUM159 cells grown on native collagen ($p = 0.0003$) and collagen pretreated with 1μM HNE ($p = 0.001$) and in HNE-treated cells grown on native collagen ($p = 0.0293$) and collagen pretreated with 10 μM HNE ($p = 0.0307$) when compared to PS. NANOG showed a similar pattern as vimentin when observing the differences between mammospheres (PS) and different cultivating surfaces. Significant increase of NANOG was observed for both non-treated and HNE-treated cells grown on native collagen ($p = 0.0055$ and $p = 0.0102$) and collagen pretreated with 1 μM HNE ($p = 0.0029$ and $p = 0.0009$), but also without differences between non-treated and HNE-treated cells grown on the same growth surface. HNE seems to be important in regulating the levels of TWIST, regardless of HNE treatment. A significant HNE concentration-dependent increase of TWIST was observed for non-treated cells grown on collagen pretreated with 1 and 10 μM HNE ($p = 0.0437$ and $p < 0.0001$). In the group of multiple HNE treatments, TWIST was increased on all collagen surfaces when compared to PS ($p < 0.001$). Interestingly, multiple HNE treatments increased TWIST levels in cells grown on native collagen ($p = 0.0003$) but decreased them significantly on collagen pretreated with 10 μM HNE ($p = 0.0014$). Among all assayed proteins, OCT4 was the only one significantly increased by multiple HNE treatments on PS ($p = 0.0128$). Additionally, growth on pretreated surfaces increases the levels of OCT4. While in non-treated cells, its levels were increased for all growth surfaces ($p < 0.0005$) in comparison to PS, in HNE-treated cells, OCT4 levels were significantly increased when cells were grown on collagen ($p = 0.0102$) and collagen pretreated with 10 μM HNE ($p = 0.0013$). Surprisingly, HNE treatment significantly decreased OCT4 on collagen pretreated with 1 μM HNE, but both of these levels were higher than on PS.

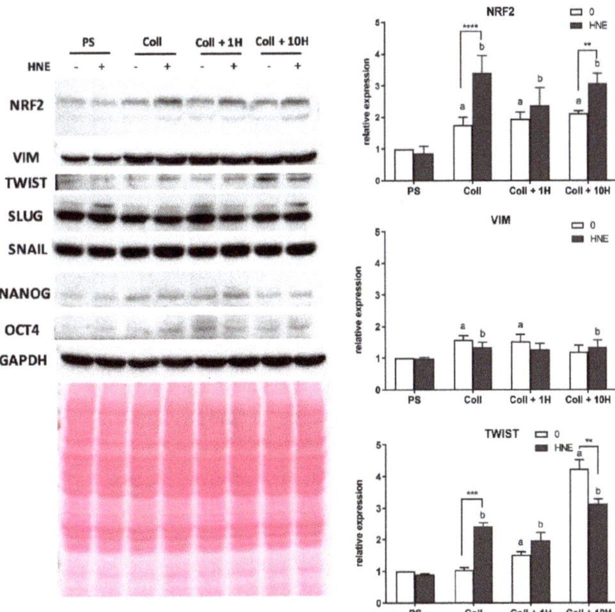

Figure 7. *Cont.*

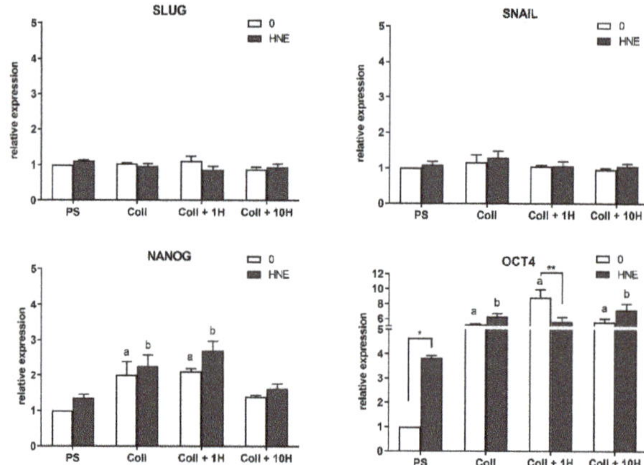

Figure 7. Western blot analyses. Representative blots and relative expression of different proteins: NRF2, VIM, TWIST, SLUG, SNAIL, NANOG, OCT4 are shown. Two-way ANOVA with Tukey's post hoc test was used to test the differences between groups: a—significantly different compared to the control on PS, at least $p < 0.05$, specified in the text; b—significantly different compared to HNE-treated PS at least $p < 0.05$, specified in the text; * $p < 0.05$, ** $p < 0.01$, *** $p < 0.001$, **** $p < 0.0001$ all control vs. HNE-treatment on the same growth surface.

4. Discussion

Cells can, to a certain extent, adapt to numerous stress conditions, and, therefore, the aim of this study was to evaluate whether oxidative stress caused by lipid peroxidation representative end-product HNE has the capability to cause specific molecular changes of tumor cells and impact the frequency of BCSC. Numerous factors may affect tumors, such as oxidative stress, which is a risk factor in tumor initiation and proliferation but can modify tumor microenvironment components, such as proteins and cells, which can further affect tumors. Additionally, a subpopulation of tumor cells, CSC, are increasingly recognized as the main factor of tumor growth and recurrence. Until now, these factors were studied separately. Our findings suggest that HNE modifications of collagen I, in combination with chronic exposure to HNE, may cause changes in the distribution of putative BCSC. Oxidative stress may cause either cell differentiation or, when chronic, an increase of BCSC population and up-regulation of EMT markers.

We have studied the influence of oxidative stress and lipid peroxidation on breast cancer cell line SUM159, modeling both the direct influence of HNE and combinations with modifications of collagen I. The microenvironment of each tumor is unique and the changes in this environment due to inflammation and oxidation processes are complex. Therefore, it is challenging to model these modifications. Oxidative stress is involved in mutagenesis, which is a driving force of (breast) cancer initiation and progression, especially in hereditary breast cancer, where the mere loss of BRCA1 increases ROS [29]. Therefore, it is not surprising that oxidative stress and lipid peroxidation biomarkers are changed in breast cancer patients [30]. HNE is also recognized as a biomarker of oxidative stress, and as such, is involved in (breast) cancer progression [31–33]. In accordance with its role are concentrations found in human plasma, where concentrations ranging from 0.1 µM to 1 µM are considered physiological, while 1 µM to 10 µM are considered as "where pathology begins" [34]. Taken that hereditary mutations in breast cancer, as well as conventional cancer treatment strategies, such as chemo- and radiotherapy, cause increases in ROS, which can, in turn, cause lipid peroxidation and HNE formation, these oxidative processes may affect numerous signaling molecules such as HNE activation of NRF2 transcription factor. In order to study the influence of ECM, we have chosen collagen I, as it can influence some of the EMT markers [15]. We show in our study that collagen may act as a protective agent on SUM159 cell viability, regardless of previous HNE modifications of the collagen. In acute HNE treatment, cell viability was affected at rather higher HNE concentrations (IC_{50} about 100 µM HNE), whereas proliferation was inhibited already at lower HNE concentrations (25 µM), thereby indicating modulation of cell growth and survival. Previously, we observed similar effects of HNE with collagen oxidized by hydroxyl radical instead of HNE [35], indicating that oxidative modifications of collagen I are an important factor when studying cell responses to different stimuli or inhibition factors.

As expected, chronic HNE treatment had a higher impact on cells. Interestingly, proliferation was generally lower in tested cultures than in acute stress, with the exception of 25 µM HNE. This decrease in the proliferation rate after 10 days could be a consequence of increased cell density. In support of this conclusion is the proliferation burst with 25 µM HNE, indicating that these cells adapted and survived the treatment, and, due to the initial decrease in proliferation, now were not spatially limited to grow. Notably, 1 µM HNE, which was considered the physiological concentration, caused differentiation, observed by a decrease in BCSC markers and an increase in hormone receptors, effects that have been described for colon cancer cells and HL-60 cells as well [36,37]. Interestingly, the BCSC marker that we show here, ALDH activity, is the enzyme that can detoxify HNE, particularly ALDH2, which is located in the mitochondrial matrix [17].

Next, we aimed to investigate the influence of chronic stress and HNE modifications of collagen on the expression and protein levels of EMT markers as well as antioxidant parameters measured by catalase activity, GSH levels, and NRF2 antioxidant transcription factor level. Interestingly, although collagen itself, regardless of HNE pretreatment, lowered ALDH, it did not influence EMT markers in the same manner. For example, fibronectin expression was unaffected by different growth surfaces nor by HNE treatment. A similar pattern of expression and protein levels was observed for vimentin, which was increasing with HNE pretreatment concentrations. SLUG was not affected by both mRNA and protein levels. Similarly, SNAIL expression pattern changes were not followed by changes in protein level. Interestingly, expression patterns of TWIST and OCT4 were not followed by protein levels, which were higher in cells on collagen, native, or HNE-pretreated, than on PS. EMT was recognized as an important factor in cancer progression because it represented a conversion between differentiated epithelial cells into migratory mesenchymal cancer cells [38]. The plasticity of CSC enabled them to follow transition traits between EMT and MET, thereby contributing to the metastatic potential of the primary tumor [39]. While many studies link EMT and cancer development and malignancy [40], the influence of oxidative stress/ROS and reactive aldehydes are simply not investigated enough [41]. Numerous factors can stimulate these transitions, and, as shown here, one of them may be chronic oxidative stress.

It was shown previously that EMT might be abolished by the addition of antioxidant curcumin, underscoring the possible role of redox signaling in this process [42]. Therefore, in addition to EMT markers, the levels of GSH, catalase activity, and ROS were measured after HNE treatment and the antioxidant transcription factor NRF2. Interestingly, while catalase activity was the highest in control mammospheres on PS, and decreased by growth on collagen, native or HNE pretreated, GSH levels were significantly increased by both HNE treatment and growth on collagen. It is not surprising that GSH levels were increased by HNE as this is the major scavenger of HNE, and the first step in HNE detoxification [43], while the thioredoxin system is inhibited by HNE and does not contribute to its detoxification [44]. Finally, and in support of GSH increase, were the levels of ROS and NRF2. In all control groups, ROS were at the same level, while the addition of HNE increased ROS, which was additionally increased by HNE pretreated collagen. Following the ROS pattern, growth on collagen increased NRF2 levels, and HNE treatments additionally increased NRF2 on native collagen and collagen pretreated with 10 µM of HNE. HNE is known to activate NRF2 by releasing it from KEAP1 inhibition, and once NRF2 is freed, it translocates to the nucleus [20]. In the nucleus, NRF2 activates transcription of antioxidant genes, among which are glutamate-cysteine ligase, catalytic subunit, and glutamate-cysteine ligase, a modifier subunit, and an enzyme which catalyzes the first step in GSH synthesis [20].

Finally, a recent study indicated that EMT is not the limiting factor for metastasis, but contributes greatly to chemoresistance [45]. Taking all the results into account, our findings indicate that under chronic stress, EMT markers remain elevated and in combination with elevated antioxidant factors such as GSH and NRF2, which can contribute to the maintenance of the BCSC phenotype and therapy resistance.

5. Conclusions

Our results suggest that chronic oxidative stress acts as a double-edged sword in supporting the BCSC phenotype. Low levels of HNE can increase differentiation markers in BCSC. In contrast, higher levels and chronic HNE presence increased GSH and NRF2, thereby increasing antioxidative protection. Concurrently, some protein EMT markers are increased, and hormone levels were decreased, thereby supporting the BCSC phenotype and its resistance to oxidative stress. Finally, a better understanding of the role of chronic oxidative stress in the modulation of the breast cancer microenvironment and its impact on breast cancer differentiation may eventually allow for the development of more effective therapeutic strategies.

Supplementary Materials: The following are available online at http://www.mdpi.com/2076-3921/8/12/633/s1, Figure S1: Dot blot of HNE-collagen I conjugates.

Author Contributions: All authors read and approved the final version of the manuscript. Conceptualization, A.Č.G. and M.B.; data curation, A.Č.G., L.M. and N.D.; formal analysis, A.Č.G., L.M., N.D., and S.S.; methodology, A.Č.G., L.M., S.S., I.P., J.V., S.Š. and C.S.; resources, A.Č.G. and M.B.; validation, L.M. and N.D.; writing—original draft, A.Č.G.; writing—review and editing, L.M. and M.B.

Funding: The research was supported by an OENB project No'14001, Croatian MSES projects, Croatian-Austrian bilateral project (WTZ HR 01/12).

Acknowledgments: The authors wish to acknowledge COST CM CM1106 Action. The authors are grateful for the help on editing English grammar to Klara Balic.

Conflicts of Interest: The authors declare no conflict of interest.

Abbreviations

HNE: 4-hydroxy-2-nonenal; CSC; Cancer stem cells; BCSC; Breast cancer stem cells; ECM; Extracellular matrix; EMT; Epithelial to mesenchymal transition; MET; Mesenchymal to epithelial transition; ROS; Reactive oxygen species; LPO; Lipid peroxidation; NRF2; Nuclear factor erythroid 2-related factor 2; KEAP1; Kelch-like ECH-associated protein 1; bFGF; Basic fibroblast growth factor; EGF; Epidermal growth factor; PS; Polystyrene; PBS; Phosphate-buffered saline; ALDH; Aldehyde dehydrogenase; MEBM; Mammary epithelial basal medium; RT; Room temperature; EDTA; Ethylenediaminetetraacetic acid; qRT- PCR; Real-time quantitative polymerase chain reaction (PCR); FN; Fibronectin; VIM; Vimentin; N CAD; N-cadherin; B2M; Beta-2-microglobulin; LDHA; Lactate dehydrogenase A; H_2O_2; Hydrogen peroxide; GSH; Glutathione; ER; Estrogen receptor; PR; Progesterone receptor

References

1. Izrailit, J.; Reedijk, M. Developmental pathways in breast cancer and breast tumor-initiating cells: Therapeutic implications. *Cancer Lett.* **2012**, *317*, 115–126. [CrossRef] [PubMed]

2. Al-Hajj, M.; Becker, M.W.; Wicha, M.; Weissman, I.; Clarke, M.F. Therapeutic implications of cancer stem cells. *Curr. Opin. Genet. Dev.* **2004**, *14*, 43–47. [CrossRef] [PubMed]

3. Balic, M.; Lin, H.; Young, L.; Hawes, D.; Giuliano, A.; McNamara, G.; Datar, R.H.; Cote, R.J. Most early disseminated cancer cells detected in bone marrow of breast cancer patients have a putative breast cancer stem cell phenotype. *Clin. Cancer Res.* **2006**, *12*, 5615–5621. [CrossRef] [PubMed]

4. Paget, S. The distribution of secondary growths in cancer of the breast. *Lancet* **1889**, *133*, 571–573. [CrossRef]

5. Gnant, M.; Mlineritsch, B.; Schippinger, W.; Luschin-Ebengreuth, G.; Pöstlberger, S.; Menzel, C.; Jakesz, R.; Seifert, M.; Hubalek, M.; Bjelic-Radisic, V.; et al. Endocrine therapy plus zoledronic acid in premenopausal breast cancer. *N. Engl. J. Med.* **2009**, *360*, 679–691. [CrossRef] [PubMed]

6. Coleman, R.E.; Gregory, W.; Marshall, H.; Wilson, C.; Holen, I. The metastatic microenvironment of breast cancer: Clinical implications. *Breast* **2013**, *22*, S50–S56. [CrossRef]

7. Liubomirski, Y.; Lerrer, S.; Meshel, T.; Rubinstein-Achiasaf, L.; Morein, D.; Wiemann, S.; Körner, C.; Ben-Baruch, A. Tumor-Stroma-Inflammation Networks Promote Pro-metastatic Chemokines and Aggressiveness Characteristics in Triple-Negative Breast Cancer. *Front. Immunol.* **2019**, *10*, 757. [CrossRef]

8. Giussani, M.; Merlino, G.; Cappelletti, V.; Tagliabue, E.; Daidone, M.G. Tumor-extracellular matrix interactions: Identification of tools associated with breast cancer progression. *Semin. Cancer Biol.* **2015**, *35*, 3–10. [CrossRef]

9. Wang, Y.J.; Bailey, J.M.; Rovira, M.; Leach, S.D. Sphere-forming assays for assessment of benign and malignant pancreatic stem cells. *Methods Mol. Biol.* **2013**, *980*, 281–290. [PubMed]

10. Balic, M.; Schwarzenbacher, D.; Stanzer, S.; Heitzer, E.; Auer, M.; Geigl, J.B.; Cote, R.J.; Datar, R.H.; Dandachi, N. Genetic and epigenetic analysis of putative breast cancer stem cell models. *BMC Cancer* **2013**, *13*. [CrossRef]

11. Celià-Terrassa, T.; Jolly, M.K. Cancer Stem Cells and Epithelial-to-Mesenchymal Transition in Cancer Metastasis. *Cold Spring Harb. Perspect. Med.* **2019**. [CrossRef] [PubMed]

12. Taube, J.H.; Herschkowitz, J.I.; Komurov, K.; Zhou, A.Y.; Gupta, S.; Yang, J.; Hartwell, K.; Onder, T.T.; Gupta, P.B.; Evans, K.W.; et al. Core epithelial-to-mesenchymal transition interactome gene-expression signature is associated with claudin-low and metaplastic breast cancer subtypes. *Proc. Natl. Acad. Sci. USA* **2010**, *107*, 15449–15454. [CrossRef] [PubMed]

13. Du, B.; Shim, J.S. Targeting epithelial-mesenchymal transition (EMT) to overcome drug resistance in cancer. *Molecules* **2016**, *21*, 965. [CrossRef] [PubMed]

14. Mimeault, M.; Batra, S.K. Altered gene products involved in the malignant reprogramming of cancer stem/progenitor cells and multitargeted therapies. *Mol. Asp. Med.* **2014**, *39*, 3–32. [CrossRef] [PubMed]

15. Wei, S.C.; Fattet, L.; Tsai, J.H.; Guo, Y.; Pai, V.H.; Majeski, H.E.; Chen, A.C.; Sah, R.L.; Taylor, S.S.; Engler, A.J.; et al. Matrix stiffness drives epithelial-mesenchymal transition and tumour metastasis through a TWIST1-G3BP2 mechanotransduction pathway. *Nat. Cell Biol.* **2015**, *17*, 678–688. [CrossRef] [PubMed]

16. Esterbauer, H.; Schaur, R.J.; Zollner, H. Chemistry and biochemistry of 4-hydroxynonenal, malonaldehyde and related aldehydes. *Free Radic. Biol. Med.* **1991**, *11*, 81–128. [CrossRef]

17. Zhong, H.; Yin, H. Role of lipid peroxidation derived 4-hydroxynonenal (4-HNE) in cancer: Focusing on mitochondria. *Redox Biol.* **2015**, *4*, 193–199. [CrossRef]

18. Guéraud, F.; Atalay, M.; Bresgen, N.; Cipak, A.; Eckl, P.M.M.; Huc, L.; Jouanin, I.; Siems, W.; Uchida, K.; Gueraud, F.; et al. Chemistry and biochemistry of lipid peroxidation products. *Free Radic. Res.* **2010**, *44*, 1098–1124. [CrossRef]

19. Barrera, G.; Pizzimenti, S.; Dianzani, M.U. 4-hydroxynonenal and regulation of cell cycle: Effects on the pRb/E2F pathway. *Free Radic. Biol. Med.* **2004**, *37*, 597–606. [CrossRef]

20. Milkovic, L.; Zarkovic, N.; Saso, L. Controversy about pharmacological modulation of Nrf2 for cancer therapy. *Redox Biol.* **2017**, *12*, 727–732. [CrossRef]

21. Balic, M.; Rapp, N.; Stanzer, S.; Lin, H.; Strutz, J.; Szkandera, J.; Daidone, M.G.; Samonigg, H.; Cote, R.J.; Dandachi, N. Novel immunofluorescence protocol for multimarker assessment of putative disseminating breast cancer stem cells. *Appl. Immunohistochem. Mol. Morphol.* **2011**, *19*, 33–40. [CrossRef] [PubMed]

22. Ginestier, C.; Hur, M.H.; Charafe-Jauffret, E.; Monville, F.; Dutcher, J.; Brown, M.; Jacquemier, J.; Viens, P.; Kleer, C.G.; Liu, S.; et al. ALDH1 is a marker of normal and malignant human mammary stem cells and a predictor of poor clinical outcome. *Cell Stem Cell* **2007**, *1*, 555–567. [CrossRef] [PubMed]

23. Palafox, M.; Ferrer, I.; Pellegrini, P.; Vila, S.; Hernandez-Ortega, S.; Urruticoechea, A.; Climent, F.; Soler, M.T.; Muñoz, P.; Viñals, F.; et al. RANK induces epithelial-mesenchymal transition and stemness in human mammary epithelial cells and promotes tumorigenesis and metastasis. *Cancer Res.* **2012**, *72*, 2879–2888. [CrossRef] [PubMed]

24. Bradford, M.M. A rapid and sensitive method for the quantitation of microgram quantities of protein utilizing the principle of protein-dye binding. *Anal. Biochem.* **1976**, *72*, 248–254. [CrossRef]

25. Góth, L. A simple method for determination of serum catalase activity and revision of reference range. *Clin. Chim. Acta* **1991**, *196*, 143–151. [CrossRef]

26. Radman Kastelic, A.; Odžak, R.; Pezdirc, I.; Sović, K.; Hrenar, T.; Čipak Gašparović, A.; Skočibušić, M.; Primožič, I. New and Potent Quinuclidine-Based Antimicrobial Agents. *Molecules* **2019**, *24*, 2675. [CrossRef]

27. Tietze, F. Enzymic method for quantitative determination of nanogram amounts of total and oxidized glutathione: Applications to mammalian blood and other tissues. *Anal. Biochem.* **1969**, *27*, 502–522. [CrossRef]

28. Srivastava, S.; Chandra, A.; Wang, L.-F.; Seifert, W.E.; DaGue, B.B.; Ansari, N.H.; Srivastava, S.K.; Bhatnagar, A. Metabolism of the Lipid Peroxidation Product, 4-Hydroxy- *trans* -2-nonenal, in Isolated Perfused Rat Heart. *J. Biol. Chem.* **1998**, *273*, 10893–10900. [CrossRef]

29. Sawczuk, B.; Maciejczyk, M.; Sawczuk-Siemieniuk, M.; Posmyk, R.; Zalewska, A.; Car, H. Salivary Gland Function, Antioxidant Defence and Oxidative Damage in the Saliva of Patients with Breast Cancer: Does the BRCA1 Mutation Disturb the Salivary Redox Profile? *Cancers* **2019**, *11*, 1501. [CrossRef]

30. Lee, J.D.; Cai, Q.; Shu, X.O.; Nechuta, S.J. The Role of Biomarkers of Oxidative Stress in Breast Cancer Risk and Prognosis: A Systematic Review of the Epidemiologic Literature. *J. Women's Health* **2017**, *26*, 467–482. [CrossRef]

31. Marquez-Quiñones, A.; Cipak, A.; Zarkovic, K.; Fattel-Fazenda, S.; Villa-Treviño, S.; Waeg, G.; Zarkovic, N.; Guéraud, F. HNE-protein adducts formation in different pre-carcinogenic stages of hepatitis in LEC rats. *Free Radic. Res.* **2010**, *44*, 119–127. [CrossRef] [PubMed]

32. Balestrieri, M.L.; Dicitore, A.; Benevento, R.; Di Maio, M.; Santoriello, A.; Canonico, S.; Giordano, A.; Stiuso, P. Interplay between membrane lipid peroxidation, transglutaminase activity, and cyclooxygenase 2 expression in the tissue adjoining to breast cancer. *J. Cell. Physiol.* **2012**, *227*, 1577–1582. [CrossRef] [PubMed]

33. Karihtala, P.; Kauppila, S.; Puistola, U.; Jukkola-Vuorinen, A. Divergent behaviour of oxidative stress markers 8-hydroxydeoxyguanosine (8-OHdG) and 4-hydroxy-2-nonenal (HNE) in breast carcinogenesis. *Histopathology* **2011**, *58*, 854–862. [CrossRef] [PubMed]

34. Zhang, H.; Forman, H.J. Signaling by 4-hydroxy-2-nonenal: Exposure protocols, target selectivity and degradation. *Arch. Biochem. Biophys.* **2017**, *617*, 145–154. [CrossRef] [PubMed]

35. Cipak, A.; Mrakovcic, L.; Ciz, M.; Lojek, A.; Mihaylova, B.; Goshev, I.; Jaganjac, M.; Cindric, M.; Sitic, S.; Margaritoni, M.; et al. Growth suppression of human breast carcinoma stem cells by lipid peroxidation product 4-hydroxy-2-nonenal and hydroxyl radical-modified collagen. *Acta Biochim. Pol.* **2010**, *57*, 165–171. [CrossRef] [PubMed]

36. Cerbone, A.; Toaldo, C.; Laurora, S.; Briatore, F.; Pizzimenti, S.; Dianzani, M.U.; Ferretti, C.; Barrera, G. 4-Hydroxynonenal and PPARgamma ligands affect proliferation, differentiation, and apoptosis in colon cancer cells. *Free Radic. Biol. Med.* **2007**, *42*, 1661–1670. [CrossRef]

37. Varnat, F.; Duquet, A.; Malerba, M.; Zbinden, M.; Mas, C.; Gervaz, P.; Ruiz i Altaba, A. Human colon cancer epithelial cells harbour active HEDGEHOG-GLI signalling that is essential for tumour growth, recurrence, metastasis and stem cell survival and expansion. *EMBO Mol. Med.* **2009**, *1*, 338–351. [CrossRef]

38. Kalluri, R.; Weinberg, R.A. The basics of epithelial-mesenchymal transition. *J. Clin. Investig.* **2009**, *119*, 1420–1428. [CrossRef]

39. Lah, T.T.; Novak, M.; Breznik, B. Brain Malignancies: Cancer Cell Trafficking in and out of the Niches. In *Seminars Cancer Biology*; Academic Press: New York, NY, USA, 2019.

40. Hass, R.; von der Ohe, J.; Ungefroren, H. Potential Role of MSC/Cancer Cell Fusion and EMT for Breast Cancer Stem Cell Formation. *Cancers* **2019**, *11*, 1432. [CrossRef]

41. Meitzler, J.L.; Konaté, M.M.; Doroshow, J.H. Hydrogen peroxide-producing NADPH oxidases and the promotion of migratory phenotypes in cancer. *Arch. Biochem. Biophys.* **2019**, *675*, 108076. [CrossRef]

42. Sun, X.-D.; Liu, X.-E.; Huang, D.-S. Curcumin reverses the epithelial-mesenchymal transition of pancreatic cancer cells by inhibiting the Hedgehog signaling pathway. *Oncol. Rep.* **2013**, *29*, 2401–2407. [CrossRef] [PubMed]

43. Castro, J.P.; Jung, T.; Grune, T.; Siems, W. 4-Hydroxynonenal (HNE) modified proteins in metabolic diseases. *Free Radic. Biol. Med.* **2017**, *111*, 309–315. [CrossRef] [PubMed]

44. Go, Y.M.; Halvey, P.J.; Hansen, J.M.; Reed, M.; Pohl, J.; Jones, D.P. Reactive aldehyde modification of thioredoxin-1 activates early steps of inflammation and cell adhesion. *Am. J. Pathol.* **2007**, *171*, 1670–1681. [CrossRef] [PubMed]

45. Zheng, X.; Carstens, J.L.; Kim, J.; Scheible, M.; Kaye, J.; Sugimoto, H.; Wu, C.C.; Lebleu, V.S.; Kalluri, R. Epithelial-to-mesenchymal transition is dispensable for metastasis but induces chemoresistance in pancreatic cancer. *Nature* **2015**, *527*, 525–530. [CrossRef]

Article

A Role for H_2O_2 and TRPM2 in the Induction of Cell Death: Studies in KGN Cells

Carsten Theo Hack [1], Theresa Buck [1], Konstantin Bagnjuk [1], Katja Eubler [1], Lars Kunz [2], Doris Mayr [3] and Artur Mayerhofer [1,*]

[1] Biomedical Center Munich (BMC), Cell Biology, Anatomy III, Ludwig-Maximilian-University (LMU), D-82152 Planegg, Germany; theo.hack@campus.lmu.de (C.T.H.); theresa.buck@lrz.uni-muenchen.de (T.B.); Konstantin.Bagnjuk@lrz.uni-muenchen.de (K.B.); Katja.Eubler@lrz.uni-muenchen.de (K.E.)

[2] Division of Neurobiology, Department Biology II, Ludwig-Maximilian-University (LMU), D-82152 Planegg, Germany; lars.kunz@biologie.uni-muenchen.de

[3] Institute of Pathology, Ludwig-Maximilian-University (LMU), D-80337 München, Germany; doris.mayr@med.uni-muenchen.de

* Correspondence: mayerhofer@bmc.med.lmu.de; Tel.: +49-89-2180-75859

Received: 21 October 2019; Accepted: 25 October 2019; Published: 29 October 2019

Abstract: Recent studies showed that KGN cells, derived from a human granulosa cell tumor (GCT), express NADPH oxidase 4 (NOX4), an important source of H_2O_2. Transient receptor potential melastatin 2 (TRPM2) channel is a Ca^{2+} permeable cation channel that can be activated by H_2O_2 and plays an important role in cellular functions. It is also able to promote susceptibility to cell death. We studied expression and functionality of TRPM2 in KGN cells and examined GCT tissue microarrays (TMAs) to explore in vivo relevance. We employed live cell, calcium and mitochondrial imaging, viability assays, fluorescence activated cell sorting (FACS) analysis, Western blotting and immunohistochemistry. We confirmed that KGN cells produce H_2O_2 and found that they express functional TRPM2. H_2O_2 increased intracellular Ca^{2+} levels and N-(p-Amylcinnamoyl)anthranilic acid (ACA), a TRPM2 inhibitor, blocked this action. H_2O_2 caused mitochondrial fragmentation and apoptotic cell death, which could be attenuated by a scavenger (Trolox). Immunohistochemistry showed parallel expression of NOX4 and TRPM2 in all 73 tumor samples examined. The results suggest that GCTs can be endowed with a system that may convey susceptibility to cell death. If so, induction of oxidative stress may be beneficial in GCT therapy. Our results also imply a therapeutic potential for TRPM2 as a drug target in GCTs.

Keywords: ovary; calcium channel; Trolox; granulosa cell tumor; cell death; mitochondria

---

1. Introduction

In a recent study, we described expression of NADPH oxidase 4 (NOX4) [1] in vivo in human granulosa cells (GCs) of ovarian follicles and in vitro in granulosa-lutein cells derived from in vitro fertilization patients. Activity of this enzyme is linked to the generation of H_2O_2 [2], which is a diffusible reactive oxygen species (ROS) and has been postulated to be an important signaling molecule within the follicle (e.g., [3]). Although precise modes of action remain to be identified, involvement in GC proliferation has been suggested by studies employing the granulosa cell tumor (GCT) line KGN [4] and a specific NOX4 blocker [2].

These results are in line with the changing view of ROS. They are no longer regarded as destructive correlates of oxidative stress only, but their importance in the regulation of cellular functions and in the maintenance of the essential redox homeostasis is being more and more recognized [5–7]. Yet, ROS in higher concentrations are indeed often associated with cell death [8–10].

Cellular actions of H_2O_2 are linked to transient receptor potential melastatin 2 (TRPM2) channel, a cation channel permeable for Ca^{2+} that is activated by oxidative stress and therefore considered to be a cellular redox sensor [11,12]. Studies in recent years have provided evidence of a role of TRPM2-mediated Ca^{2+} influx in physiological and pathophysiological functions, such as insulin release by pancreatic β-cells, pro-inflammatory cytokine production in immune cells, endothelial permeability and cell death [11]. Cell death is the most outstanding and common consequence of TRPM2 channel activation, and has been described in several publications (e.g., [8,13,14]). The exact mechanism of TRPM2 activation by H_2O_2 is still a subject of ongoing research; however, there are well described inhibitors such as N-(p-Amylcinnamoyl)anthranilic acid (ACA) that are widely being used in studies on TRPM2 [15–17].

Information about TRPM2 in ovarian cells is sparse. However, data mining of recently published single-cell RNA sequencing data has revealed that this channel is expressed in human GCs in situ [18]. To our knowledge, TRPM2 expression in GCT has not been explored yet.

In the present study we examined KGN, a model for GCT [4,19]. KGN cells express NOX4 and generate H_2O_2 [2]. We found that they also express TRPM2, which can be activated by H_2O_2 and facilitate an influx of Ca^{2+}, followed by mitochondrial fragmentation and cell death. Immunohistochemical analysis of tissue microarrays (TMAs) revealed that both NOX4 and TRPM2 were expressed by all GCT samples we examined. Our findings suggest that induction of oxidative stress in GCT may result in cell death. Furthermore, the results implicate a therapeutic potential of TRPM2 as a possible drug target.

2. Materials and Methods

2.1. KGN Cell Culture

Procedures have recently been described [2]. The patented KGN cell line was obtained from RIKEN BioResource Center [4] with permission by T. Yanase. KGN cells were cultured in Dulbecco's Modified Eagle's Medium (DMEM)/Ham's F12 medium (Life Technologies, Paisley, UK) supplemented with penicillin (100 U/mL), streptomycin (100 µg/mL) (Biochrom, Berlin, Germany) and 10% fetal calf serum (FCS) (Capricorn Scientific, Ebsdorfergrund, Germany) at 37 °C and with 5% CO_2. For stimulation experiments, 20 µM Trolox (Santa Cruz Biotechnology, Dallas, TX, USA) or 100 µM/1 mM H_2O_2 (Sigma-Aldrich, St. Louis, MO, USA) was diluted in DMEM/Ham's F12 medium (Life Technologies; colorless medium without phenol red was used for live cell fluorescence imaging to reduce background autofluorescence).

2.2. Reverse Transcription PCR

RNeasy Plus Micro Kit (Qiagen, Hilden, Germany) was used to isolate RNA. Concentration and purity were measured as described [2]. Superscript II (Invitrogen, Carlsbad, CA, USA) and random 15-mer primers (metabion international, Munich, Germany) were used for reverse transcription (RT). Oligonucleotide primers for amplification of NOX4 were described previously [2]. For TRPM2, we used primers with the following sequences: 5′–AGGCTGAACTCTAACCTGCAC–3′ (forward) and 5′–GGAGGAGGGTCTTGTGGTTC–3′ (reverse) (yielding a 103 bp fragment). Negative controls were performed by replacing cDNA with RNA (-RT) or water (H_2O). Amplicon identity was verified by agarose gel electrophoresis, consecutive cDNA extraction with Wizard SV Gel and PCR Clean-up System (Promega, Madison, WI, USA) and sequence analysis (GATC, Konstanz, Germany).

2.3. Western Blotting

Protein isolation and Western blotting were performed as previously described [20]. KGN cells were lysed using RIPA buffer plus protease and phosphatase inhibitors (Thermo Fisher Scientific, Waltham, USA). A total of 7 µg (NOX4) or 20 µg (cleaved caspase 3, clCASP3) protein per lane was loaded on a 10% (NOX4) or 12% (clCASP3) SDS gel and run (NOX4: 20 min at 100 V + 70 min at 120 V;

clCASP3: 20 min at 75 V + 40 min at 150 V). After blotting (NOX4: 55 min at 100 V; clCASP3: 60 min at 100 V) and blocking with 5% non-fat dry milk (Roth, Karlsruhe, Germany) in Tris-buffered saline with Tween 20 (5 mM Tris, 100 mM NaCl, 0.05% Tween 20, pH 7.5), rabbit anti-NOX4 polyclonal antiserum (1:1000, #7927, ProSci, Fort Collins, CO, USA) or rabbit anti-clCASP3 monoclonal antibody (1:1000, #9664, Cell Signaling Technology, Danvers, MA, USA) were administered to detect these proteins. As a loading control, mouse anti-β-actin monoclonal antibody (1:10000, #A5441, Sigma-Aldrich) was used. HRP-conjugated goat anti-rabbit and goat anti-mouse secondary antibodies (Jackson ImmunoResearch Europe, Cambridgeshire, UK) were used to visualize specific binding. Band intensities were determined using the FIJI software [21].

2.4. Immunohistochemistry

Immunohistochemistry was performed using TMAs assembled from anonymized archival material. All patients were treated surgically at the same institution (Department of Gynaecology, University of Munich, Germany) and diagnosed at the Institute for Pathology, LMU, Munich, Germany. The diagnoses were confirmed by an experienced gynaecopathologist (D.M.). Tissue biopsies ($n = 73$) were taken from representative regions of larger paraffin-embedded tumor samples and arrayed into a new recipient paraffin block using MTA-1 (Manual Tissue Arrayer) from Beecher Instruments, Sun Prairie, WI, USA. Staining procedures were conducted as previously described [22]. In brief, sections were deparaffinized, antigens were unmasked and endogenous peroxidase activity was blocked, followed by incubation in 10% goat serum, diluted in PBS, to prevent unspecific binding. Polyclonal rabbit antisera raised against human NOX4 (1:500, #7927, ProSci, Fort Collins, CO, USA) or against human TRPM2 (1:100, #HPA035260, Sigma-Aldrich) were used to identify these proteins in the TMAs. Specific binding was detected by biotinylated goat anti-rabbit secondary antibody and Vectastain Elite ABC kit (Vector Laboratories, Burlingame, CA, USA). For negative controls, incubation with normal rabbit serum instead of the primary antiserum was performed. Sections were counterstained with haematoxylin and visualized using a Zeiss Axioplan microscope with the Achroplan 63x/0.80 objective (Carl Zeiss Microscopy, Jena, Germany) and a Jenoptik camera (Progres Gryphax Arktur; Jenoptik, Jena, Germany).

2.5. Measurement of H_2O_2

The generation of H_2O_2 was measured using an Amplex Red Hydrogen Peroxide/Peroxidase Assay Kit (Life Technologies, Eugene, OR, USA) as described previously [2,23]. In brief, KGN cells were seeded in black 96-well plates (1.5×10^4 cells/well, $n = 6$) and cultured overnight. Amplex Red reagent (10-acetyl-3,7-dihydroxyphenoxazine) was used in a final concentration of 5.0 μM and fluorescence levels were measured at 544 nm excitation/590 nm emission in a fluorometer (FLUOstar Omega, BMG LABTECH, Ortenberg, Germany) for 115 min at 37 °C. Data points were normalized according to the starting point value. To compare H_2O_2 concentrations in the supernatant, 3×10^5 cells were seeded on a 60-mm (diameter) cell culture dish the day before stimulation. After 72 h of stimulation with 20 μM Trolox or serum-free medium only, supernatants were collected, centrifuged and measured with the Amplex Red Kit according to the manufacturer's instructions ($n = 3$).

2.6. Cell Viability Assay, Confluence Measurement and Cell Counting

Cell viability was estimated by measuring cellular ATP content (the indicator for metabolically active cells) using CellTiter-Glo Luminescent Cell Viability Assay (Promega, Madison, WI, USA) as described previously [2,24]. KGN cells were seeded on a white 96-well microtiter plate (5.0×10^3 cells/well, $n = 12$) one day prior to stimulation, then exposed to 20 μM Trolox or serum-free medium for 72 h. After removal of the supernatant and washing with PBS, wells were filled with a 1:1 mixture of CellTiter-Glo reagent and DMEM/Ham's F12 without phenol red (100 μl/well), mixed on a plate shaker and incubated for 10 min at room temperature. Luminescence was measured by a luminometer (FLUOstar Omega; BMG LABTECH). Confluence was analyzed with the JuLIBr Live cell movie analyzer (NanoEnTek, Waltham, MA, USA) for a period of 72 h. For determination of cell numbers, KGN cells were counted using the CASY Cell Counter (OLS OMNI Life Science, Bremen, Germany).

2.7. Fluorescence-Activated Cell Sorting (FACS) Analysis

FITC-conjugated annexin V (ALX-209-256-T100, Enzo Life Sciences, Farmingdale, NY, USA) and SYTOX Red Dead Cell Stain (S34859, Invitrogen) were used to examine the occurrence of apoptosis. KGN cells were incubated in colorless DMEM/Ham's F12 medium for 24 h with or without 100 μM H_2O_2, or for 72 h with or without 20 μM Trolox. They were trypsinized, washed with PBS and incubated with annexin V-FITC (2.5 μg/mL) according to the manufacturer's instructions. SYTOX Red Dead Cell Stain—a nucleic acid stain labeling cells with damaged membranes—was added (1:2000), and cells were analyzed using the BD FACSCanto (Becton, Dickinson and Company, Franklin Lakes, NJ, USA). Annexin V-FITC signal was obtained using a 488 nm excitation laser and a 530/30 bandpass (BP) filter. For the SYTOX Red signal, a 633-nm laser and a 660/20 BP filter were used, with 20,000 events recorded for each treatment. Signals were analyzed using BD FACSDiva Software (version 8.0.1, Becton, Dickinson and Company). Cells show single staining with annexin V in an early stage of apoptosis (intact membrane), while they are double-positive for annexin V and nucleic acid stains in late apoptosis (cf. [25,26]). The apoptotic/late apoptotic (L/A) fraction of the analyzed cells comprise early and late stage apoptosis.

2.8. Calcium Imaging

For all calcium imaging experiments, KGN cells were incubated with 5 μM Fluoforte Reagent (Enzo Life Sciences)—a fluorescent dye detecting intracellular Ca^{2+}—in DMEM/Ham's F12 without phenol red for 30 min at 37 °C and 5% CO_2 in ibidi dishes optimized for microscopy (μ-dish 35 mm, ibidi, Gräfelfing, Germany). After washing with colorless medium, fluorescence was monitored every 5 s using a wide-field microscope (microscope: Axio Observer.Z1; light-source: Colibri.2; camera: Axiocam 506 mono; objective: Plan-Apochromat 20x/0.8 Ph2 M27; software: ZEN 2.6; Carl Zeiss Microscopy) with a 450–490 nm BP excitation and 500–550 nm BP emission filter (F46-002; AHF analysentechnik, Tübingen, Deutschland). A continuous flow of medium with or without stimulant was generated by a peristaltic pump (Instech Laboratories, Plymouth Meeting, PA, USA) linked to needles that were placed under the surface of the medium, close to the observed cells. H_2O_2 was utilized in a higher concentration (1 mM) to address diluting effects. For the blocking experiments, KGN cells were incubated with either ACA (20 μM, Sigma-Aldrich) or DMSO (solvent control) for 4 h prior to and during measurements. Stimulation with 0.05‰ trypsin (Biochrom, Berlin, Germany) served as a positive control (cf. [27,28]). FIJI software was used to obtain fluorescence intensities of the regions of interest (ROIs) and to optimize the images and videos provided in the Supplementary Materials. Background fluorescence was subtracted from the raw data and results were normalized to the starting point values. Images and videos showed fluorescence intensity based on a pseudo-color scale from black/red (low Ca^{2+}) to yellow/white (high Ca^{2+}).

2.9. Mitochondrial Imaging

KGN cells were incubated with 100 nM MitoTracker Green FM (Molecular Probes, Eugene, OR, USA) for 30 min at 37 °C and 5% CO_2 in microscopy-optimized cell culture dishes (μ-dish 35 mm, ibidi). The staining solution was prepared in DMEM/Ham's F12 without phenol red. To examine changes of the mitochondrial structure, cells were treated with 100 μM H_2O_2 after staining and washing. Fluorescence was recorded with a wide-field microscope (microscope: Axio Observer.Z1; light-source: Colibri.2; camera: Axiocam 506 mono; objective: Plan-Apochromat 63x/1.40 Oil Ph 3 M27; software: ZEN 2.6; Carl Zeiss Microscopy). A 450–490 nm BP (excitation) and a 500–550 nm BP (emission) were used (F46-002; AHF analysentechnik). In a second approach, stimulation with H_2O_2

for 4 h in colorless medium prior to staining and imaging was performed to rule out phototoxicity due to multiple imaging as a reason for mitochondrial fragmentation.

To evaluate the effect of H_2O_2 (100 µM) on mitochondria, 239 control cells and 122 treated cells in two dishes each for both groups were analyzed after 4 h of treatment. Examples for KGN cells with elongatedor fragmented mitochondrial networks are given in the corresponding figure.

2.10. Statistics

GraphPad Prism 6.0 Software (GraphPad Software, San Diego, CA, USA) was used to perform unpaired *t*-tests (two-tailed) for comparisons of H_2O_2, ATP and cell counts. Control and Trolox-treatments were performed in parallel ($n = 3$), derived samples were run next to each other on the gels and results (band intensities) of clCASP3 Western blots were analyzed using paired *t*-test.

3. Results

3.1. KGN Cells Express H_2O_2 Generating NOX4

Expression of NOX4, an enzyme known to generate H_2O_2, was detected by Western blot (68 kDa) and RT-PCR (160 bp) (Figure 1A). The Amplex Red Hydrogen Peroxide Assay ($n = 6$) provided evidence for basal H_2O_2 production and release by untreated KGN cells, resulting in increasing levels in the supernatant over time (Figure 1B).

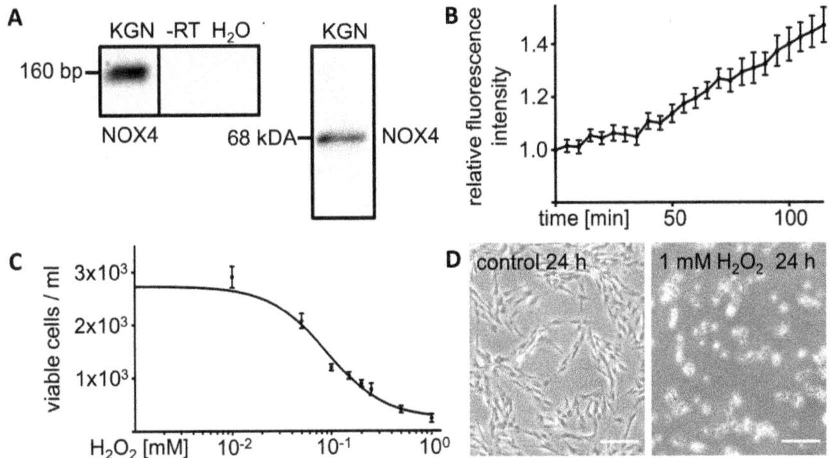

Figure 1. H_2O_2 production and release by cultured KGN cells and effects of exogenous H_2O_2. (**A**) NOX4 RT-PCR analysis and Western blot of cultured KGN cells show single bands of 160 bp and 68 kDa, respectively. Controls using RNA (-RT) or H_2O instead of cDNA (H_2O) were negative. (**B**) Hydrogen peroxide assay of untreated KGN cells showed increasing H_2O_2 levels in the supernatant over a time period of 2 h ($n = 6$). Signal intensities were normalized to start point values. Bars indicate SEM. (**C**) Exogenously added H_2O_2 reduced cell viability in a dose dependent manner. Cell counts after treatment of KGN cells with different concentrations of H_2O_2 for 24 h ($n = 2$–5 for each concentration) are shown with an interpolated sigmoidal standard curve ($r^2 = 0.9361$). Bars indicate SEM. (**D**) Images of KGN cells treated with H_2O_2 (1 mM) for 24 h compared to untreated control cells. Scale bars indicate 200 µm.

3.2. Exogenous H_2O_2 Kills KGN Cells in a Dose Dependent Manner

To examine effects of H_2O_2, KGN cells were exposed to H_2O_2 at different concentrations for 24 h ($n = 2$ to 5 for each concentration). We observed a dose dependent reduction of cell viability (cell counting, live cell imaging). Cell numbers decreased with a calculated EC$_{50}$ of 89.7 μM (72.16–111.5 μM; interpolated sigmoidal standard curve: $f(x) = 23800 + \frac{272929-23800}{1+10^{(1.953-x)(-1.507)}}$; $r^2 = 0.9361$) (Figure 1C). Figure 1D shows a live cell image of KGN cells treated with H$_2$O$_2$ (1 mM) for 24 h compared to controls, revealing the damaging effects of exogenous H$_2$O$_2$.

3.3. Trolox Promotes Survival of KGN Cells in Serum-Free Medium

Culturing KGN cells in serum-free medium for 72 h led to a drop in confluence after an initial 2.6 ± 0.2 fold increase for the first 40 h (mean). This decrease was prevented by the addition of Trolox (20 μM), a water-soluble derivate of vitamin E. Trolox is well known for its antioxidative activity [29,30] and kept KGN cells prospering until the end of the measurement. Pictures taken by a live cell imaging system show the difference between treated and control cells. While KGN cells looked healthy within the first part of the observation under both conditions, they detached after 72 h under control conditions, but were further propagated with Trolox (Figure 2A). The H$_2$O$_2$ concentration in the supernatant of KGN cells after 72 h in serum-free medium was significantly ($n = 3$, $p < 0.0001$, t-test) reduced by Trolox (Figure 2B). Cell counts ($n = 4$, $p < 0.0001$, t-test) (Figure 2C) and viability assay ($n = 12$, $p < 0.0001$, t-test) (Figure 2D) after 72 h gave further evidence for the positive effects of Trolox on KGN cell survival in serum-free medium.

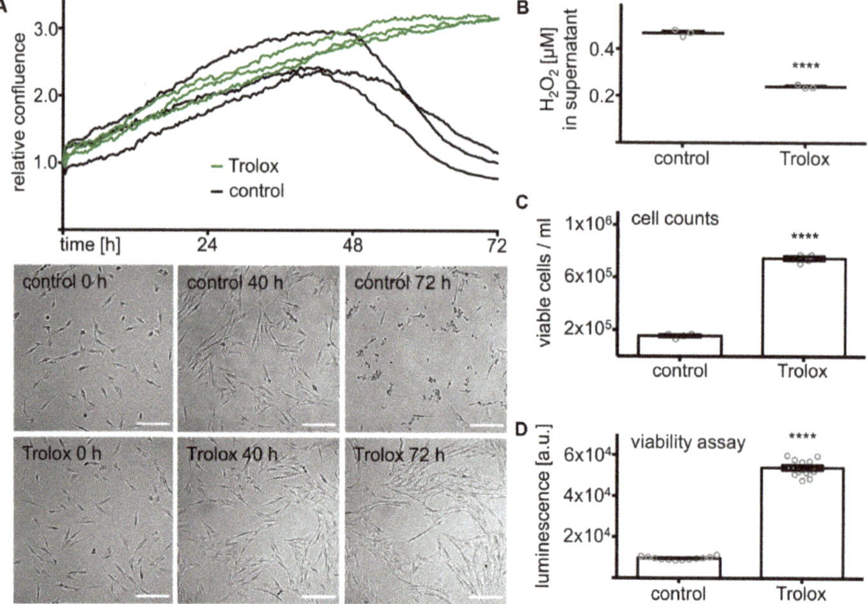

Figure 2. Antioxidant Trolox antagonized the effects of endogenous H$_2$O$_2$. (**A**) Changes in confluence of KGN cells over the course of 72 h for cells treated with Trolox (20 μM) compared to serum-free medium only ($n = 3$). Results were normalized to the respective start value. Images at different time points are shown below. Scale bars indicate 200 μm. (**B**) H$_2$O$_2$ in the supernatant after treatment with Trolox (20 μM) for 72 h, measured by a hydrogen peroxide assay, was significantly lower compared to ontrols ($n = 3$, **** $p < 0.0001$). Means and SEM, as well as individual results are given. (**C**) Cell counts relative to the average untreated controls significantly increased ($n = 4$, **** $p < 0.0001$). Means and SEM, as well as individual results are shown. (**D**) ATP viability assay-generated luminescence signaling was significantly higher in Trolox-treated cells ($n = 12$, **** $p < 0.0001$) after 72 h of treatment. Means and SEM, as well as individual results (circles), are presented.

3.4. Effects of Cultivation in Serum-Free Medium and Exogenous H_2O_2 on Markers of Apoptosis and Necrosis

FACS analysis of KGN cells co-stained with FITC-conjugated annexin V and the nucleic acid stain SYTOX Red Dead Cell Stain revealed an 11.4-fold increase in the apoptotic/late apoptotic (L/A) fraction, while the necrotic fraction (N) only doubled in cells treated with H_2O_2 (100 µM) for 24 h compared to the serum-free medium control (Figure 3A,B). Culturing KGN cells in serum-free medium resulted in a 4.1-fold higher L/A cell fraction after 72 h compared to 24 h, whereas the N fraction reduced by 23% (Figure 3A,C).

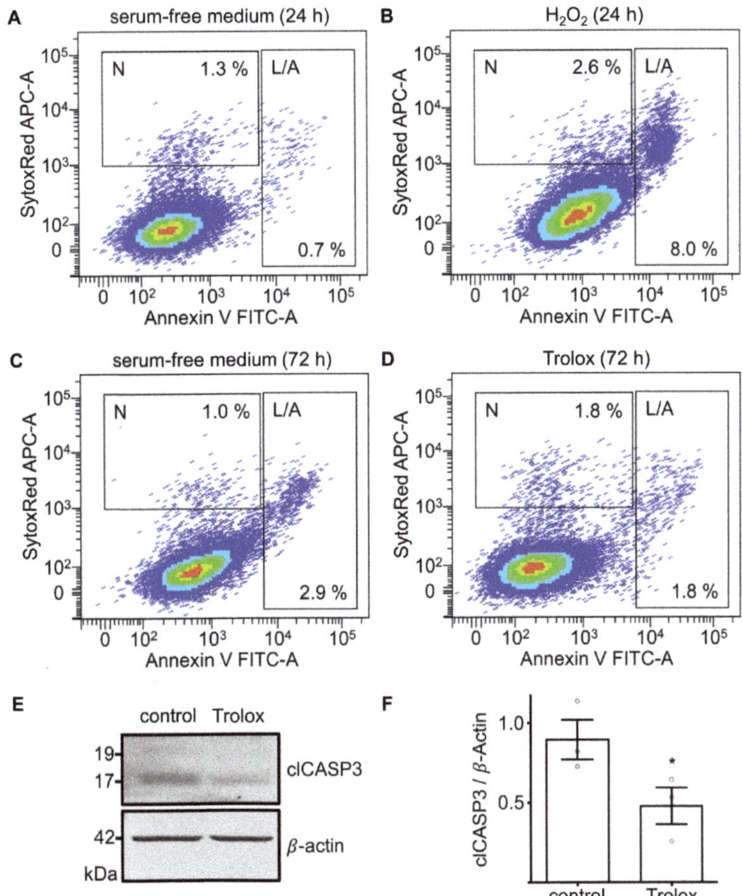

Figure 3. Effects of serum-free medium, exogenous H_2O_2 and Trolox on markers of apoptosis and necrosis. (**A–D**) FACS analysis of KGN cells co-stained with annexin V and SYTOX Red Dead Cell Stain. N indicates necrotic (single stained with SYTOX Red), apoptotic and late apoptotic (L/A) cells (single stained with annexin V or double-stained). Unstained cells in the left lower quadrant were viable. Percentage values of the N and L/A fractions are shown for each treatment: (**A**) Serum-free medium for 24 h; (**B**) H_2O_2 (100 µM) in serum-free medium for 24 h; (**C**) serum-free medium for 72 h; (**D**) Trolox (20 µM) in serum-free medium for 72 h; (**E**) Western blot membrane with clCASP3 and β-actin bands; (**F**) clCASP3 levels in KGN cells after treatment with Trolox (20 µM) for 72 h are compared with control cells in serum-free medium. clCASP3 relative to β-actin was significantly lower in treated cells ($n = 3$, * $p < 0.05$). Means and SEM and individual results are shown.

3.5. Treatment with Trolox Reduces Markers of Apoptosis

Addition of Trolox (20 µM) to the serum-free medium reduced the L/A fraction after 72 h in serum-free medium by 37.9% (Figure 3C,D), but not the N fraction, which actually increased. Western blot analysis confirmed the effect on apoptosis by showing a significant reduction in clCASP3 ($n = 3$, $p = 0.0225$, paired *t*-test) (Figure 3E,F).

3.6. KGN Cells Express Functional TRPM2

Expression of TRPM2 channel, a H_2O_2-responsive Ca^{2+}-permeable cation channel, was detected by RT-PCR (Figure 4A) and sequencing. To examine functionality of TRPM2, changes in intracellular Ca^{2+} levels were imaged. Stimulation with H_2O_2 (1 mM) caused a transient increase in Ca^{2+} levels in three independent measurements, which occurred with a delay of more than 1 min and quickly disappeared after terminating the stimulation. Repeated stimulation was possible (Figure 4B,C). The Ca^{2+} increase was blocked by treatment with the TRPM2 inhibitor ACA (20 µM) [15–17] applied for 4 h prior to and during the measurement. Cellular response to the positive control (trypsin) was not affected (Figure 4D). The H_2O_2-derived signal was obtained in the control experiments with the solvent (Figure 4E). Videos are provided in the Supplementary Materials (Videos S1–S3). Blocking experiments and according controls were repeated in four independent measurements each.

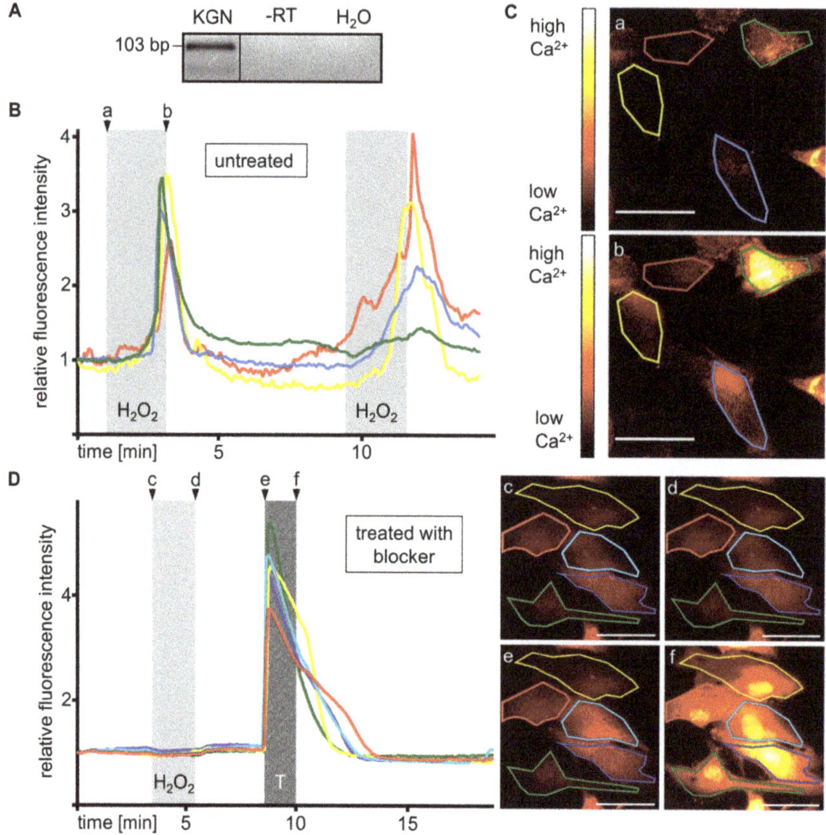

Figure 4. *Cont.*

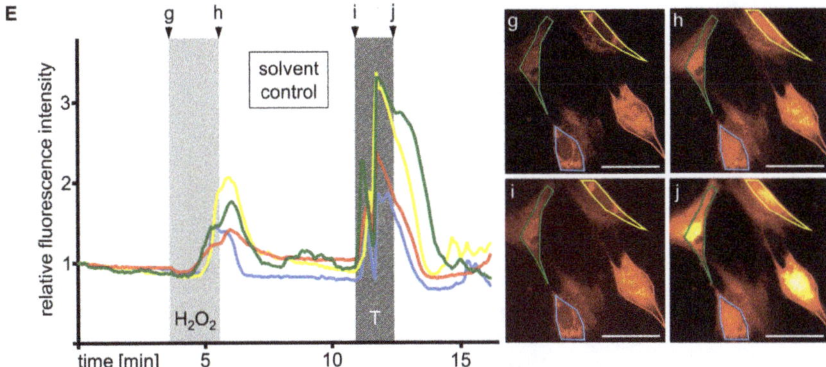

Figure 4. KGN cells express functional TRPM2. (**A**) TRPM2 RT-PCR shows a band at 103 bp. Controls with RNA (-RT) or H_2O instead of cDNA (H_2O) were negative. (**B**) Addition of H_2O_2 (1 mM) increased the fluorescence signal of the four individual KGN cells shown, which were loaded with the Ca^{2+}-sensitive dye Fluoforte. Background signals were subtracted and fluorescence is shown relative to the respective start value of each region of interest (ROI). (**C**) Fluorescence images, taken before (**a**) and after (**b**) the first stimulation with H_2O_2. (**D**) Treatment with the inhibitor (ACA; 20 µM), 4 h prior to and during the measurement, blocked the Ca^{2+} increase upon stimulation with H_2O_2, but not with 0.05‰ trypsin (T). Images (**c–f**) represent the indicated time points. (**E**) The H_2O_2-derived Ca^{2+} increase was obtained in the DMSO control and thus ruled out solvent effects. Images (**g–j**) represent the indicated time points. The pseudo-color scale shown in (**c**) applies for all live cell images. Colored frames mark the cells represented in the corresponding graphs. Scale bars indicate 50 µm.

3.7. Exogeneous H_2O_2 Causes Mitochondrial Fragmentation

Monitoring mitochondria of KGN cells during treatment with H_2O_2 (100 µM) revealed fragmentation over time (Figure 5A). Stimulation for 4 h prior to staining and imaging was performed to rule out phototoxicity due to multiple imaging during time series as a reason for fragmentation. Mitochondria were fragmented in this approach (Figure 5B), and comparison of H_2O_2-treated cells to the control cells revealed a vast difference in the portion of KGN cells presenting mitochondrial fragmentation. One out of 239 (0.4%) untreated and 50 out of 122 (41.0%) treated cells showed fragmentation (Figure 5C).

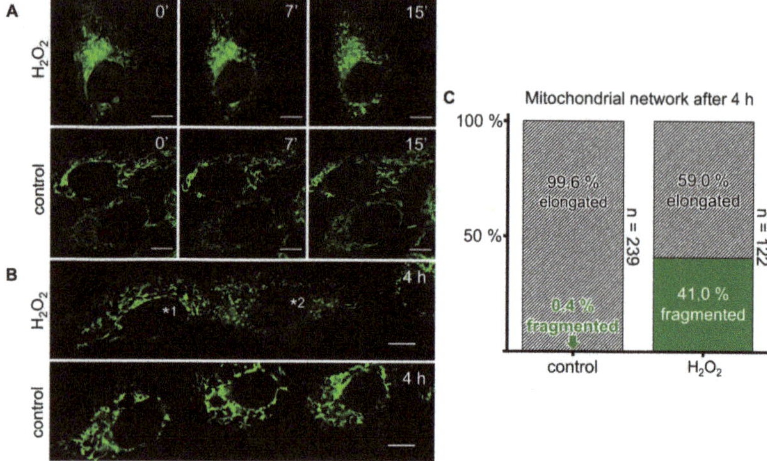

Figure 5. H_2O_2 causes mitochondrial fragmentation. (**A**) MitoTracker Green FM-based live cell fluorescence images show mitochondria of KGN cells treated with H_2O_2 (100 µM) at different time points, compared to medium-only controls. (**B**) KGN cells stained and imaged after 4 h of treatment with H_2O_2 (100 µM) or medium only. [*1] indicates a cell presenting an elongated mitochondrial network, [*2] indicates an example of a fragmented mitochondrion. (**C**) Portion of cells with fragmented mitochondria after 4 h, as determined by counting 239 control and 122 treated KGN cells. Mitochondrial fragmentation was markedly increased in H_2O_2-treated cells. Scale bars (**A**,**B**) indicate 10 µm.

3.8. Primary GCT Express NOX4 and TRPM2

As the KGN cell line serves as a well-established in vitro model for GCTs, we analyzed NOX4 and TRPM2 expression in 73 GCT samples using TMAs. Immunohistochemical analysis revealed that all of the tumors expressed both NOX4 and TRPM2 (Figure 6). Both proteins were detected in GCT cells and showed a generally homogenous distribution, but intensities varied between different tumors.

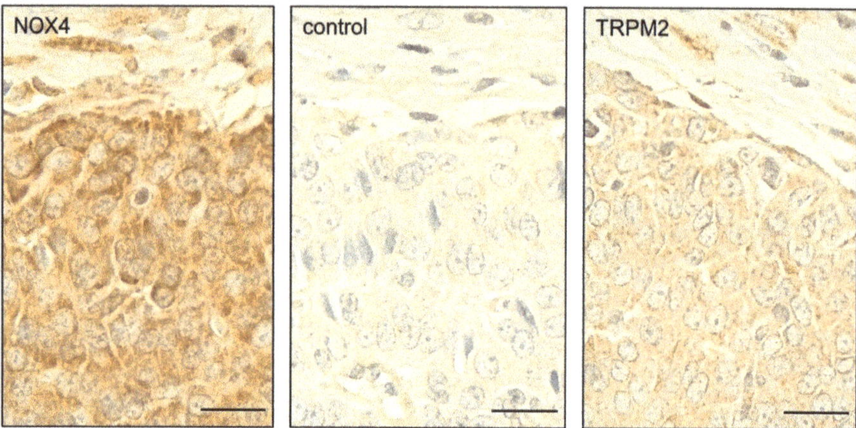

Figure 6. Granulosa cell tumors (GCTs) express NOX4 and TRPM2. Immunohistochemical staining of NOX4 and TRPM2 in one of the 73 tumor samples analyzed. NOX4 and TRPM2 were detected in GCT cells, normal rabbit serum control was negative. Scale bars indicate 20 µm.

4. Discussion

Our results show that H_2O_2 from both exogenous and endogenous sources is able to induce cell death in KGN cells. The majority of endogenous H_2O_2 is most likely generated by NOX4, as previously shown [2]. NOX4 expression in KGN cells was confirmed by RT-PCR and Western blot. Trolox, a typical ROS scavenging antioxidant [29,30], reduced endogenously produced H_2O_2 in the supernatant and rescued the cells, as shown by confluence measurements, cell counting and ATP cell viability assay.

There is ample evidence implying that H_2O_2 activates the ROS-gated cation channel TRPM2, and that the consecutive Ca^{2+} influx may cause cell death (e.g., [8–10]). Single cell RNA sequencing data in a recent publication [18] revealed expression of TRPM2 in GCs of human follicles in situ. We showed that TRPM2 is expressed in the GCT-derived tumor cell line KGN and calcium imaging suggested its functionality. Intracellular Ca^{2+} levels increased upon H_2O_2 stimulation and disappeared after termination of the treatment. The observed delay of more than 1 min provided an additional indication of TRPM2 involvement, since its activation by H_2O_2 is reported to be slow and take up to minutes [11].

To provide further evidence for the functionality of TRPM2 in KGN cells, we utilized ACA to inhibit TRPM2 in the calcium imaging experiment. ACA completely blocked the Ca^{2+} influx upon H_2O_2 stimulation, pinpointing TRPM2 as the channel responsible for the signal observed in untreated cells. As we found in our experiments, extracellular application of ACA was reported previously to completely block the H_2O_2-induced increase of intracellular Ca^{2+} in TPRM2-expressing cells at a concentration of 20 µM [15]. Yet, ACA is also a phospholipase A_2 (PLA_2) blocker. Although the inhibitory action on TRPM2 was reported to be independent of effects on PLA_2 [15], detrimental consequences of PLA_2 inhibition cannot be ruled out, especially in long-term treatments. We therefore did not perform additional experiments.

In accordance with previous studies reporting apoptosis upon H_2O_2 stimulation in primary GCs [31] and other cells [8–10], we detected a distinct induction of apoptosis/late apoptosis by H_2O_2 in KGN cells. Although annexin V and nucleic acid stain double-stained cells are often referred to as late apoptotic [25,26], necrotic cells might be double-positive as well [32]. Given that changes in the necrotic cell fraction do not match the changes in the annexin V signals, our results provide evidence that exogenous H_2O_2 induces apoptosis, although other cell death forms might be involved as well.

Fetal calf serum (FCS) was reported to feature a total anti-oxidant capacity (TAC) equaling 360 ± 40 µM Trolox [33], which implicates a minor role of endogenously produced H_2O_2 in media supplemented with FCS. Even 10% of FCS, as usual in our culture medium, exceeds the TAC of 20 µM Trolox. For different experiments we therefore used serum-free medium. Under these conditions, H_2O_2 produced by cells may reach concentrations that might be high enough to activate TRPM2, especially in their immediate surroundings, although the overall concentration within the supernatant remains relatively low compared to the concentration used for exogenous stimulation with H_2O_2. In accordance, results of the FACS analysis and the clCASP3 Western blot revealed an induction of apoptosis by long-term cultivation in serum-free medium. Reduction of markers for apoptosis by the ROS scavenger Trolox and the differences in H_2O_2 levels in the supernatants suggest that endogenous H_2O_2 might play an important role in the induction of apoptosis. These results are in line with the ability of another antioxidant, N-acetyl-L-cysteine, to reduce H_2O_2-induced apoptotic cell death in human melanocytes [8].

In their review on mitochondrial dynamics and apoptosis, Suen et al. [34] discussed the role of mitochondrial fragmentation and pointed out that while there are many different conditions wherein this can be observed, it is always involved in apoptosis and appears before caspase activation. To further examine the mechanism underlying the H_2O_2 effects on KGN cells, we performed live cell imaging of mitochondria labeled with MitoTracker Green FM. We observed rapid and massive fragmentation of these organelles upon H_2O_2 stimulation. This may reflect another manifestation of an activated apoptosis machinery and be a consequence of Ca^{2+}-dependent phosphorylation of dynamin-related protein 1 (Drp1) [35], a critical player in mitochondrial fragmentation [36]. The exact mechanisms in KGN cells and the question of whether other forms of cell death are involved remain to be studied.

To explore in vivo the relevance of our cellular results, we studied the expression of NOX4, a typical source for endogenous H_2O_2 [1,2], and TRPM2, a possible target for H_2O_2 [11,12], in primary GCT. Immunohistochemical analysis of TMAs revealed that both NOX4 and TRPM2 were expressed in all 73 tumors analyzed in this study. We noticed that signal intensities varied between the different GCTs. However, as we studied archival material, we reasoned that variations in sample preparations or storage could not be ruled out. Therefore, we did not attempt to further evaluate these differences.

Investigations of other tumor's entities indicated that NOX4-derived ROS may limit tumor progression (liver carcinoma [37]) and that TRPM2 overexpression may enhance induction of cell death by H_2O_2 (neuroblastoma [14]). TRPM2 confers susceptibility of different tumors to H_2O_2-mediated neutrophil cytotoxicity, thereby limiting metastasis [38]. Interestingly, inflammatory neutrophil infiltration is mediated by TRPM2 activation, resulting in chemokine production by monocytes [39].

In line with these observations, our immunohistochemical analysis of 73 GCTs and the results of experiments with KGN cells implicate that GCTs can be endowed with a relevant system that may convey susceptibility to cell death. Our in vitro-studies provide evidence that induction of oxidative stress may be beneficial in GCT therapy and that there is a therapeutic potential for TRPM2 as a drug target. Whether the new insights of our study are indeed of relevance in vivo remains to be shown.

Supplementary Materials: The following are available online at http://www.mdpi.com/2076-3921/8/11/518/s1: Video S1. Increasing intracellular Ca^{2+} levels upon stimulation with H_2O_2. Fluorescence images of the experiment represented in Figure 4B,C were compiled to a time-lapse video. The movie shows reversible increases in intracellular Ca^{2+} levels after addition of 1 mM H_2O_2 to KGN cells. Video S2. Time-lapse movie of the experiment represented in Figure 4D. This video shows levels of intracellular Ca^{2+} in cells treated with the TRPM2 blocker ACA. Ca^{2+} levels do not increase upon stimulation with 1 mM H_2O_2; yet, they increase upon stimulation with trypsin. Video S3. Obtained H_2O_2 effect in the solvent control. Time-lapse video of the experiment represented in Figure 4E. The movie shows that the Ca^{2+} response upon stimulation with H_2O_2 is obtained in the DMSO control. The second increase in Ca^{2+} derives from stimulation with trypsin.

Author Contributions: Conceptualization, C.T.H., A.M. and L.K.; methodology, C.T.H., K.E. and K.B.; validation, C.T.H., A.M., K.B., T.B. and L.K.; formal analysis, C.T.H.; investigation, C.T.H., K.B. and T.B.; resources, A.M., L.K. and D.M.; data curation, C.T.H.; writing—original draft preparation, C.T.H. and A.M.; writing—review and editing, C.T.H., A.M., K.B., T.B., L.K., K.E. and D.M.; visualization, C.T.H.; supervision, A.M. and L.K.; project administration, A.M. and L.K.; funding acquisition, A.M., L.K. and D.M.

Funding: This research was funded by DFG, MA1080/26-1 and MA1080/19-2 (to A.M.), KU1282/7-1 (to L.K.), MA 4790/4-1 (to D.M.).

Acknowledgments: We gratefully acknowledge the expert technical help of Kim Dietrich, Carola Herrmann and Astrid Tiefenbacher. We especially thank Toshihiko Yanase, Fukuoka University, Japan, for the permission to use KGN cells. Furthermore, we gratefully acknowledge the Core Facility Flow Cytometry at the Biomedical Center, Ludwig-Maximilians-Universität München, for analyzing samples and providing equipment and expertise. This work was done in partial fulfilment of the requirements of a Dr. med. degree at LMU (CTH).

Conflicts of Interest: The authors declare no conflict of interest. The funders had no role in the design of the study, in the collection, analyses or interpretation of data, in the writing of the manuscript or in the decision to publish the results.

Ethics Approval and Consent to Participation: The ethical committee of the LMU has approved the study (project 390-15).

References

1. Guo, S.; Chen, X. The human Nox4: Gene, structure, physiological function and pathological significance. *J. Drug Target* **2015**, *23*, 888–896. [CrossRef] [PubMed]

2. Buck, T.; Hack, C.T.; Berg, D.; Berg, U.; Kunz, L.; Mayerhofer, A. The NADPH oxidase 4 is a major source of hydrogen peroxide in human granulosa-lutein and granulosa tumor cells. *Sci. Rep.* **2019**, *9*, 3585. [CrossRef] [PubMed]

3. Shkolnik, K.; Tadmor, A.; Ben-Dor, S.; Nevo, N.; Galiani, D.; Dekel, N. Reactive oxygen species are indispensable in ovulation. *Proc. Natl. Acad. Sci. USA* **2011**, *108*, 1462–1467. [CrossRef] [PubMed]

4. Nishi, Y.; Yanase, T.; Mu, Y.; Oba, K.; Ichino, I.; Saito, M.; Nomura, M.; Mukasa, C.; Okabe, T.; Goto, K.; et al. Establishment and characterization of a steroidogenic human granulosa-like tumor cell line, KGN, that expresses functional follicle-stimulating hormone receptor. *Endocrinology* **2001**, *142*, 437–445. [CrossRef]

5. Rhee, S.G. Cell signaling. H2O2, a necessary evil for cell signaling. *Science* **2006**, *312*, 1882–1883. [CrossRef]

6. Gough, D.R.; Cotter, T.G. Hydrogen peroxide: A Jekyll and Hyde signalling molecule. *Cell Death Dis.* **2011**, *2*, e213. [CrossRef]

7. Schröder, K. NADPH oxidase-derived reactive oxygen species: Dosis facit venenum. *Exp. Physiol.* **2019**, *104*, 447–452. [CrossRef]

8. Kang, P.; Zhang, W.; Chen, X.; Yi, X.; Song, P.; Chang, Y.; Zhang, S.; Gao, T.; Li, C.; Li, S. TRPM2 mediates mitochondria-dependent apoptosis of melanocytes under oxidative stress. *Free Radic. Biol. Med.* **2018**, *126*, 259–268. [CrossRef]

9. Yang, K.T.; Chang, W.L.; Yang, P.C.; Chien, C.L.; Lai, M.S.; Su, M.J.; Wu, M.L. Activation of the transient receptor potential M2 channel and poly(ADP-ribose) polymerase is involved in oxidative stress-induced cardiomyocyte death. *Cell Death Differ.* **2006**, *13*, 1815–1826. [CrossRef]

10. Jones, B.E.; Lo, C.R.; Liu, H.; Pradhan, Z.; Garcia, L.; Srinivasan, A.; Valentino, K.L.; Czaja, M.J. Role of caspases and NF-kappaB signaling in hydrogen peroxide- and superoxide-induced hepatocyte apoptosis. *Am. J. Physiol. Gastrointest. Liver Physiol.* **2000**, *278*, G693–G699. [CrossRef]

11. Jiang, L.H.; Yang, W.; Zou, J.; Beech, D.J. TRPM2 channel properties, functions and therapeutic potentials. *Expert Opin. Ther. Targets* **2010**, *14*, 973–988. [CrossRef] [PubMed]

12. Takahashi, N.; Kozai, D.; Kobayashi, R.; Ebert, M.; Mori, Y. Roles of TRPM2 in oxidative stress. *Cell Calcium* **2011**, *50*, 279–287. [CrossRef] [PubMed]

13. Miller, B.A. The role of TRP channels in oxidative stress-induced cell death. *J. Membr. Biol.* **2006**, *209*, 31–41. [CrossRef] [PubMed]

14. An, X.; Fu, Z.; Mai, C.; Wang, W.; Wei, L.; Li, D.; Li, C.; Jiang, L.H. Increasing the TRPM2 Channel Expression in Human Neuroblastoma SH-SY5Y Cells Augments the Susceptibility to ROS-Induced Cell Death. *Cells* **2019**, *8*, 28. [CrossRef] [PubMed]

15. Kraft, R.; Grimm, C.; Frenzel, H.; Harteneck, C. Inhibition of TRPM2 cation channels by N-(p-amylcinnamoyl)anthranilic acid. *Br. J. Pharmacol.* **2006**, *148*, 264–273. [CrossRef] [PubMed]

16. Li, F.; Abuarab, N.; Sivaprasadarao, A. Reciprocal regulation of actin cytoskeleton remodelling and cell migration by Ca^{2+} and Zn^{2+}: Role of TRPM2 channels. *J. Cell. Sci.* **2016**, *129*, 2016–2029. [CrossRef] [PubMed]

17. Tamano, H.; Nishio, R.; Morioka, H.; Furuhata, R.; Komata, Y.; Takeda, A. Paraquat as an Environmental Risk Factor in Parkinson's Disease Accelerates Age-Related Degeneration Via Rapid Influx of Extracellular Zn^{2+} into Nigral Dopaminergic Neurons. *Mol. Neurobiol.* **2019**, *56*, 7789–7799. [CrossRef]

18. Zhang, Y.; Yan, Z.; Qin, Q.; Nisenblat, V.; Chang, H.M.; Yu, Y.; Wang, T.; Lu, C.; Yang, M.; Yang, S.; et al. Transcriptome Landscape of Human Folliculogenesis Reveals Oocyte and Granulosa Cell Interactions. *Mol. Cell* **2018**, *72*, 1021–1034. [CrossRef]

19. Li, J.; Bao, R.; Peng, S.; Zhang, C. The molecular mechanism of ovarian granulosa cell tumors. *J. Ovarian Res.* **2018**, *11*, 13. [CrossRef]

20. Bagnjuk, K.; Stöckl, J.B.; Fröhlich, T.; Arnold, G.J.; Behr, R.; Berg, U.; Berg, D.; Kunz, L.; Bishop, C.; Xu, J.; et al. Necroptosis in primate luteolysis: A role for ceramide. *Cell Death Discov.* **2019**, *5*, 67. [CrossRef]

21. Schindelin, J.; Arganda-Carreras, I.; Frise, E.; Kaynig, V.; Longair, M.; Pietzsch, T.; Preibisch, S.; Rueden, C.; Saalfeld, S.; Schmid, B.; et al. Fiji: An open-source platform for biological-image analysis. *Nat. Methods* **2012**, *9*, 676–682. [CrossRef] [PubMed]

22. Blohberger, J.; Kunz, L.; Einwang, D.; Berg, U.; Berg, D.; Ojeda, S.R.; Dissen, G.A.; Fröhlich, T.; Arnold, G.J.; Soreq, H.; et al. Readthrough acetylcholinesterase (AChE-R) and regulated necrosis: Pharmacological targets for the regulation of ovarian functions? *Cell Death Dis.* **2015**, *6*, e1685. [CrossRef] [PubMed]

23. Blohberger, J.; Buck, T.; Berg, D.; Berg, U.; Kunz, L.; Mayerhofer, A. L-DOPA in the human ovarian follicular fluid acts as an antioxidant factor on granulosa cells. *J. Ovarian Res.* **2016**, *9*, 62. [CrossRef] [PubMed]

24. Saller, S.; Kunz, L.; Dissen, G.A.; Stouffer, R.; Ojeda, S.R.; Berg, D.; Berg, U.; Mayerhofer, A. Oxytocin receptors in the primate ovary: Molecular identity and link to apoptosis in human granulosa cells. *Hum. Reprod.* **2010**, *25*, 969–976. [CrossRef]

25. Telford, W.; Tamul, K.; Bradford, J. Measurement and Characterization of Apoptosis by Flow Cytometry. *Curr. Protoc. Cytom.* **2016**, *77*, 9–49. [CrossRef]

26. Fathi, E.; Farahzadi, R.; Valipour, B.; Sanaat, Z. Cytokines secreted from bone marrow derived mesenchymal stem cells promote apoptosis and change cell cycle distribution of K562 cell line as clinical agent in cell transplantation. *PLoS ONE* **2019**, *14*, e0215678. [CrossRef]

27. Eubler, K.; Herrmann, C.; Tiefenbacher, A.; Kohn, F.M.; Schwarzer, J.U.; Kunz, L.; Mayerhofer, A. Ca(2+) Signaling and IL-8 Secretion in Human Testicular Peritubular Cells Involve the Cation Channel TRPV2. *Int. J. Mol. Sci.* **2018**, *19*, 2829. [CrossRef]

28. Azimi, I.; Bong, A.H.; Poo, G.X.H.; Armitage, K.; Lok, D.; Roberts-Thomson, S.J.; Monteith, G.R. Pharmacological inhibition of store-operated calcium entry in MDA-MB-468 basal A breast cancer cells: Consequences on calcium signalling, cell migration and proliferation. *Cell Mol. Life Sci.* **2018**, *75*, 4525–4537. [CrossRef]

29. Scott, J.W.; Cort, W.M.; Harley, H.; Parrish, D.R.; Saucy, G. 6-Hydroxychroman-2-Carboxylic Acids—Novel Antioxidants. *J. Am. Oil Chem. Soc.* **1974**, *51*, 200–203. [CrossRef]

30. Howard, A.C.; McNeil, A.K.; McNeil, P.L. Promotion of plasma membrane repair by vitamin E. *Nat. Commun.* **2011**, *2*, 597. [CrossRef]

31. Nakahara, T.; Iwase, A.; Nakamura, T.; Kondo, M.; Bayasula; Kobayashi, H.; Takikawa, S.; Manabe, S.; Goto, M.; Kotani, T.; et al. Sphingosine-1-phosphate inhibits H2O2-induced granulosa cell apoptosis via the PI3K/Akt signaling pathway. *Fertil. Steril.* **2012**, *98*, 1001–1008. [CrossRef] [PubMed]

32. Wallberg, F.; Tenev, T.; Meier, P. Analysis of Apoptosis and Necroptosis by Fluorescence-Activated Cell Sorting. *Cold Spring Harb. Protoc.* **2016**, *2016*, pdb.prot087387. [CrossRef] [PubMed]

33. Lewinska, A.; Wnuk, M.; Slota, E.; Bartosz, G. Total anti-oxidant capacity of cell culture media. *Clin. Exp. Pharmacol. Physiol.* **2007**, *34*, 781–786. [CrossRef] [PubMed]

34. Suen, D.F.; Norris, K.L.; Youle, R.J. Mitochondrial dynamics and apoptosis. *Genes Dev.* **2008**, *22*, 1577–1590. [CrossRef] [PubMed]

35. Han, X.J.; Lu, Y.F.; Li, S.A.; Kaitsuka, T.; Sato, Y.; Tomizawa, K.; Nairn, A.C.; Takei, K.; Matsui, H.; Matsushita, M. CaM kinase I alpha-induced phosphorylation of Drp1 regulates mitochondrial morphology. *J. Cell Biol.* **2008**, *182*, 573–585. [CrossRef]

36. Fonseca, T.B.; Sanchez-Guerrero, A.; Milosevic, I.; Raimundo, N. Mitochondrial fission requires DRP1 but not dynamins. *Nature* **2019**, *570*, E34–E42. [CrossRef]

37. Crosas-Molist, E.; Bertran, E.; Sancho, P.; Lopez-Luque, J.; Fernando, J.; Sanchez, A.; Fernandez, M.; Navarro, E.; Fabregat, I. The NADPH oxidase NOX4 inhibits hepatocyte proliferation and liver cancer progression. *Free Radic. Biol. Med.* **2014**, *69*, 338–347. [CrossRef]

38. Gershkovitz, M.; Caspi, Y.; Fainsod-Levi, T.; Katz, B.; Michaeli, J.; Khawaled, S.; Lev, S.; Polyansky, L.; Shaul, M.E.; Sionov, R.V.; et al. TRPM2 Mediates Neutrophil Killing of Disseminated Tumor Cells. *Cancer Res.* **2018**, *78*, 2680–2690. [CrossRef]

39. Yamamoto, S.; Shimizu, S.; Kiyonaka, S.; Takahashi, N.; Wajima, T.; Hara, Y.; Negoro, T.; Hiroi, T.; Kiuchi, Y.; Okada, T.; et al. TRPM2-mediated Ca^{2+} influx induces chemokine production in monocytes that aggravates inflammatory neutrophil infiltration. *Nat. Med.* **2008**, *14*, 738–747. [CrossRef]

 *antioxidants*

Article

Combination of PDT and NOPDT with a Tailored BODIPY Derivative

Loretta Lazzarato [1,†], Elena Gazzano [2,†], Marco Blangetti [3], Aurore Fraix [4], Federica Sodano [1], Giulia Maria Picone [2], Roberta Fruttero [1], Alberto Gasco [1], Chiara Riganti [2,*] and Salvatore Sortino [4,*]

[1] Department of Drug Science and Technology, University of Torino, 10125 Torino, Italy; loretta.lazzarato@unito.it (L.L.); federica.sodano@unito.it (F.S.); roberta.fruttero@unito.it (R.F.); alberto.gasco@unito.it (A.G.)

[2] Department of Oncology, University of Torino, 10126 Torino, Italy; elena.gazzano@unito.it (E.G.); giulia.picone@edu.unito.it (G.M.P.)

[3] Department of Chemistry, University of Torino, 10125 Torino, Italy; marco.blangetti@unito.it

[4] Department of Drug Sciences, University of Catania, 95125 Catania, Italy; fraix@unict.it

* Correspondence: chiara.riganti@unito.it (C.R.); ssortino@unict.it (S.S.); Tel.: +39-011-6705-857 (C.R.); +39-095-7385-079 (S.S.)

† These authors contributed equally to this work.

Received: 9 October 2019; Accepted: 5 November 2019; Published: 7 November 2019

Abstract: The engineering of photosensitizers (PS) for photodynamic therapy (PDT) with nitric oxide (NO) photodonors (NOPD) is broadening the horizons for new and yet to be fully explored unconventional anticancer treatment modalities that are entirely controlled by light stimuli. In this work, we report a tailored boron-dipyrromethene (BODIPY) derivative that acts as a PS and a NOPD simultaneously upon single photon excitation with highly biocompatible green light. The photogeneration of the two key species for PDT and NOPDT, singlet oxygen ($^1O_2$) and NO, has been demonstrated by their direct detection, while the formation of NO is shown not to be dependent on the presence of oxygen. Biological studies carried out using A375 and SKMEL28 cancer cell lines, with the aid of suitable model compounds that are based on the same BODIPY light harvesting core, unambiguously reveal the combined action of $^1O_2$ and NO in inducing amplified cancer cell mortality exclusively under irradiation with visible green light.

Keywords: photodynamic therapy; singlet oxygen; nitric oxide; light; combination therapy

---

1. Introduction

Photodynamic therapy (PDT) for cancer involves the combined use of nontoxic photosensitizers (PSs) and visible light of appropriate wavelength [1,2]. Upon light excitation, PSs reach the lowest excited singlet state ($^1PS^*$) and, after intersystem crossing (ISC), their lowest-energy excited triplet state ($^3PS^*$). Due to their long lifetime (µs-ms), the $^3PS^*$ can dissipate most of their energy via quenching with nearby molecular oxygen, potentially giving rise to two different reactions. The first (Type I) reaction) involves an initial electron transfer to $^3O_2$ and the consequent formation of superoxide anion (O_2^-), from which hydrogen peroxide (H_2O_2) arises. This, in turn, can eventually generate the highly reactive and toxic hydroxyl radical (OH•). These species are collectively called reactive oxygen species (ROS). The second (Type II reaction) entails energy transfer from $^3PS^*$ to $^3O_2$ with the production of another ROS, highly toxic singlet oxygen ($^1O_2$), which is the key species in PDT [2,3]. The local irradiation of a tumor after the systemic administration of PS can, therefore, lead to a burst of cytotoxic ROS in a confined area, triggering cell death with high spatial accuracy, however, the success of PDT is strictly related to the presence of O_2, which is necessary for the production of ROS. Solid tumors have poor blood supply in their bulk and consequently have a low O_2 concentration. This is the reason for the scant efficacy of PDT against these kinds of malignancies. A few strategies have recently been proposed to overcome this problem [4]. In this frame, one of the most appealing approaches is based on the use of nitric oxide (NO). NO is an endogenous messenger that is ubiquitous in mammalian tissues and cells. This ephemeral free radical, which has a half-life of ca. 5 s and diffusion radius of ca. 200 μm in tissues, is involved not only in maintaining and regulating important physiological processes, such as cell proliferation, mitochondrial oxidative metabolism, antimicrobial defences, and vasodilation, but also plays crucial roles in an extensive number of different diseases, including cancer [5,6]. It has been shown that low (nM) NO concentrations promote cancer growth while high (μM) concentrations are toxic and reduce cancer progression [7,8]. Other factors, besides concentration, can influence the effects of NO on tumor growth. These factors include the duration of NO exposure and cellular sensitivity in particular [7,9]. The toxic effects of NO can be both direct and indirect. The former is related to the capacity of NO to react with the transition metals in some biomolecules that are essential for cellular life (e.g., the iron-sulfur centers in proteins and iron-containing enzymes). The indirect effects are caused by the ability of NO to react with O_2 or $O_2^-\bullet$, affording reactive nitrogen oxide species (RNOS) that can oxidize, nitrate and nitrosate a variety of biological targets, such as ribonucleotide reductase and proteins that contain iron-sulfur clusters, such as aconitase and the components of mitochondrial respiration complexes [7,8]. Peroxynitrite ($ONOO^-$) is one of these types of RNOS and is highly toxic. It is formed in the rapid reaction of NO with $O_2^-\bullet$ ($k = 6,7 \times 10^9$ M^{-1} s^{-1}) and is both a potent oxidant and nitrating agent. In the physiological environment, $ONOO^-$ gives rise to a pair of $OH\bullet$ and $NO_2\bullet$ radicals, via homolyses, that can combine to yield nitrate. In addition, it can interact with carbon dioxide (CO_2), to produce the carbonate radical ($CO_3^-\bullet$), which is another strong oxidant [10].

Therefore, it is not surprising that NO donor-based therapy has recently received particular attention as a potential clinical treatment for cancer. It is centred on the use of NO donors, namely products that can release NO under physiological conditions [11,12]. However, the main limitation of this approach is the lack of precise spatial-temporal control of the NO amount that is to be released which, as described above, is fundamental for a positive therapeutic outcome. For these reasons, there is currently great interest in compounds that are able to release NO under the action of light, namely NO photodonors (NOPDs). In these structures, NO is "caged", by a covalent bond, into a photoactivatable scaffold that harnesses the excitation energy to break bonds and, consequently, either liberate NO [13–20] or compounds that, in turn, spontaneously release it [21–26]. In these cases, NO release is limited to the irradiated area and fine temporal control of its release is achieved by simply tuning the duration and intensity of the irradiation. NOPDs are especially appealing in the field of light-controlled NO anticancer photodynamic therapy (NOPDT) as it may be a means to overcome the limitations of PDT in hypoxic tumors [4]. The combination of PS and NOPDT is an ideal, innovative and unconventional methodology thanks to the important properties that $^1O_2$ and NO share which include: (i) no resistance being developed towards these species, (ii) multitarget activity, and (iii) confinement of their action to a restricted region of space due to their short lifetimes and diffusion radius [27].

We have recently reported compound **1** (Scheme 1), in which a boron dipyrromethene (BODIPY) scaffold is covalently joined to cupferron, a C-diazeniumdiolate that is able to spontaneously release NO [26]. Under the action of biocompatible green light, **1** affords NO following the release of the cupferron moiety. In order to pursue the recent idea of incorporating a NOPD and a PS within the same molecular skeleton [28], we have devised the 2,6-diiododerivative, **2** (Scheme 1).

In this case, the presence of the two iodine substituents is expected to encourage the population of the triplet state, which is fundamental for ROS generation via type I and type II mechanisms (vide supra) without, hopefully, hampering NO photorelease properties. In this paper, we report the synthesis, and the photochemical and photophysical characterization of compound **2** and the validation of the cytotoxicity, against human melanoma A375 and SKMEL28 cell lines, of this compound, the non-iodinate analogue, **1**, and the 8-OCH₃ substituted BODIPYs, **3** and **4**, which were chosen as suitable model compounds for **1** and **2**, respectively. The nature of the reactive species involved in the antiproliferative effects of these compounds is discussed with reference to a detailed study of their photochemical profiles.

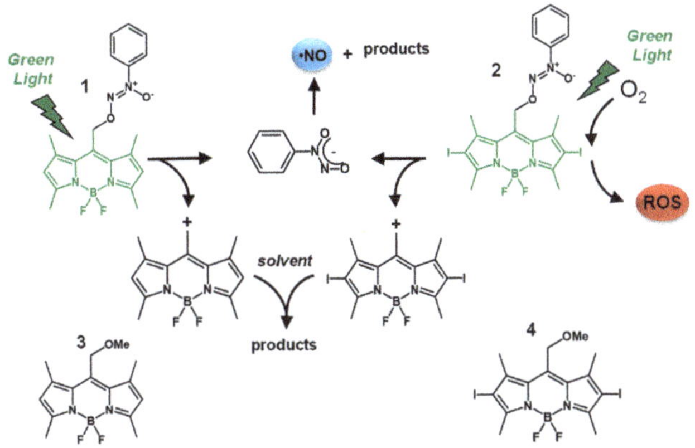

Scheme 1. Mechanism for the photorelease of NO from compound **1**, and for the nitric oxide (NO) and reactive oxygen species (ROS) generation attributed to compound **2**. The molecular structures of the related model compounds, **3** and **4**, are also shown.

2. Materials and Methods

2.1. Materials

All reagents (Sigma-Aldrich) were of high commercial grade and were used without further purification. All solvents used (from Carlo Erba) were of spectrophotometric grade.

2.2. Synthetic Procedures

All reactions involving air-sensitive reagents were performed under nitrogen in oven-dried glassware using the syringe-septum cap technique. All solvents were purified and degassed before use. Chromatographic separation was achieved under pressure using Merck silica gel 60 and flash-column techniques. Reactions were monitored by thin-layer chromatography (TLC) on 0.25 mm silica-gel-coated aluminium plates (Merck 60 F254) (Merck KGaA, Darmstadt, Germany) using UV light (254 nm) as the visualising agent. Unless otherwise specified, all reagents were used as received without further purification. Compounds **1** and **3** were synthesised according to the previously reported procedure [26,29].

2.2.1. Synthesis of 2-((4,4-difluoro-2,6-diiodo-1,3,5,7-tetramethyl-4-bora-3a,4a-diaza-s-indacen-8-yl) methoxy)-1-phenyldiazene oxide 2

A stirred solution of compound **1** (40 mg, 0.10 mmol) in EtOH (35 mL) was treated with iodine (63.5 mg, 0.25 mmol) and then with iodic acid (42.1 mg, 0.25 mmol) in 1 mL of water. The reaction mixture was stirred at 80 °C for 3 h. Once this was completed, as shown by TLC analysis, the mixture was cooled to room temperature and the solvent was removed under reduced pressure. The residue was dissolved in dichloromethane (DCM), and then washed with water (2 × 20 mL) and brine (2 × 20 mL). The combined organic layers were dried over sodium sulphate. Purification via silica gel chromatography, eluting with 1/1 petroleum ether/DCM, gave **2** as a dark red crystalline solid (51 mg, 78%). M.p. = 195.7–196.8 °C (with dec.); ^{1}H NMR (300 MHz, CDCl$_3$): δ 7.91 (d, J = 7.2 Hz, 2H), 7.58–7.44 (m, 3H), 5.70 (s, 2H), 2.64 (s, 6H), 2.56 (s, 6H); ^{13}C NMR (75 MHz, CDCl$_3$): δ 158.4, 144.0, 143.2, 132.9, 131.9, 131.6, 129.3, 121.3, 87.7, 66.6, 18.8, 16.6; ESI-MS [M + Na]$^+$: m/z 673.2; ESI-HRMS(+):

m/z 672.95491 [M + Na]$^+$, $C_{20}H_{19}BF_2I_2N_4O_2Na$ requires 672.95506; HPLC purity >95% (CH_3CN/H_2O TFA 0.1% 90:10 (v/v), flow = 1.0 mL/min, t_R = 8.7 min) at 226, 254 and 550 nm.

2.2.2. Synthesis of 4,4-difluoro-2,6-diiodo-8-methoxymethyl-1,3,5,7-tetramethyl-4-bora-3a,4a-diaza-s-indacene **4**

A solution of compound **3** (0.230 mmol) in EtOH (30 mL) was treated with iodine (0.570 mmol) and then with iodic acid (0.570 mmol) in 2 mL of water. The reaction mixture was stirred for 2 h at r.t., and the solvent was removed under reduced pressure. The residue was dissolved in EtOAc, and then washed with water (2 × 20 mL) and brine (2 × 20 mL). The combined organic layers were dried over sodium sulphate. Purification by silica gel chromatography, eluting with 95/5 petroleum ether/EtOAc, gave the title product, **4**, as a dark red solid (96 mg, 77%). M.p. = 186.4–187.0 °C (with dec.); ^{1}H NMR (300 MHz, $CDCl_3$): δ 4.54 (s, 2H), 3.48 (s, 3H), 2.61 (s, 6H), 2.46 (s, 6H); ^{13}C NMR (75 MHz, $CDCl_3$): δ 157.4, 143.9, 135.3, 133.1, 86.8, 65.6, 58.9, 17.8, 16.4; ESI-MS [M-H]$^-$: m/z 543.2; HPLC purity >95% (CH_3CN/H_2O TFA 0.1% 90:10 (v/v), flow = 1.0 mL/min, t_R = 8.7 min) at 226, 254 and 550 nm.

2.3. Instrumentation

^{1}H NMR and ^{13}C NMR spectra were recorded on a Bruker Avance 300 (Bruker, Billerica, MA, USA) at room temperature at 300 and 75 MHz, respectively, and calibrated using $SiMe_4$ as the internal reference. Chemical shifts (δ) are given in parts per million (ppm) and coupling constants (J) in Hertz (Hz). The following abbreviations are used to designate the multiplicities: s = singlet, d = doublet, t = triplet, q = quartet, m = multiplet, and br = broad. Low-resolution mass spectra were recorded on a Micromass Quattro microTM API (Waters Corporation, Milford, MA, USA). High-resolution mass spectra were recorded on a Bruker Bio Apex Fourier transform ion cyclotron resonance (FT-ICR) mass spectrometer equipped with an Apollo I ESI source, a 4.7 T superconducting magnet, and a cylindrical infinity cell (Bruker Daltonics, Billerica, MA, USA). HPLC analyses were performed using a HP 1200 chromatograph system (Agilent Technologies, Palo Alto, CA, USA) equipped with a quaternary pump (model G1311A, Agilent Technologies, Palo Alto, CA, USA), a membrane degasser (G1322A, Agilent Technologies, Palo Alto, CA, USA), and a multiple wavelength UV detector (MWD, model G1365D, Agilent Technologies, Palo Alto, CA, USA) integrated into the HP1200 (Agilent Technologies, Palo Alto, CA, USA) system. Data analysis was performed using a HP ChemStation system (Agilent Technologies). The sample was eluted in a Merck LiChrospher C18 end-capped column (250 × 4.6 mm ID, 5 μm) (Merck KGaA, Darmstadt, Germany). The injection volume was 20 μL (Rheodyne, Cotati, CA, USA). The mobile phase consisted of acetonitrile 0.1% TFA (solvent A) and 0.1% TFA (solvent B) at flow rate = 1.0 mL/min. For the photoproduct distribution studies, irradiation was performed using a white LED 90W lamp (Led Engin, San Jose, CA, USA) equipped with a green filter BP-540 80HT (transmission range 515–565 nm, T >90%, Schneider, Kreuznach, Germany) as the light source. Irradiance was measured using a HD2302.0 Delta Ohm lightmeter (Delta Ohm, Caselle di Selvazzano (PD), Italy) equipped with a Delta Ohm LP471RAD light probe (Delta Ohm, Caselle di Selvazzano (PD), Italy).

UV-Vis absorption and fluorescence emission spectra were recorded on a JascoV-560 spectrophotometer (Jasco, Easton, MD, USA) and a Spex Fluorolog-2 (mod. F-111) spectrofluorimeter (Horiba, Kyoto, Japan), respectively, in air-equilibrated solutions, using quartz cells with a path length of 1 cm. Fluorescence lifetimes were recorded with the same fluorimeter, which was equipped with a Time-Correlated Single Photon Counting (TCSPC) Triple Illuminator (Horiba, Kyoto, Japan). The samples were irradiated by a pulsed diode excitation source (Nanoled) at 455 nm and the decays were monitored at 500 nm. The system allowed fluorescence lifetimes to be measured from 200 ps. The multiexponential fit of the fluorescence decay was obtained using the following equation:

$$I(t) = \Sigma \alpha_i \exp^{(-t/\tau_i)}$$

Absorption spectral changes were monitored by irradiating the sample in a thermostated quartz cell (1 cm path length, 3 mL capacity) under gentle stirring, using a continuum laser with $\lambda exc = 532$ nm, ca. 50 mW and a beam diameter of ca. 1.5 mm.

The direct monitoring of NO release in solution was performed via amperometric detection (World Precision Instruments, Sarasota, FL, USA), with a ISO-NO meter, equipped with a data acquisition system, and was based on the direct amperometric detection of NO with a short response time (<5 s) in a 1 nM to 20 μM sensitivity range. The analogue signal was digitalized using a four-channel recording system and transferred to a PC. The sensor was accurately calibrated by mixing standard solutions of $NaNO_2$ with 0.1 M H_2SO_4 and 0.1 M KI, according to the reaction:

$$4H^+ + 2I^- + 2NO_2^- \rightarrow 2H_2O + 2NO + I_2$$

Irradiation was performed in a thermostated quartz cell (1 cm path length, 3 mL capacity) using the continuum laser at $\lambda exc = 532$ nm. NO measurements were carried out under stirring with the electrode positioned outside the light path in order to avoid NO signal artefacts that could be caused by photoelectric interference with the ISO-NO electrode.

$^1O_2$ emission was registered using a Fluorolog-2 (Model, F111) spectrofluorimeter that was equipped with a NIR-sensitive liquid nitrogen cooled photomultiplier, which excited the air-equilibrated samples with a CW laser at 532 nm.

Laser Flash Photolysis

All solutions were excited with the second harmonic of a Nd-YAG Continuum Surelite II-10 laser (532 nm, 6 ns FWHM) (Continuum, Santa Clara, CA, USA), using quartz cells with path length 1.0 cm. The excited solutions were analyzed using Luzchem Research mLFP-111 apparatus (Luzchem Research Inc., Gloucester, ON, Canada) with an orthogonal pump and probe configuration. The probe source was a ceramic xenon lamp coupled to quartz fibre optic cables. The laser pulse and the mLFP-111 system were synchronized using a Tektronix TDS 3032 digitizer (Tektronix, Beaverton, OR, USA), which was operated in pretrigger mode. The signals from a compact Hamamatsu photomultiplier (Hamamatsu Photonics K.K., Hamamatsu, Japan) were initially captured by the digitizer and then transferred to a personal computer, which was controlled by Luzchem Research software (Luzchem Research Inc., Gloucester, ON, Canada) operating in the National Instruments LabView 5.1 (National Instruments, Austin, TX, USA) environment. The solutions were deoxygenated via bubbling with a vigorous and constant flux of pure nitrogen (previously saturated with solvent). The solution temperature was 295 ± 2 K. The energy of the laser pulse was measured at each shot with a SPHD25 Scientech pyroelectric meter (Scientech Inc., Boulder, CO, USA).

2.4. Biological Experiments

2.4.1. Cell Cultures

Culture media were supplied by Invitrogen Life Technologies (Carlsbad, CA, USA), and plasticware for the cell cultures was obtained from Falcon (Becton Dickinson, Franklin Lakes, NJ, USA). The protein content of the cell monolayers was assessed using the BCA kit from Sigma Chemical Co. (St. Louis, MO, USA). Unless otherwise specified, all reagents were from Sigma Chemical Co. The compounds were dissolved in dimethyl sulfoxide (DMSO). The stock solution was diluted in culture medium to reach the final concentration. The concentration of DMSO in the culture medium under each experimental condition was less than 0.1%.

Human melanoma A375 and SKMEL28 cells (ATCC, Manassas, VA, USA) were cultured in DMEM medium that was supplemented with 10% fetal bovine serum and 1% penicillin-streptomycin. Cell cultures were maintained in a humidified atmosphere at 37 °C and 5% CO_2. When indicated, cells were incubated for 4 h with the compounds and then exposed for 30 min to the light emitted by a white LED 90W lamp with an irradiance of 1.75 mW/cm^2 and a bandpass filter of 540 nm $\pm$ 40 nm (IFG BP 540-80 HT), in DMEM medium without phenol red at room temperature. Non-irradiated cells were maintained in a dark room for 30 min in DMEM medium without phenol red at room temperature. After this period, cells were left for 19.5 h in the incubator before the experimental procedures, described below, were performed.

2.4.2. NO Photorelease in Cells

After the cells were incubated and irradiated, as described in the "Cell" section, 1 mL of cell supernatant was collected and centrifuged for 10 min at 13,000× g. The presence of nitrite in the reaction mixture was then determined using the Griess assay [30], and a Synergy HT microplate reader (Bio-Tek Instruments, Winooski, VT, USA).

2.4.3. Cytotoxicity

The cytotoxic effect of the compounds was measured, using a Synergy HT microplate reader (Bio-Tek Instruments, Winooski, VT, USA), as previously described [31], as the leakage of lactate dehydrogenase (LDH) into the extracellular medium. Both intracellular and extracellular LDH were measured, and then extracellular LDH activity (LDH out) was calculated as a percentage of the total (intracellular + extracellular) LDH activity (LDH tot) in the dish. For the calculation of IC_{50}, i.e., the concentration of each compound that decreased the cell viability by 50%, cells were incubated with either 0, 0.1, 0.25, 0.5, 1, 2.5, 5, 10, 25, 50, 75, or 100 µM of the single compounds, and were either maintained in the dark or irradiated for 30 min. Viable cells were measured after 24 h using the ATPlite kit (PerkinElmer, Waltham, MA, USA). Chemiluminescence was read using a Synergy HT microplate reader. The mean relative luminescence units (RLUs) of the untreated cells was considered to be 100%, and the RLUs of the other experimental conditions were expressed as percentage of viable cells vs. untreated cells. IC_{50} was calculated using GraphPad Prism (v.6.01) software.

2.4.4. Statistical Analysis

All the data in the text and figures are provided as means ± SD. The results were analysed via a one-way analysis of variance (ANOVA) and Tukey's test. $p < 0.05$ was considered significant.

3. Results and Discussion

3.1. Photochemical and Spectroscopic Properties

BODIPY derivative **2** is highly soluble in the H_2O:MeOH (20:80 v/v) solution, and its absorption features are dominated by the BODIPY chromophore in the visible region (Figure 1A). This compound is very stable under these conditions at room temperature in the dark for several days. By contrast, clear photodegradation is observed upon irradiation with green light at 532 nm (Figure 1). It is worth noting that neither the photolysis nor rate profile were affected by the presence of oxygen (see inset Figure 1A), suggesting that the primary photodegradation pathway proceeds from a short-lived excited state (i.e., the lowest singlet). The HPLC analysis of the irradiated solution revealed the presence of two solvent substitution products that lack the cupferron moiety, accompanied by nitrosobenzene (Figure 1B). These findings account for a photodegradation mechanism that is similar to the one that has already been observed for the non-iodinate analogue **1** [26], and that involves the heterolytic rupture of the C-O bond, with the consequent formation of a BODIPY-methyl carbocation via solvent substitution (see Scheme 1). Furthermore, the presence of nitrosobenzene as a photolysis product suggests that NO is quickly lost from the photodecaged cupferron. NO release was monitored via its real-time detection by an ultrasensitive NO electrode, which directly detects NO concentration using an amperometric technique. The results illustrated in Figure 1C provide clear evidence that compound **2** is stable in the dark, but that it generates NO under irradiation with Vis green light.

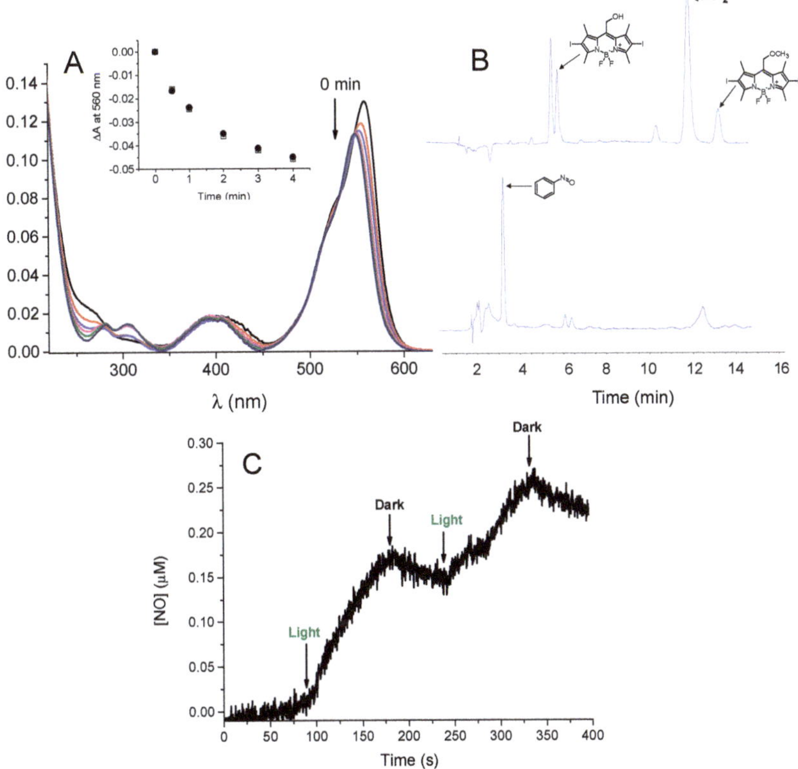

Figure 1. (**A**) Changes in absorption spectra observed upon exposure of an H_2O:MeOH (20:80, v/v) air-equilibrated solution of **2** to λexc = 532 nm from 0 to 5 min. T = 25 °C. The inset shows the kinetic profile monitored at λ = 560 nm in aerated ($\bullet$) and N2-saturated solutions ($\square$). (**B**) HPLC trace of the photolyzed solution detected at λ = 560 nm (top) and λ = 306 nm (bottom). The main products observed were identified by comparison of their retention time with standards. (**C**) NO release profile observed upon 532 nm light irradiation (200 mW) of a H_2O:MeOH (20:80 v/v) solution of **2**.

As expected, the presence of the iodine substituents in compound **2** dramatically amplified the ISC process leading to a drastic reduction in fluorescence emission and lifetime (more than two orders of magnitude compared with **1**) and the effective population of the lowest excited triplet state, in contrast to non-iodinate **1**. Laser flash photolysis with nanosecond time resolution is a powerful tool with which to examine the spectroscopic and kinetic features of excited triplets of many PSs, since these transient species exhibit very intense absorption in the visible region, and lifetimes in the order of microseconds. Figure 2A shows the transient absorption spectrum of compound **2** observed 0.1 µs after the laser pulse. It consists of two bands, around 440 nm and 650 nm, and bleaching in the region that corresponds to the ground state absorption of the BODIPY chromophore, which is in line with the typical triplet state absorption of BODIPY derivatives [32]. The triplet state of **2** decays monoexponentially (inset Figure 2A) with a lifetime in the microsecond time regime in N_2-saturated solution and is quenched by molecular oxygen with a diffusional quenching constant. Accordingly, a negligible triplet signal was observed in the optically matched solution of **1** (see Figure 2A). As outlined above, triplet quenching by molecular oxygen can lead to either $^1O_2$ or O_2^- production, according to the type of mechanism. $^1O_2$ was detected directly by measuring its typical phosphorescence in the near-IR spectral window, upon excitation with green light. Figure 2B shows that the typical luminescence signal of $^1O_2$, with maximum at ca. 1270 nm, is only one observed in the case of compound **2**.

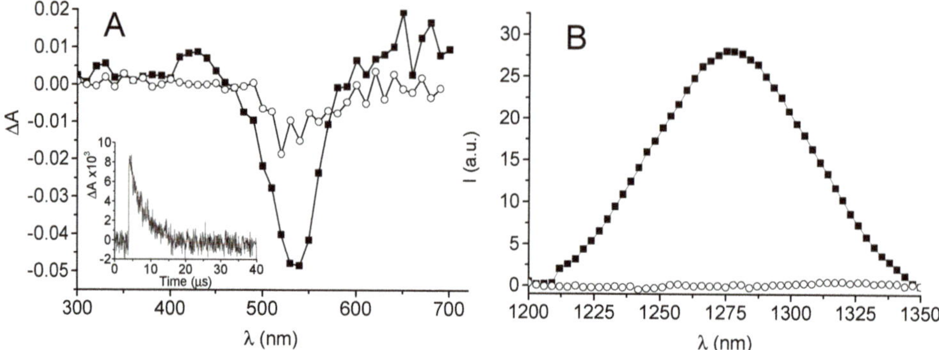

Figure 2. (**A**) Transient absorption spectra observed 0.1 µs after 532 nm laser excitation ($E_{532} \approx 10$ mJ/pulse) of optically matched N_2-saturated solutions of **1** (○) and **2** (●) in H_2O:MeOH (20:80 v/v). The inset shows the decay traces monitored at 440 nm and the related first-order fitting of **2**. $R^2 = 0.89$. (**B**) $^1O_2$ luminescence detected upon 532 nm light excitation of optically matched solutions of **1** (○) and **2** (■). D_2O:MeOD (20:80 v/v).

The spectroscopic and photochemical scenario described above, therefore, demonstrates that the cupferron decaging, and subsequent NO release, occurs from the excited singlet state and that $^1O_2$ is the main quenching product from the lowest excited triplet state, according to Scheme 2.

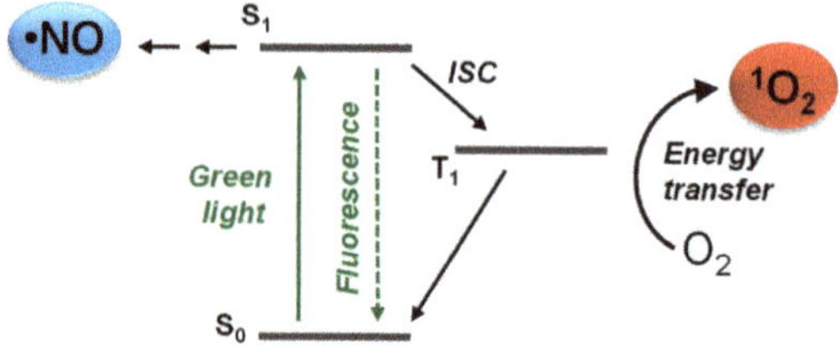

Scheme 2. Schematic illustration of the main photoprocesses that lead to the generation of NO and $^1O_2$ from compound **2**.

3.2. Biological Experiments

3.2.1. NO Photorelease in Cells

The NO photorelease demonstrated by **1** and **2** was also investigated in A375 and SKMEL28, two melanoma cell lines. After 4 h of incubation with the products at three different concentrations, the cells were irradiated for 30 min with the light emitted by a white LED 90 W lamp with an irradiance of 1.75 mW/cm^2 and a bandpass filter 540 nm ± 40 nm. NO was measured as nitrite (its main oxidation product) as determined in the supernatants by Griess assay. Similar experiments were carried out on model compounds **3** and **4** which, unsurprisingly, failed to show any nitrite generation (data not shown). The data reported in Figure 3 clearly demonstrate that products **1** and **2** were able to produce nitrite when irradiated in cells. In particular, they afforded significant amounts of nitrite in the two cell lines at 2.5 to 5 µM concentrations, i.e., in a concentration range that is compatible with the possible cytotoxic and antitumor effects of NO-derived reactive species [7–9]. The amount of nitrite produced by **1** and **2** was similar in the two cell lines. Individual cell lines can differ somewhat in how enzymes metabolize NO and its stable derivatives; superoxide dismutase neutralises O_2^- and limits the generation of $ONOO^-$, while nitrate reductase increases the conversion of nitrate into nitrite and amplifies the levels of nitrite that are detected using the Griess method. The cell independency shown in nitrite production, here, strongly suggests that NO generation is dependent on the photochemical properties of the compounds. **1** is a more efficient nitrite producer than **2** in the 2.5–5 µM concentration range. Since we have ascertained that NO photorelease takes place from the lowest excited singlet state of both **1** and **2**, the lower efficiency found for **2** reflects the effective ISC that occurs in this compound, which competes with the cupferron decaging and therefore impacts upon the NO produced.

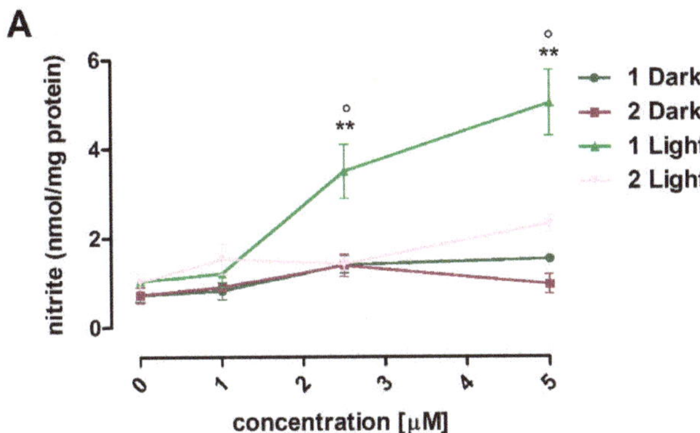

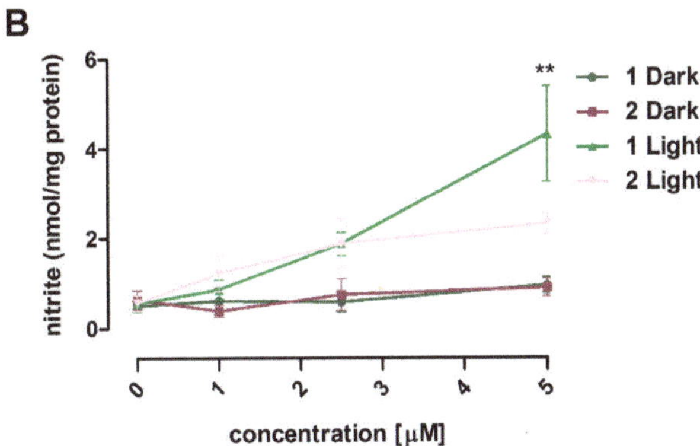

Figure 3. Nitric oxide release in the extracellular medium of A375 (panel **A**) and SKMEL28 (panel **B**) melanoma cells incubated with compounds **1** and **2** and either maintained in the dark or irradiated at 1.75 mW/cm^2 in the 500–580 nm range (green light) for 30 min. Measurements were performed in triplicate and data are presented as means ± SEM ($n = 3$). Vs untreated cells (ctrl), ** $p < 0.01$ and Vs compound 1 (light), ° $p < 0.05$.

3.2.2. Cytotoxicity against Human Melanoma A375 and SKMEL28 Cells

A375 and SKMEL28 cells were incubated with 1, 2.5, and 5 µM of compounds **1** and **2** and were then either left in the dark or irradiated for 30 min with green light. Cytotoxicity was evaluated 24 h after the completion of irradiation by measuring the release of LDH, a sensitive index of membrane damage due to oxidative [31] and nitrosative stress [33], which correlates well with cell death [34]. The results shown in Figure 4 demonstrate that both products are nontoxic in the dark, but that they show a good profile of activity, which is finely dependent on concentration, under irradiation. While the cytotoxicity of **2** is similar in the two cell lines, **1** was more toxic in A375 than in SKMEL28 cells. According to NO photogeneration and the lack of $^1O_2$ photosensitisation, the toxicity observed in the case of **1** is most likely principally due to NO toxicity and not mediated by $ONOO^-$. The differential cytotoxicity of **1**, in particular in the 1–2.5 μM range, in the A375 and SKMEL28 cells may be due to the fact that the former are more dependent on efficient mitochondrial oxidative respiration than the latter; inhibiting mitochondrial respiration with KCN reduces cell viability in A375 cells more than in SKMEL28 cells [35]. Since NO is a proven inhibitor of complexes I, III, and IV [7–9], as well as of aconitase, in this range of concentration, the cytotoxicity caused by **1** is most likely due to the impairing of mitochondrial respiration elicited by the released NO.

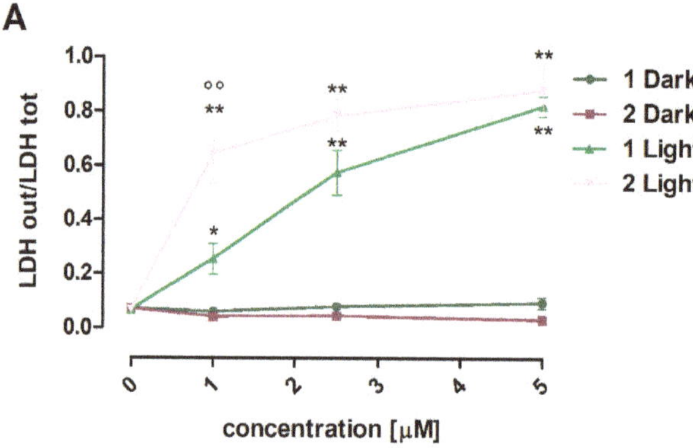

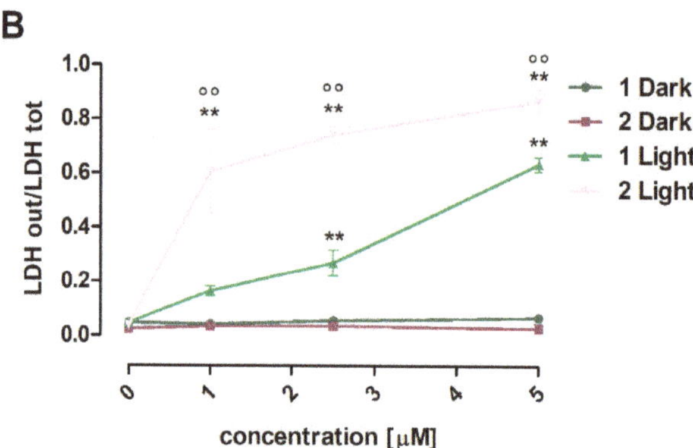

Figure 4. Cytotoxicity of compounds **1** and **2** towards A375 (panel **A**) and SKMEL28 (panel **B**) melanoma cells that were either maintained in the dark or irradiated at 1.75 mW/cm^2 in 500–580 nm range (green light) for 30 min. Measurements were performed in triplicate and data are presented as means ± SEM (n = 3). Vs untreated cells (ctrl), * $p < 0.05$ and ** $p < 0.01$ and Vs compound 1 (light), °° $p < 0.01$.

Moreover, the activity profile observed for compound **2** against both cell lines shows higher cytotoxicity than **1**. In order to establish whether this higher cytotoxicity is due to $^1O_2$, NO or the combined action of the two species, cytotoxicity experiments were also performed with the model compounds **3** and **4**. The results reported in Figure 5 show that **3** fails to show toxicity in the dark and upon irradiation. This is not surprising as this compound does not produce NO and does not generate

$^1O_2$. By contrast, compound **4** shows toxicity under green light irradiation at concentrations higher than 2 µM. As we found that **4** generates $^1O_2$ with similar efficiency to **2**, the photodynamic action observed can be exclusively attributed to this transient species. A comparison of the data in Figures 4 and 5 provides quite a clear scenario of the biological effects. In fact, compound **2** already exhibits relevant levels of phototoxicity at a concentration of 1 µM in both cell lines, unlike model compound **4**, which is inactive at the same concentration. These data, together with the phototoxicity observed for compound **1**, which exclusively produces NO, suggest that the phototoxic effects mediated by **2** are clearly a result of the combined action of $^1O_2$ and NO. We can infer that the higher cytotoxicity that **2** shows, in particular in the 1–2 µM concentration range, as compared with **4** at these concentrations can be explained by the different reactive species generated by the compounds upon irradiation, i.e., $^1O_2$ and NO from **2**, only $^1O_2$ from **4**. This comparison allows us to separate the proportions of the cytotoxic effects of NO and those of $^1O_2$.

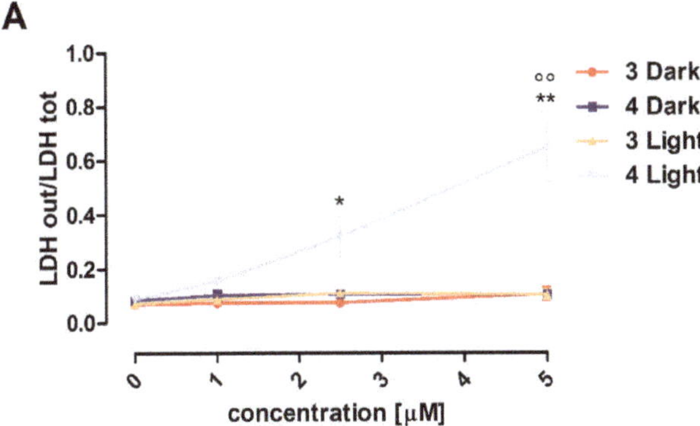

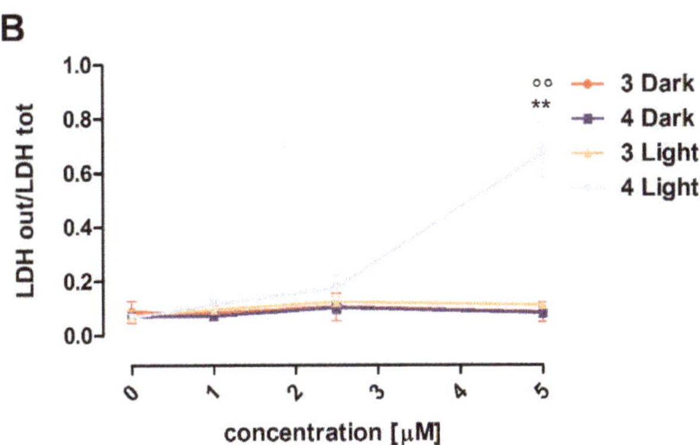

Figure 5. The cytotoxicity of compounds **3** and **4** towards A375 (panel **A**) and SKMEL28 (panel **B**) melanoma cells that were either maintained in the dark or irradiated at 1.75 mW/cm^2 in 500–580 nm range (green light) for 30 min. Measurements were performed in triplicate and data are presented as means ± SEM (*n* = 3). Vs untreated cells (ctrl), * *p* < 0.05 and ** *p* < 0.01 and Vs compound **3** (light), °° *p* < 0.01.

The results obtained from the LDH release assays were further confirmed by the calculation of IC$_{50}$ values (Table 1). The IC$_{50}$ values for all of the compounds were higher than the maximum concentration tested and compatible with a good solution of the compounds, i.e., 100 μM, indicating that the compounds did not exert peculiar toxicity if not irradiated. Upon irradiation, the IC_{50} of **1**, **2**, and **4** was significantly lower. In particular, the IC_{50} of **2** was lower than the toxicity of **1**, which is in line with the ability of **2** to release two toxic species, i.e., NO and $^1O_2$, and the ability of **1** to release NO only. The IC_{50} of the respective model compounds, **3** and **4**, were higher. The IC_{50} of **3** was similar to that measured in the dark, since this compound does not produce either NO or $^1O_2$ upon irradiation. The IC_{50} of **4**, which generates $^1O_2$ similarly to **2**, but not NO, is lower than the IC_{50} of **2** upon irradiation. These data confirm that the maximal phototoxicity is produced by the simultaneous release of RNS and ROS, as occurs in compound **2**.

Table 1. IC_{50} (μM) of compounds per cell line analyzed.

Cell Lines	1 (Dark)	2 (Dark)	3 (Dark)	4 (Dark)	1 (Light)	2 (Light)	3 (Light)	4 (Light)
A375	>100	>100	>100	>100	8.6 ± 1.2 ***	4.6 ± 0.4 ***,°	>100	7.3 ± 1.6 ***
SKMEL28	>100	>100	>100	>100	10.1 ± 2.5 ***	5.9 ± 0.7 ***,°	>100	8.1 ± 0.7 ***

The viability of A375 and SKMEL28 melanoma cells, either maintained in the dark or irradiated at 1.75 mW/cm^2 in 500–580 nm range (green light) for 30 min, in the presence of each compound incubated at 0, 0.1, 0.25, 0.5, 1, 2.5, 5, 10, 25, 50, 75, and 100 μM. Viable cells were measured with a chemiluminescence-based assay after 24 h. Measurements were performed in triplicate and data are presented as means ± SEM (n = 3). IC_{50} were calculated with the GraphPad Prism (v.6.01) software. Vs untreated cells (ctrl), *** p < 0.001 and Vs compound **1** (light), ° p < 0.05.

4. Conclusions

In conclusion, we have reported a new BODIPY derivative, **2**, that is able to release NO and $^1O_2$ under the action of green light. The photogeneration of these transient cytotoxic species has been unambiguously demonstrated by their direct detection. With the aid of analogue derivative **1**, which exclusively photogenerates NO, and two suitable model compounds based on the same BODIPY core, we have provided evidence that the in vitro cytotoxic activity observed upon green light irradiation of **2** against human melanoma cell lines is due to the combined action of NO and $^1O_2$, whereas only NO is involved in the antiproliferative effects induced by **1**. Our results clearly demonstrate that the possibility of releasing both NO and $^1O_2$ in a spatially and temporally controlled manner is the most powerful tool for killing tumor cells, while sparing non-transformed, non-irradiated healthy tissues. These data are especially promising as they have been obtained against melanoma cells, which are known for their high resistance to chemotherapy and targeted therapy [36]. Interestingly, NO photogeneration is not affected by the presence of oxygen. These findings, therefore, provide further confirmation that strategic approaches that are based on the photoregulated release of NO and its combination with ROS are very appealing as potential tools for the innovative treatment of tumors. Moreover, our compounds are active in vitro in low micromolar concentrations, like most chemotherapeutic drugs. The BODIPY derivatives used in the present study do not show preferential accumulation within a specific organelle, as demonstrated previously by [16]. However, the BODIPY scaffold can be chemically conjugated with mitochondria, endoplasmic reticulum, and lysosome-vectorising moieties. Work on this topic is currently ongoing in our group. Using this approach, we plan to concentrate the simultaneous release of ROS and RNS within a single compartment, and thus produce higher selective damage. Indeed, it is known that RNS exerts antitumor effects by inhibiting mitochondrial metabolism, inducing ER stress, and regulating lysosomal functions when released at micromolar concentrations [11]. The targeted release of RNS and ROS within a specific compartment will boost the cytotoxic potential of our hybrids. Our present work is a proof of concept that demonstrates the opportunity provided by a technique that can achieve the same antitumor efficacy of classical chemotherapeutic drugs, avoid the problem of the onset of chemoresistance and undesired side effects, while also exploiting the intratumor destruction produced by oxidative and nitrosative damage upon irradiation with visible light.

Author Contributions: Conceptualisation, R.F., A.G., C.R., and S.S.; methodology, L.L., E.G., and M.B.; formal analysis, L.L., E.G., and A.F.; investigation, E.G., A.F., F.S., and G.M.P.; writing—original draft preparation, E.G.,

A.F., and F.S.; writing—review and editing, L.L., R.F., A.G., C.R., and S.S.; supervision, L.L., R.F., A.G., C.R., and S.S.

Funding: We thank AIRC Project IG-19859 and the University of Turin "Ricerca Locale" (grant LAZL_RILO_18_01 to L.L.) for financial support. This article is based on work from COST Action 17104 STRATAGEM, supported by the COST association (European Cooperation in Science and Technology).

Acknowledgments: We thank Dale Lawson, the Department of Drug Science and Technology, for language revision and Costanzo Costamagna, the Department of Oncology, University of Turin, for technical assistance.

Conflicts of Interest: The authors declare no conflict of interest.

References

1. Wainwright, M. *Photosensitizers in Biomedicine*; John Wiley & Sons Ltd.: West Sussex, UK, 2009.
2. Ormond, A.B.; Freeman, H.S. Dye sensitizers for photodynamic therapy. *Materials* **2013**, *6*, 817–840. [CrossRef] [PubMed]
3. Castano, A.P.; Demidova, T.N.; Hamblin, M.R. Mechanisms in photodynamic therapy: Part one—Photosensitizers, photochemistry and cellular localization. *Photodiagnosis Photodyn. Ther.* **2004**, *1*, 279–293. [CrossRef]
4. Li, X.; Kwon, N.; Guo, T.; Liu, Z.; Yoon, J. Innovative strategies for hypoxic-tumor photodynamic therapy. *Angew. Chem. Int. Ed.* **2018**, *57*, 11522–11531. [CrossRef] [PubMed]
5. Kervin, J.F., Jr.; Lancaster, J.R., Jr.; Feldman, P.L. Nitric oxide: A new paradigm for second messengers. *J. Med. Chem.* **1995**, *38*, 4343–4362.
6. Gross, S.S.; Wolin, M.S. Nitric oxide: Pathophysiological mechanisms. *Annu. Rev. Physiol.* **1995**, *57*, 737–769. [CrossRef]
7. Kroncke, K.-D.; Fehsel, K.; Kolb-Bachofen, V. Nitric oxide: Cytotoxic versus cytoprotection—How, why, and where? *Nitric Oxide* **1997**, *1*, 107–120. [CrossRef]
8. Wink, D.A.; Mitchell, J.B. Chemical biology of nitric oxide: Insights into regulatory, cytotoxic, and cytoprotective mechanism of nitric oxide. *Free Radic. Biol. Med.* **1998**, *25*, 434–456. [CrossRef]
9. Ridnour, L.A.; Thomas, D.D.; Donzelli, S.; Espey, M.G.; Roberts, D.D.; Wink, D.A.; Isenberg, J.S. The biphasic nature of nitric oxide responses in tumor biology. *Antioxid. Redox Signal.* **2006**, *8*, 1329–1337. [CrossRef]
10. Ferrer-Sueta, G.; Campolo, N.; Trujillo, M.; Bartesaghi, S.; Carballal, S.; Romero, N.; Alvarez, B.; Radi, R. Biochemistry of peroxynitrite and protein tyrosine nitration. *Chem. Rev.* **2018**, *118*, 1330–1408. [CrossRef]
11. Huerta, S.; Chilka, S.; Bonavida, B. Nitric oxide donors: Novel cancer therapeutics. *Int. J. Oncol.* **2008**, *33*, 909–927. [CrossRef]
12. Huang, Z.; Fu, J.; Zhang, Y. Nitric-oxide donor-based cancer therapy: Advances and prospects. *J. Med. Chem.* **2017**, *60*, 7617–7635. [CrossRef] [PubMed]
13. Suzuki, T.; Nagae, O.; Kato, Y.; Nakagawa, H.; Fukuhara, K.; Miyata, N. Photoinduced nitric oxide release from nitrobenzene derivatives. *J. Am. Chem. Soc.* **2005**, *127*, 11720–11726. [CrossRef] [PubMed]
14. Kitamura, K.; Kawaguchi, M.; Ieda, N.; Miyata, N.; Nakagawa, H. Visible light-controlled nitric oxide release from hindered nitrobenzene derivatives for specific modulation of mitochondrial dynamics. *ACS Chem. Biol.* **2016**, *11*, 1271–1278. [CrossRef] [PubMed]
15. Caruso, E.B.; Petralia, S.; Conoci, S.; Giuffrida, S.; Sortino, S. Photodelivery of nitric oxide from water soluble platinum nanoparticles. *J. Am. Chem. Soc.* **2007**, *129*, 480–481. [CrossRef] [PubMed]
16. Parisi, C.; Failla, M.; Fraix, A.; Rolando, B.; Gianquinto, E.; Spyrakis, F.; Gazzano, E.; Riganti, C.; Lazzarato, L.; Fruttero, R.; et al. Fluorescent nitric oxide photodonors based on BODIPY and rhodamine antennae. *Chem. Eur. J.* **2019**, *47*, 11080–11084. [CrossRef]
17. Ford, P.C. Polychromophoric metal metal complexes for generating the bioregulatory agent nitric oxide by single- and two-photon excitation. *Acc. Chem. Res.* **2008**, *41*, 190–200. [CrossRef]
18. Ford, P.C. Photochemical delivery of nitric oxide. *Nitric Oxide* **2013**, *34*, 56–64. [CrossRef]
19. Ieda, N.; Hotta, Y.; Miyata, N.; Kimura, K.; Nakagawa, H. Photomanipulation of vasodilation with a blue-light controllable nitric oxide releaser. *J. Am. Chem. Soc.* **2014**, *136*, 7085–7091. [CrossRef]
20. Sortino, S. Light-controlled nitric oxide delivering molecular assemblies. *Chem. Soc. Rev.* **2010**, *39*, 2903–2913. [CrossRef]
21. Makings, L.R.; Tsien, R.Y. Caged nitric oxide. *J. Biol. Chem.* **1994**, *269*, 6282–6285.

22. Ruane, P.H.; Bushan, K.M.; Pavlos, C.M.; D'Sa, R.A.; Toscano, J.P. Controlled photochemical release of nitric oxide from O_2-benzyl-substituted diazeniumdiolates. *J. Am. Chem. Soc.* **2002**, *124*, 9806–9811. [CrossRef] [PubMed]

23. Behara, K.K.; Rajesh, Y.; Venkatesh, Y.; Pinninti, B.R.; Mandal, M.; Singh, N.D.P. Cascade photocaging of diazeniumdiolates: A novel strategy for one and two photon triggered uncaging with real time reporting. *Chem. Commun.* **2017**, *53*, 9470–9473. [CrossRef] [PubMed]

24. Luo, X.; Wu, J.; Lv, T.; Lai, Y.; Zhang, H.; Lu, J.-J.; Zhang, Y.; Huang, Z. Synthesis and evaluation of novel O_2-derived diazeniumdiolates as photochemical and real-time monitoring nitric oxide delivery agents. *Org. Chem. Front.* **2017**, *4*, 2445–2449. [CrossRef]

25. Stroppel, A.S.; Paolillo, M.; Ziegler, T.; Feil, R.; Stafforst, T. Npom-protected NONOate enables light-triggered NO/cGMP signalling in primary vascular smooth muscle cells. *ChemBioChem* **2018**, *19*, 1312–1318. [CrossRef]

26. Blangetti, M.; Fraix, A.; Lazzarato, L.; Marini, E.; Rolando, B.; Sodano, F.; Fruttero, R.; Gasco, A.; Sortino, S. A nonmetal-containing nitric oxide donor activated with single-photon green light. *Chem. Eur. J.* **2017**, *23*, 9026–9029. [CrossRef] [PubMed]

27. Fraix, A.; Sortino, S. Combination of PDT photosensitizers with NO-photodonors. *Photochem. Photobiol. Sci.* **2018**, *17*, 1709–1727. [CrossRef] [PubMed]

28. Parisi, C.; Failla, M.; Fraix, A.; Rescifina, A.; Rolando, B.; Lazzarato, L.; Cardile, V.; Graziano, A.C.E.; Fruttero, R.; Gasco, A.; et al. A molecular hybrid producing simultaneously singlet oxygen and nitric oxide by single photon excitation with green light. *Bioorg. Chem.* **2019**, *85*, 18–22. [CrossRef]

29. Krumova, K.; Cosa, G. Bodipy dyes with tunable redox potentials and functional groups for further tethering: Preparation, electrochemical, and spectroscopic characterization. *J. Am. Chem. Soc.* **2010**, *132*, 17560–17569. [CrossRef]

30. Chegaev, K.; Fraix, A.; Gazzano, E.; Abd-Ellatef, G.E.; Blangetti, M.; Rolando, B.; Conoci, S.; Riganti, C.; Fruttero, R.; Gasco, A.; et al. Light-regulated NO release as a novel strategy to overcome doxorubicin multidrug resistance. *ACS Med. Chem. Lett.* **2017**, *8*, 361–365. [CrossRef]

31. Riganti, C.; Costamagna, C.; Bosia, A.; Ghigo, D. The NADPH oxidase inhibitor apocynin (acetovanillone) induces oxidative stress. *Toxicol. Appl. Pharmacol.* **2006**, *212*, 179–187. [CrossRef]

32. Rachford, A.A.; Ziessel, R.; Bura, T.; Retailleau, P.; Castellano, F.N. Boron dipyrromethene (Bodipy) phosphorescence revealed in [Ir(ppy)2(bpy-CtC-Bodipy)]. *Inorg. Chem.* **2010**, *49*, 3730–3736. [CrossRef] [PubMed]

33. Riganti, C.; Costamagna, C.; Doublier, S.; Miraglia, E.; Polimeni, M.; Bosia, A.; Ghigo, D. The NADPH oxidase inhibitor apocynin induces nitric oxide synthesis via oxidative stress. *Toxicol. Appl. Pharmacol.* **2008**, *228*, 277–285. [CrossRef] [PubMed]

34. Aldieri, E.; Fenoglio, I.; Cesano, F.; Gazzano, E.; Gulino, C.; Scarano, D.; Attanasio, A.; Mazzucco, G.; Ghigo, D.; Fubini, B. The role of iron impurities in the toxic effects exerted by short multiwalled carbon nanotubes (MWCNT) in murine Aalveolar Macrophages. *J. Toxicol. Environ. Health Part A* **2013**, *76*, 1056–1071. [CrossRef] [PubMed]

35. Corazao-Rozas, P.; Guerreschi, P.; Jendoubi, M.; André, F.; Jonneaux, A.; Scalbert, C.; Garçon, G.; Malet-Martino, M.; Balayssac, S.; Rocchi, S.; et al. Mitochondrial oxidative stress is the Achille's heel of melanoma cells resistant to Braf-mutant inhibitor. *Oncotarget* **2013**, *4*, 1986–1998. [CrossRef] [PubMed]

36. Kalal, B.S.; Upadhya, D.; Pai, V.R. Chemotherapy resistance mechanisms in advanced skin cancer. *Oncol. Rev.* **2017**, *11*, 326. [CrossRef]

Article

Protein Carbonylation in Patients with Myelodysplastic Syndrome: An Opportunity for Deferasirox Therapy

Alba Rodríguez-García [1], María Luz Morales [1], Vanesa Garrido-García [1], Irene García-Baquero [1], Alejandra Leivas [1], Gonzalo Carreño-Tarragona [1], Ricardo Sánchez [1], Alicia Arenas [1], Teresa Cedena [1], Rosa María Ayala [1], José M. Bautista [2], Joaquín Martínez-López [1,3,*,†] and María Linares [1,2,†]

[1] Department of Hematology, 16473 Hospital Universitario 12 de Octubre, Hematological Malignancies Clinical Research Unit H120-CNIO, 28041 Madrid, Spain; albarodriguezgarcia@hotmail.com (A.R.-G.); mlmorales17@gmail.com (M.L.M.); gvaneg@hotmail.com (V.G.-G.); irene.g.baquero@gmail.com (I.G.-B.); alejandraleial@gmail.com (A.L.); gonzalocarreno.gomeztarragona@gmail.com (G.C.-T.); ricardsanchez.hdoc@gmail.com (R.S.); arenasalic@gmail.com (A.A.); mariateresa.cedena@salud.madrid.org (T.C.); rayaladiaz12@gmail.com (R.M.A.); mlinares@ucm.es (M.L.)
[2] Department of Biochemistry and Molecular Biology and Research Institute Hospital 12 de Octubre, Universidad Complutense de Madrid, Ciudad Universitaria, 28040 Madrid, Spain; jmbau@vet.ucm.es
[3] Department of Medicine, Universidad Complutense de Madrid, Ciudad Universitaria, 28040 Madrid, Spain
* Correspondence: jmarti01@med.ucm.es; Tel.: +34-9177922611
† These authors have contributed equally to this work.

Received: 11 September 2019; Accepted: 22 October 2019; Published: 24 October 2019

Abstract: Control of oxidative stress in the bone marrow (BM) is key for maintaining the interplay between self-renewal, proliferation, and differentiation of hematopoietic cells. Breakdown of this regulation can lead to diseases characterized by BM failure such as the myelodysplastic syndromes (MDS). To better understand the role of oxidative stress in MDS development, we compared protein carbonylation as an indicator of oxidative stress in the BM of patients with MDS and control subjects, and also patients with MDS under treatment with the iron chelator deferasirox (DFX). As expected, differences in the pattern of protein carbonylation were observed in BM samples between MDS patients and controls, with an increase in protein carbonylation in the former. Strikingly, patients under DFX treatment had lower levels of protein carbonylation in BM with respect to untreated patients. Proteomic analysis identified four proteins with high carbonylation levels in MDS BM cells. Finally, as oxidative stress-related signaling pathways can modulate the cell cycle through p53, we analyzed the expression of the p53 target gene p21 in BM cells, finding that it was significantly upregulated in patients with MDS and was significantly downregulated after DFX treatment. Overall, our results suggest that the fine-tuning of oxidative stress levels in the BM of patients with MDS might control malignant progression.

Keywords: myelodysplastic syndromes; carbonylation; oxidative stress; deferasirox

---

1. Introduction

Myelodysplastic syndromes (MDS) are a heterogeneous group of hematopoietic stem cell disorders characterized by the presence of immature myeloid precursors (blasts) and dysplastic hematopoiesis in the bone marrow (BM). A current "hot" topic in MDS research is oxidative stress and its potential effects on cell biology, DNA damage and carcinogenesis [1,2]. Whereas the origin of MDS disease is now better understood, high levels of reactive oxygen species (ROS) and consequent oxidative damage in hematopoietic cells have been reported in patients with this disease [3–6], but the consequences are less clear. Regulation of intracellular ROS levels is known to be key for maintaining the balance between self-renewal, proliferation, and differentiation of progenitor cells and a loss of this control can lead to diseases characterized by BM failure [7,8]. The potential effects of ROS on hematopoietic cells are particularly relevant because they are acutely vulnerable to oxidative damage associated with the accumulation of free radicals [9].

Studies have shown that BM cells from patients with MDS have increased levels of intracellular peroxide and decreased levels of the antioxidant glutathione (GSH), as compared with normal cells [10–12]. Importantly, patients with MDS and with high ROS or low GSH levels, and a high superoxide/peroxide ratio, have a lower overall survival [10]. The effects of increased intracellular ROS production are well recognized and include direct damage to biomolecules and/or dysregulation of ROS-dependent signaling pathways [8,13,14]. In the context of blood cells, an interplay has been reported between oxidative damage of DNA in CD34+ cells and subsequent increased oxidation levels in precursor cells such as blasts or erythroid precursors [4]. Also, oxidative stress correlates with DNA hypermethylation in patients with MDS [11] and other pathological conditions [15].

Proteins are important targets of ROS [16], and oxidation can lead to aggregation, polymerization, unfolding or conformational changes that cause structural or functional loss. Although several oxidative modifications to proteins are possible, most involve the formation of carbonyl groups, which are irreversibly introduced into proteins [17–19]. Protein carbonylation can occur by several pathways, but the two main contributors are: (i) direct metal-catalyzed oxidation of specific amino acid residues (lysine, arginine, proline and threonine), and (ii) secondary reactions of nucleophilic amino acid side-chains with ROS-induced lipid peroxidation products such as 4-hydroxynonenal (HNE) [17,19]. These modifications likely have important roles in cell signaling [18,20].

As a specific marker of oxidative damage, assessing protein carbonylation is useful for understanding the role of oxidative stress in several disease [19]. A common method of protein carbonylation detection involves derivatization by 2, 4-dinitrophenylhydrazine (DNPH) to form a stable dinitrophenylhydrazone (DNP) product, which is subsequently detected by specific antibodies to visualize carbonyl groups bound to tissues and cells. This method can therefore provide information both on the distribution of in vivo oxidative damage, and the identification of their protein targets by complementary methods such as mass spectrometry (MS) [18,21,22].

Signaling pathways activated by oxidative stress can control the cell cycle via the p53 gene [23]. Indeed, the p53 target gene p21, a cyclin-dependent kinase inhibitor, is involved in the cell response to genotoxic stress [16] and has a role in the regulation of senescence by controlling proliferation [24]. Defective activation of p21 can lead to leukemogenesis [25,26]. Along this line, it has been suggested that during the first phase of MDS, the ineffective maturation of progenitor hematopoietic cells is associated with an increase in apoptosis, whereas in more advanced stages, BM cells switch to a proliferative phenotype with negligible apoptotic control [27].

Many patients with MDS require frequent blood transfusions and can develop transfusion-dependent iron-overload [28], which in turn can lead to an increase in the generation of ROS [29]. Accordingly, iron chelator therapy has been proposed to address primary oxidative stress in MDS [28,30,31]. Indeed, Neukirchen et al. [32] presented data showing a survival benefit for low-risk MDS patients receiving iron chelation therapy. One such treatment related to oxidative stress that has been used in patients with MDS is the iron chelator deferasirox (DFX) [33]. DFX seems to constrain ROS damage by activating transcription factors and mitochondrial biogenesis [34], and by inhibiting the *de novo* generation of free radicals through the suppression of the active redox forms of iron [35]. Protein carbonylation can be reduced/eliminated with DFX, which can trap Fe-III ions by forming a 1:2 octahedral complex and preventing their reduction [36].

In this report, we demonstrate for the first time the therapeutic benefit of DFX to inhibit protein carbonylation associated with MDS. Moreover, we provide direct evidence of four proteins with increased carbonylation in the BM of patients. Lastly, our results suggest that the oxidative damage in MDS is related, at least in part, to signaling pathways regulating the cell cycle through p53/p21, which are moderated by DFX treatment, overall highlighting the important role of DFX in the control of the oxidative stress response in MDS.

2. Materials and Methods

BM samples were obtained from patients diagnosed with MDS (n = 21, median age 75 years; range 57–90 years) who were grouped according to the World Health Organization classification of tumors of hematopoietic and lymphoid tissues (2008), and to the International Prognostic Scoring System [37]. The main characteristics of the patients are summarized in Table 1.

The control group consisted of individuals (n = 13, median age 69 years; range 29–94 years) with no signs of the disease in BM no infiltration of the disease in BM. The study was approved by the Comité Ético de Investigación Clínica of the Instituto de Investigación Biomédica of the Hospital 12 de Octubre, and all patients signed an informed consent after having understood all issues involved in the study participation, in accordance with the guidelines described in the Declaration of Helsinki, Convention of the Council of Europe on Human Rights and Biomedicine, Universal Declaration of the United Nations Educational, Scientific and Cultural Organization on the human genome and human rights and requirements established in Spanish legislation in the field of biomedical research, the protection of personal data and bioethics.

Immunohistochemistry analysis was performed in MDS (n = 9), control (n = 7) and DFX-treated (n = 6) samples, as previously described [19], with some modifications. BM smears, previously fixed in methanol, were refreshed in phosphate buffered saline/1% bovine serum albumin, and endogenous peroxidase was inactivated by incubation for 5 min with 3% hydrogen peroxide. Antigen unmasking and recovery was performed by heat-mediated antigen retrieval with citrate buffer. Protein carbonylation was detected by sample derivatization with DNPH (ref. 04732) and specific detection of the DNP derivatives by indirect peroxidase staining [19] using an anti-DNP rabbit antibody (1/200 dilution; ref. D9656) (both from Sigma-Aldrich, St. Louis, MO, USA). An immunohistochemistry assay using an anti-4-HNE rabbit antibody (1:100 dilution, ref. ab46545) (Abcam, Cambridge, UK) was also performed. Smears were counterstained with Carazzi's hematoxylin solution. Finally, samples were dehydrated in an ethanol series (100%, 96%), cleared in xylol and coverslipped with DPX mounting medium (Sigma-Aldrich). Samples were visualized and imaged using a Nikon Eclipse 80i microscope (Nikon, Tokyo, Japan) equipped with a Nikon digital camera. All cells were counted from 4 randomly chosen fields and averaged.

Primary cultures used for OxyBlot assays were obtained from BM samples of patients with MDS (n = 8) and control subjects (n = 6). Mononuclear cells were separated by density gradient centrifugation through Ficoll (GE Healthcare, Chicago, IL, USA). Subsequently, CD34+ cells were selected using the CD34+ Microbead Kit and magnetic cell separation (MiniMACS, Miltenyi Biotec, Bergisch Gladbach, Germany). As previously described [38], cells were cultured in enriched Stem Spam medium with erythropoietin (0.5 U/mL, which was increased after six days), stem cell factor (100 ng/mL), IL-3 (10 ng/mL), lipoproteins (40 µg/mL), dexamethasone (1 µM), glutamine (2 mM) and 1% penicillin-streptomycin, with the aim of promoting erythroid precursor development. All cell culture products were purchased from Stem Cell Technologies (Vancouver, BC, Canada) with the exception of lipoproteins (Sigma-Aldrich, St. Louis, MO, USA). To characterize erythroblasts, after 10 days of culture, cells were stained with Wright's stain and analyzed by flow cytometry on a BD FacsAria Fusion cytometer (BD Biosciences, Franklin Lakes, NJ, USA). The gating strategy was based on dead/live cells and doublet discrimination. When possible, a minimum of 10,000 events of the population of interest was analyzed. The erythroblast antigen expression profile was evaluated by labeling with the fluorescently-conjugated antibodies CD71-APC and CD36-PE. FCS Express 6 (De NovoSoftware, Glendale, CA, USA) was used for data analysis (Figure A1).

Table 1. MDS patients' characteristics.

Demographic Data				Clinical Features								Performed Analysis			
Patient (No)	Sex (M/F)	Age (years)	MDS Subtypes(WHO)	Hb (g/dL)	ANC $(10^9/L)$	PTLs $(10^9/L)$	Blasts (%)	Cytogenetics (FISH)	Risk Groups (IPSS)	Iron Chelation Therapy	Ferritin (ng/m L)	A	B	C	D
1	M	83	RCMD-RS	10.5	3	250	1	N	Low	No	779	✔	✔		✔
2	F	75	CMML	12.3	4.6	163	2	N	Low	No	45		✔		
3	M	84	RCMD	12.8	1.1	99	2	N	Low	No	88	✔			
4	M	59	RCMD-RS	9.4	1.5	177	1	N	Low	No	582	✔	✔	✔	
5	M	75	RAEB-1	14	1.7	164	6	N	Int-1	No	168	✔			
6	M	83	RAEB-1	11.4	1.2	80	9	Mon (7, 20). Del (5q, 12p. 17q)	Low	No	180	✔			✔
7	F	84	RCMD	11.4	2.9	144	4	N	Low	No	136		✔	✔	
8	M	81	RCMD-RS	9.9	3.2	363	1	N	Low	No	437		✔		
9	F	73	RAEB-2	11.5	3.4	440	15	N	Int-1	No	1123		✔		
10	F	90	RCMD	9.1	4.4	166	4	N	Int-2	No			✔		
11	F	87	RARS	8.8	4.2	306	1	N	Low	No	1547	✔			
12	F	70	RAEB-1	10.6	1	133	5	N	Low	No	24	✔			
13	M	82	RCMD-RS	9.9	3.2	363	1	N	Low	No	319	✔			
14	F	81	RARS	10.9	4.3	165	2	N	Low	No	157	✔			
15	M	66	RARS	10.3	4.4	414	2	N	Low	Yes	1137.6	✔	✔		✔
16	M	70	RARS	6.7	6	313	2	Del (20q)	Low	Yes	884	✔			✔
17	F	66	RAEB-2	10.2	9	161	16	Del (5q)	Int-1	Yes	163				✔
18	M	71	RCMD-RS	7.8	7.8	185	4	N	Int-1	Yes	643	✔			
19	M	71	RCMD	8.2	1.8	31	1	N	Low	Yes	206.2	✔			
20	M	57	RCMD	7.5	8	22	<1	N	Int-1	Yes	767	✔			
21	F	82	RAEB-1	11.3	8.6	26	7	Del (5q)	Int-1	Yes	1179	✔			

No = sample number; M = male; F = female; RCMD-RS = refractory cytopenia with multilineage dysplasia and ringed sideroblasts; CMML = chronic myelomonocytic leukemia; RCMD = refractory cytopenia with multilineage dysplasia; RAEB-1 = refractory anemia with excess of blasts type 1; RAEB-2 = refractory anemia with excess of blasts type 2; RARS = refractory anemia with ringed sideroblasts; Hb = hemoglobin; ANC = absolute neutrophil count; PTLs = platelets; IPSS =International Prognostic Scoring System; Low = low-risk; Int-1 = intermediate-1-risk; Int-2 = intermediate-2-risk; *N* = normal; Mon (7, 20) =, monosomy of chromosomes 7 and 20; Del(5q, 12p, 17q) = deletion of chromosome 5,12,7. Performed analyses methods: A = immunohistochemistry; B = one-dimensional OxyBlot; C = two-dimensional OxyBlot; D = quantitative polymerase chain reaction.

After 10 days in culture, cells were lysed in a buffer containing 50 mM Tris pH 8, 50 mM NaCl, 3% (*w/v*) CHAPS and protease inhibitors. Total protein lysates were denatured at 100 °C for 3 min in the presence of 6% sodium dodecyl sulfate (SDS) and then derivatized as previously described [18], with some modifications. One volume of a solution containing 10 mM DNPH in 10% trifluoroacetic acid was added to the samples followed by incubation for 10 min at 25 °C. The mixture was neutralized by adding one volume of stop solution (2 M Tris, 30% glycerol and 15% β-mercaptoethanol). Samples were precipitated and resuspended in a buffer containing 7 M urea, 2 M thiourea, 3% (*w/v*) CHAPS, 0.5% (*w/v*) MEGA-10, 0.5% (*w/v*) LPC, and 10 mM dithioerythritol. Protein quantification was performed with the Bradford 1× Dye Reagent Quick Start ™ reagent from Bio-Rad (Hercules, CA, USA) and the absorbance at 595 nm was read on an EPOCH plate reader (BioTek, Highland Park, VT, USA) using Gen5 2.0 software (BioTek).

After quantification, a one-dimensional (1D)-OxyBlot was performed with 4 µg of protein samples electrophoresed through a denaturing 10% SDS-polyacrylamide gel and transferred to polyvinylidene fluoride (PVDF) membranes. Membranes were incubated overnight at 4 °C with an anti-DNP primary antibody (as in immunohistochemistry assay, but at 1/10,000 dilution) followed by an anti-rabbit-IgG HRP-linked secondary antibody (1/5000 dilution; ref. 7074, Cell Signaling Technologies, Danvers, MA, USA). Glyceraldehyde-3-phosphate dehydrogenase (antibody Ref. AM4300, Ambion/ThermoFisher, Waltham, MA, USA) was used as a loading control. Chemiluminescense was used to visualize proteins according to the ECL Prime Western Kit Blotting Detection Reagent Amersham protocol from Sigma-Aldrich, on the ChemiDoc™ MP Imaging System (Bio-Rad). Images were captured and analyzed with the ImageLab 5.0 program (Bio-Rad).

For 2D-OxyBlot and preparative gel electrophoresis, four gels ($n = 2$ from MDS samples; $n = 2$ from control samples), with 100 µg protein per sample, were run in parallel under identical conditions and transferred to PVDF membranes. Membranes were incubated with anti-DNP primary and anti-rabbit-IgG HRP-linked secondary antibodies, as described for the 1D-OxyBlot. Chemiluminescense was performed as above with exposure on Curix RP2 PLUS film (AGFA, Mortsel, Belgium). Due to the limitation of available protein, an additional SDS-polyacrylamide gel containing a representative pool of the samples (400 µg) was run in parallel and stained with Colloidal Coomassie brilliant blue G250 to display total proteins and for subsequent protein identification by MS analysis.

Spots of interest were manually excised from the 2D preparative gel, in-gel reduced, alkylated, and then digested with trypsin, as described elsewhere [18]. Proteins were identified by matching the trypsin-digested peptide mass against the SwissProt database (SwissProt 553231 sequences; 197953409 residues) using MASCOT 1.9 (http://www.matrixscience.com) through the Global Protein Server (v3.5) from Applied Biosystems (Foster City, CA, USA). The parameters for the peptide mass fingerprinting search were as follows: modification on cysteine residues by carbamidomethylation was set as fixed modification; methionine oxidation was considered as a variable modification; the maximum number of missed tryptic cleavages was one; peptide mass tolerance was set to 50 ppm and monoisotopic masses were considered. In some cases, taxonomy was restricted to human. All of the identified proteins fulfilled the criterion of being significant ($P < 0.05$) based on the MOWSE scoring scheme. MS data have been deposited to the ProteomeXchange Consortium via the PRIDE partner repository [39], with the identifier code PXD013609. To corroborate that the signal level was due exclusively to a higher level of carbonylation and not to a higher protein expression, specific antibodies for the identified proteins were used.

Analysis of p21 mRNA expression was performed by quantitative PCR (qPCR). Synthesis of cDNA from BM-extracted RNA of MDS ($n = 5$), control ($n = 8$) and DFX-treated ($n = 3$) samples was performed with the High Capacity cDNA Reverse Transcription Kit (ThermoFisher) on the Veriti 96 Well Thermal Cycler platform (Applied Biosystems), maintaining a 1:1 ratio of reverse transcriptase buffer and RNA. Gene expression levels of p21 (ref. Hs00355782_m1) and the constitutive gene β-glucuronidase (Gus) (ref. Hs00939627_m1) gene were measured following the TaqMan® Gene Expression Assay protocol provided by ThermoFisher. As a positive control, samples with known transcriptional p21 activation were used as a reference. All samples were tested in triplicate and the changes in gene expression were calculated using the comparative ΔCT method [40]. The Gus gene served as an endogenous control for any slight variation in the initial RNA concentration, the total RNA quality, and the conversion and the efficiency of the reverse transcription reaction. Finally, expression levels were normalized to control samples.

Data are presented as the mean ± SEM. Comparisons between two groups were performed using the parametric Student's t-test or the non-parametric Mann-Whitney U test, as appropriate. Multiple group comparisons were performed using the parametric analysis of variance (ANOVA) or the non-parametric Kruskal-Wallis test. Differences were considered to be statistically significant at $P \leq 0.05$. All analyses were performed using GraphPad Prism 5.01 software (GraphPad Software Inc., La Jolla, CA, USA).

3. Results

3.1. Protein Carbonylation is Increased in Myeloid Series from MDS Patients and is Decreased in Patients Treated with DFX

Immunohistochemistry analysis of BM samples revealed higher oxidative stress, measured by protein carbonylation, in MDS patients than in controls or in DFX-treated MDS patients (Figure 1a). Carbonylation immunostaining was mostly cytoplasmic and was evidently more intense in the myeloid series. Quantitative analysis, measured as the percentage of positive stained cells of different randomly chosen fields, showed that the number of positively stained cells was significantly higher ($P \leq 0.01$) in MDS samples than in equivalent controls. In addition, BM samples from DFX-treated MDS patients had a significantly lower number of positively stained cells than untreated patients ($P \leq 0.01$), whereas there was no statistically different change in the number of positive cells between samples from controls and DFX-treated MDS patients ($P = 0.82$) (Figure 1b). A negative control sample without primary antibody verified that staining was exclusively due to the presence of carbonyl groups. It should be noted that although DNPH is the most commonly used probe for this approach, it lacks the ability to characterize the type of carbonyl modification [17].

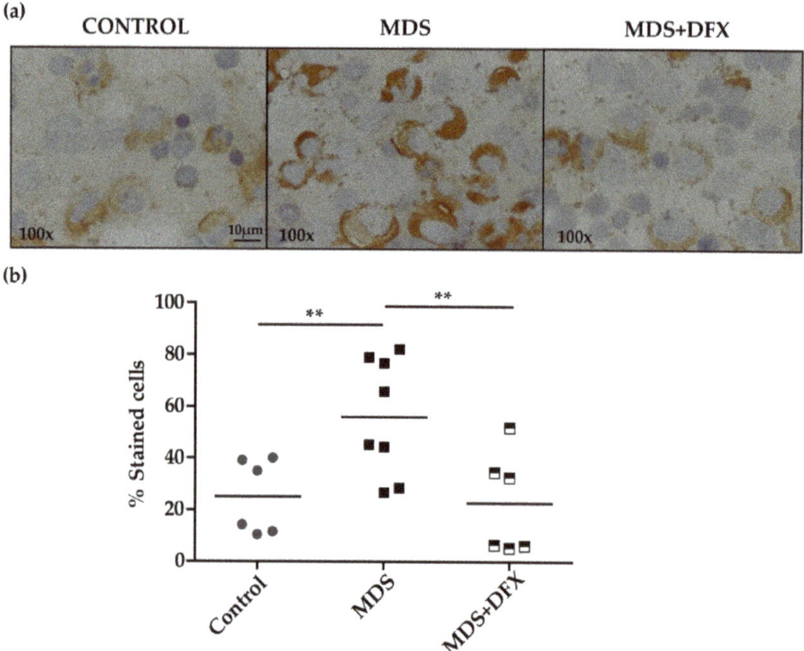

Figure 1. Immunodetection of carbonyl groups in derivatized bone marrow samples using an anti-DNP primary antibody. (**a**) Representative images (100× magnification) of bone marrow smears from controls

and MDS patients treated or not with DFX (MDS and MDS+DFX, respectively), derivatized with 2,4-dinitrophenylhydrazine. (**b**) Quantification of carbonylation-positive cells detected by immunohistochemistry: $n = 14$: control ($n = 6$), MDS ($n = 8$), and MDS+DFX ($n = 6$); mean ± SEM, **$P \leq 0.01$.

To examine for the presence of 4-HNE adducts, we performed an immunohistochemistry assay with MDS, control and MDS+DFX BM samples (Figure 2a). Quantitative analysis showed no statistically differences in the 4-HNE staining between these groups ($P = 0.29$) (Figure 2b).

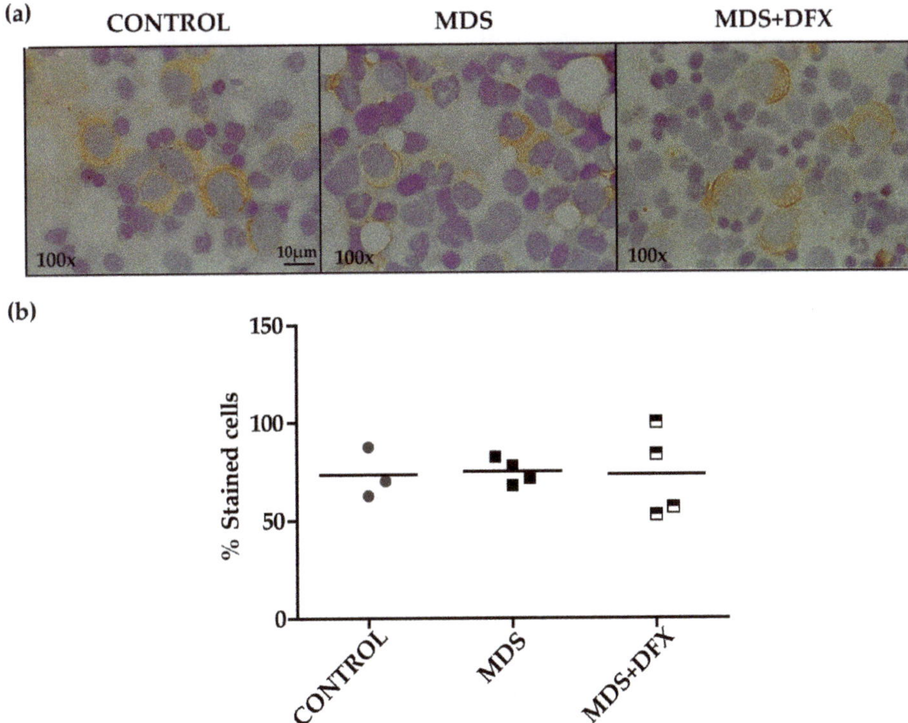

Figure 2. Immunodetection of 4-hydroxynonenal (HNE) in bone marrow samples. (**a**) Representative images (100× magnification) of bone marrow smears from controls, MDS and DFX-treated MDS patients (MDS+DFX), detected by anti 4-HNE immunostaining. (**b**) Quantification of 4-HNE-positive cells detected by immunohistochemistry: $n = 11$: control ($n = 3$), MDS ($n = 4$), and MDS+DFX ($n = 4$); mean ± SEM.

3.2. Protein Carbonylation is Increased in MDS Erythroid Precursors and is Decreased by DFX Treatment

To address the possibility of oxidative stress in MDS erythroid precursors, we examined the pattern of carbonylated proteins in cultured erythroblasts (Figure A1) from the three groups. We performed 1D-OxyBlot analysis of total protein lysates from CD34+ cells cultured from MDS and control BM samples. Results showed a significant increase ($P \leq 0.01$) in protein carbonylation in MDS over control samples (Figure 3a,b and Figure A2). Closer inspection of the 1D-OxyBlots showed strongly staining carbonylated protein species in MDS erythroblasts ranging in molecular weight from 40 to 250 kDa, with a major band at 100 kDa. We also observed a particularly highly carbonylated ~40 kDa band present in all samples, which was stronger in MDS lysates (Figure 3a). As expected, we observed that carbonylated protein staining was considerably weaker in erythroblasts cultured from DFX-treated MDS BM (Figure 3c). Ex vivo treatment with DFX also produced a reduction in protein carbonylation (Figure A3).

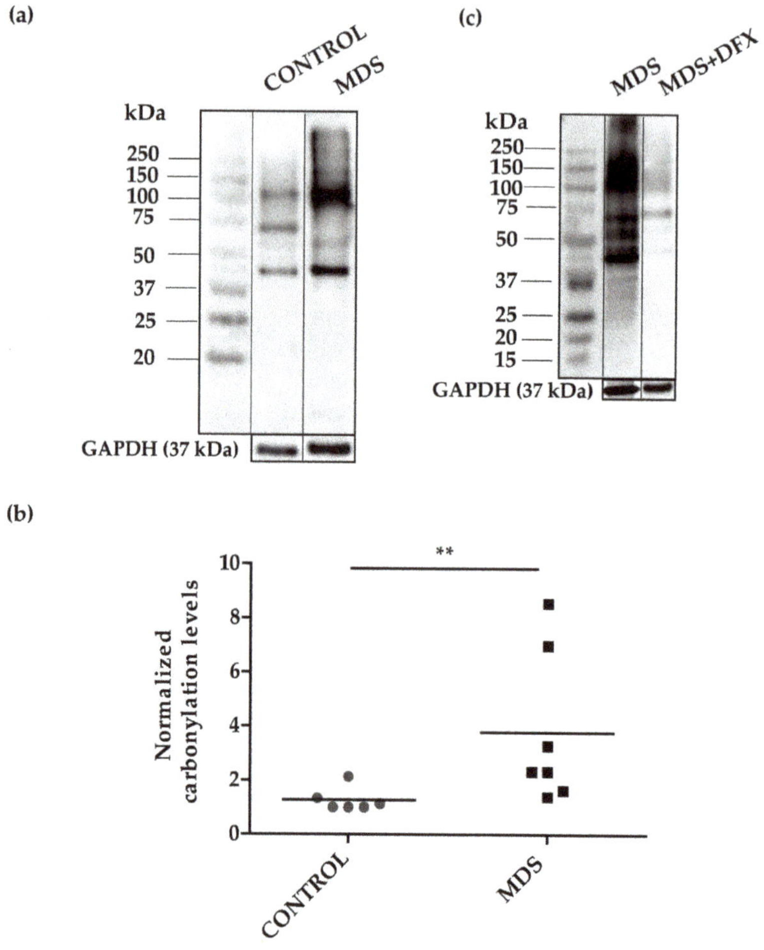

Figure 3. Protein carbonylation in erythroid precursor extracts. (**a**) Representative 1D-OxyBlot from cultured CD34+ cells of control (C) and MDS samples derivatized with DNPH and detected by anti-DNP immunostaining. (**b**) Quantification of normalized carbonylation levels by 1D-OxyBlots. $n = 13$: control ($n = 6$), and MDS ($n = 7$); mean ± SEM, ** $P \leq 0.01$. (**c**) 1D-OxyBlot of protein extracts from cultured CD34+ cells from MDS and DFX-treated MDS BM samples. A protein standard is also shown for reference.

We next aimed to identify the proteins responsible for the overall increased carbonylation in MDS erythroblasts. To do this, we performed 2D-OxyBlot analysis (Figure 4a) of the carbonylated proteins in cultured erythroblasts. A clear difference in the pattern of carbonylated proteins could be observed in the 2D-OxyBlots between MDS and control samples (Figure 4a). Although a greater number of carbonylated proteins in MDS samples than in controls might be expected, only a small number of proteins appeared to be responsible for the increased levels of total protein carbonylation. Consistent with the results from 1D- and 2D-OxyBlots, proteins ranging in size from 37 to 250 kDa had the highest carbonylation levels (Figure 4b).

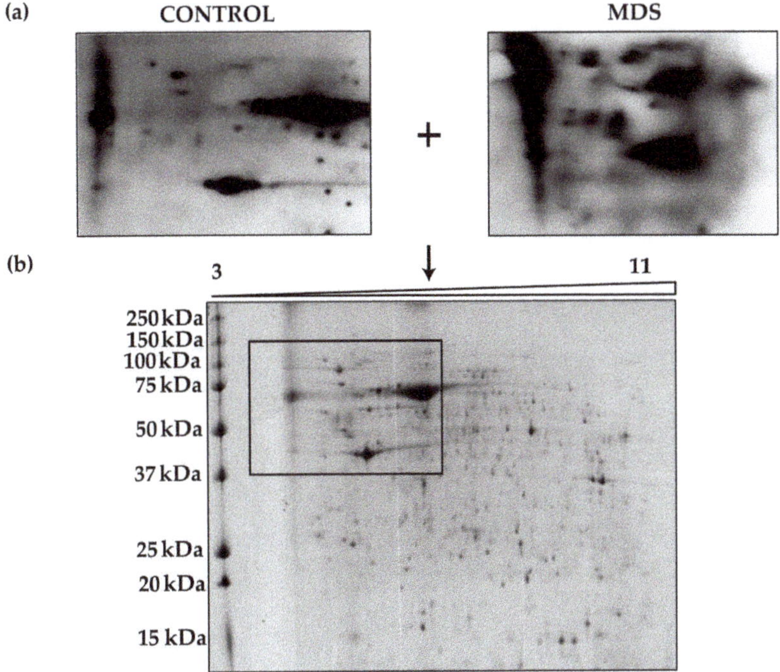

Figure 4. Protein carbonylation in erythroid precursors analyzed by 2D-Oxyblot (**a**) Representative 2D-OxyBlots of protein extracts from cultured CD34+ cells of representative control and MDS samples derivatized with DNPH and stained with an anti-DNP antibody. (**b**) Preparative 2D gel from pooled samples derivatized with DNPH and detected by Coomassie staining.

Four proteins with higher carbonylation levels in the MDS samples were excised from the preparative gel. MS analysis identified the proteins as: cytoplasmic actin, zinc finger protein 846 (ZNF846), 14-3-3 protein zeta/delta, and L-lactate dehydrogenase (LDH) A chain (Table 2). These four carbonylated proteins in MDS cells are related mainly to cell cytoskeleton, transcriptional regulation and metabolism.

Table 2. Identification of carbonylated proteins.

Protein ID	Accesion No.	No. of Peptides Matched/Searched	%Coverage	%Score	Nominal Mass (Mr)/Pi
Actin, cytoplasmic 1	sp|P60709|ACTB_HUMAN	21/65	49	147	42,052/5.29
Zinc finger protein 846	sp|Q147U1|ZN846_HUMAN	12/65	26	68	62,109/9.21
14-3-3 protein zeta/delta	sp|P63104|1433Z_HUMAN	12/65	47	90	27,899/4.73
L-lactate dehydrogenase A chain	sp|P00338|LDHA_HUMAN	14/65	45	104	36,950/8.44

Protein ID = protein ID from NCBInr database; Accession no. = accession numbers from NCBInr database; No. of peptides matched/searched = number of matched peptides versus total number of peptides; %Coverage = coverage of the matched peptides in relation to the full-length sequence; %Score = probability-based MOWSE score; Nominal mass (Mr)/Pi = theoretical nominal mass (Mr) and isoelectric point (pI) from the NCBInr database.

3.3. Upregulation of p21 in MDS is Controlled by DFX

To study signaling pathways potentially activated by oxidative stress and involved in MDS pathogenesis [25], and their possible modulation by DFX, we analyzed the expression of the p53 target gene, p21, in BM samples. Results of qPCR analysis showed that the expression of p21 was considerably and significantly higher ($P \leq 0.001$) in BM from MDS patients than from control samples

(Figure 5a). Interestingly, p21 expression was modestly but significantly lower ($P \leq 0.05$) in MDS samples after DFX treatment (Figure 5b), thus demonstrating an association between oxidative stress in MDS and p21 expression. Figure A4 summarizes the values obtained sample-by-sample when the same patients were analyzed by several techniques.

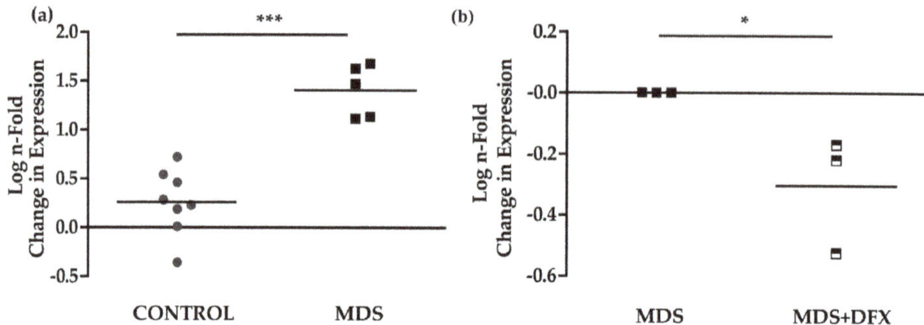

Figure 5. p21 mRNA expression. (**a**) Comparison of normalized p21 mRNA levels between samples from controls (C) ($n = 8$) and MDS patients (MDS) ($n = 5$); mean ± SEM ***$P \leq 0.001$. (**b**) p21 mRNA levels from MDS patients treated with DFX (MDS+DFX) ($n = 3$) normalized to p21 levels before DFX treatment (MDS) ($n = 3$); mean ± SEM, *$P \leq 0.05$.

4. Discussion

Oxidative stress and its effects on cell biology, DNA damage and carcinogenesis are important in the pathogenesis of MDS [9,10]. In an attempt to identify possible effectors of oxidative damage, here we provide the first analysis, to our knowledge, of protein carbonylation in hematopoietic lineages from patients with MDS. This was, however, a challenging endeavor due to the limited number of patients with this pathology, the restricted amount of sample available for research and the difficulty in the expansion of cell progenitors. Analyzing BM samples by proteomics necessitates a large amount of sample, which likely explains the limited number of proteomic studies performed for this pathology.

Immunohistochemical analysis revealed an increase in protein carbonylation in BM samples from patients with MDS, mostly in cells of the myeloid series. This was probably not triggered by lipid peroxidation, as reflected in the similarities in 4-HNE staining between samples from patients and controls. As iron levels are elevated in MDS patients (Table 1), metal-catalyzed oxidation should be a major contributor to protein damage in myeloid cells. Proteomic analysis also revealed an increase in protein carbonylation in erythroid precursors, which could be related to the increased DNA damage in MDS CD34+ cells [4,41]. Indeed, a recent study demonstrated an increase in DNA damage in MDS patients, which was reverted by treatment with an iron chelator [6].

The present study demonstrates for the first time that protein carbonylation in BM samples from MDS patients is decreased by DFX treatment, pointing to an important role of iron overload and the potential therapeutic benefit of iron chelators, as has been described in patients with low-risk MDS [30].

Previous assessments of oxidative stress in BM of patients with MDS using antioxidant biomarkers showed that all BM cell types have an imbalance in ROS levels that is related to lower overall survival [6,10,42]. This phenotype was also reverted by iron chelation [6,42], suggesting that amelioration of the oxidative stress effects by DFX treatment is beneficial for these patients. Indeed, an improvement in the pathogenesis of MDS has previously been reported after DFX therapy [20,33].

Interestingly, the findings from 2D-OxyBlot analysis suggest that perhaps only a few proteins are responsible for the bulk of total protein carbonylation in MDS samples. Four carbonylated proteins were successfully identified by MS analysis and are discussed below.

Actin is extremely susceptible to carbonylation [43]. This process seems to occur at an extent of oxidative insult higher than that causing the oxidation of some critical amino acids residues and causes the disruption of the cytoskeleton. Indeed, the increase in carbonylated actin found in a number of medical conditions is associated with severe oxidative modifications leading to functional impairment [44]. Actin carbonylation also occurs in malignant transformation [43] and, accordingly, our data suggest that this modification could be a marker of MDS pathogenesis.

Lactate dehydrogenase, a redox-active enzyme, reversibly converts pyruvate to lactate during anaerobic glycolysis [45]. Interestingly, in the context of oxidative stress, cells rely heavily on anaerobic glycolysis for ATP production [46]. Diseased cells upregulate anaerobic glycolytic enzymes – particularly LDH – to increase energy in the form of ATP to maintain homeostasis [47,48]. LDH oxidation results in a loss of its catalytic activity due to the modification of functional protein residues [47,49], and LDH carbonylation has been associated with decreased activity in disorders related to oxidative stress imbalance [50]. Our data suggest that oxidative damage to LDH observed in MDS could be a key switch for the cell dysplasia observed in this disease.

The 14-3-3 proteins are ubiquitous proteins involved in myriad processes, including the regulation of metabolism, signal transduction, cell cycle control through the p53/p21 pathway [51], apoptosis, protein trafficking, transcription, stress responses, and malignant transformation [52]. They are known to regulate the activity of a broad array of targets via protein-protein interactions, modulating signaling pathways related to redox regulation [53,54]. In addition, cysteine residues in 14-3-3 proteins are redox sensors which, in the oxidized state, impact their biological activity [55]. These oxidized post-translational modifications in 14-3-3 proteins are known disease markers in atherosclerosis [56] and neurological disorders [52], and could also serve as indicators of MDS.

Finally, ZNF846 is involved in transcriptional regulation through its association with DNA binding sites, to control the expression of multiple genes, and also acts as an RNA polymerase II transcription factor coadjutant by virtue of its sequence-specific DNA binding activity [57]. Oxidative modification of ZNF846, via zinc finger cysteine thiols, leads to the release of zinc molecules from the binding site. This results in the loss of zinc finger protein function related to DNA-binding and also in an increase of free zinc that may stimulate and interfere with cellular signaling cascades [58,59]. These processes could contribute to multiple cellular dysfunctions involved in the pathogenesis and progression of MDS. Therefore, further exploration of this marker as a potential key factor in MDS-associated diseases is warranted. Interestingly, a regulatory transcription factor-binding site for p53 in the ZNF846 gene promoter has been found in the GeneCards database (www.genecards.org).

Signaling pathways activated by oxidative stress can control the cell cycle via p53 [23]. As expected, we found that the mRNA expression of p21 was significantly increased in MDS patients with respect to controls, and was significantly reduced by DFX treatment. Thus, mechanistically, the amelioration of oxidative stress by DFX seems to involve, at least in part, the p53/p21 pathway, which is a major signaling pathway activated by oxidative stress and altered in MDS [24]. To our knowledge, this is the first study that demonstrates the mitigating effect of DFX on p21 expression in the context of MDS. Overall, our data suggest that treatment of MDS with DFX could be effective in (1) inhibiting/reversing protein carbonylation and its harmful downstream consequences and (2) restoring the altered signaling pathways associated with oxidative stress.

5. Conclusions

Patients with MDS present a higher level of protein carbonylation with respect to control peers, along with an altered p53/p21 signaling pathway. The identification of four main carbonylated proteins might suggest a role for these oxidation-sensitive proteins in the pathogenesis of MDS. Finally, our data on DFX treatment in patients indicates that the inhibition of protein carbonylation associated with enhanced p21 signaling is possible in MDS.

Author Contributions: A.R.-G. participated in the conceptualization of the study, collected the data and methodology, performed the formal analysis and wrote the initial draft of the manuscript. M.L.M. contributed to performing the analysis and discussed methology and results. A.A. helped to discuss methology. A.L. and G.C.-T. helped to characterize cells from cultures. V.G.-G., I.G.-B., R.S. and T.C. provided resources (samples used in the study). R.M.A. and J.M.B. participated in the conceptualization of the study, discussed methodology, provided resources and financial support. J.M.-L. and M.L. conceptualized, administered and supervised the study. All authors helped to write the manuscript and approved the final manuscript.

Funding: This research was funded by the Research Institute Hospital 12 de Octubre (i+12), Institute of Health Carlos III, the CRIS foundation and MINECO grant to J.M.B. BIO2016-77430R. M.L. had a postdoctoral fellowship from the Spanish Ministry of Economy and Competitiveness (FPDI-2013-16409) and holds a grant from the Spanish Society of Haematology and Hemotherapy.

Acknowledgments: We are indebted to all the patients who donated bone marrow and for their collaboration. We thank the Proteomics Unit of the Complutense University of Madrid (ProteoRed, PRB2-ISCIII, supported by grant PT13/0001) for performing the proteomic analysis.

Conflicts of Interest: The authors declare no conflict of interest.

Appendix A

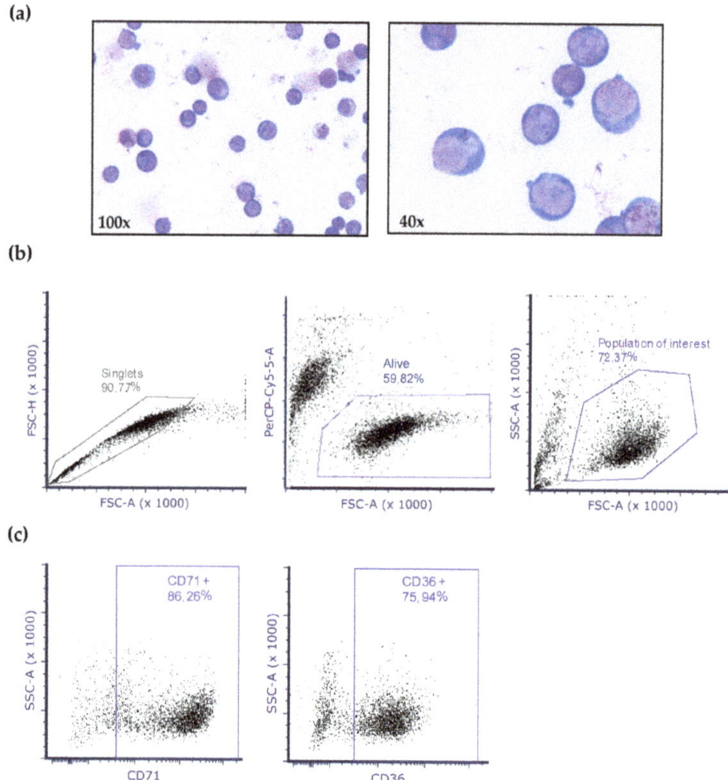

Figure A1. Characterization of erythroblasts in cell culture. (**a**) After 10 days of culture, cells were stained with Wright´s stain and 98% of cells were identified as erythroblasts. (**b–c**) Representative dot plots from flow cytometry analysis of erythroblast cultures. (**b**) Analysis and gating strategy of erythroblast cells. (**c**) Cells were analyzed after 10 days in culture for CD71 and CD36 expression, which is specific for erythroblasts.

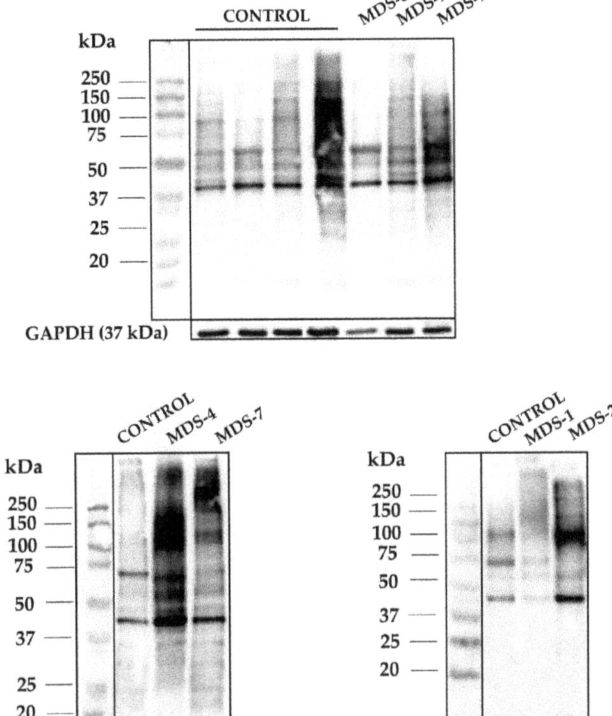

Figure A2. 1D-OxyBlots from all cultured CD34+ cells of control and MDS samples derivatized with DNPH and detected by anti-DNP immunostaining. A protein standard is shown for reference.

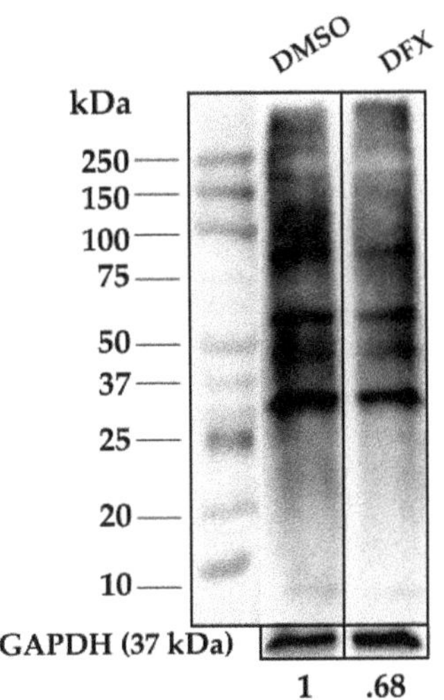

Figure A3. Representative 1D-OxyBlot from CD34+ cells treated with 50 μM DFX for 24 h, *n* = 2. Carbonylation levels were normalized to GAPDH levels (loading control).

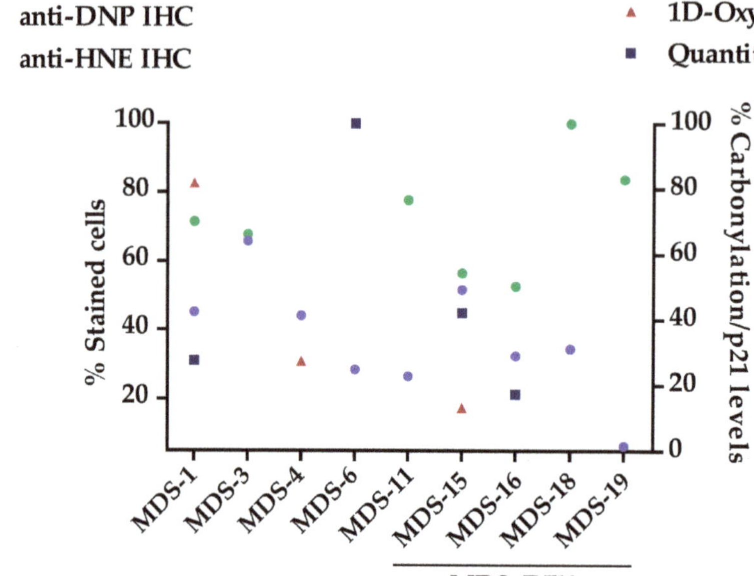

Figure A4. 2D plot sample-by-sample of patients analyzed by several techniques. Correlation of the percentage of positive stained cells for DNP and HNE antibodies and the percentage of carbonylation/p21 levels from MDS patients (MDS-1, 3, 4, 6) and MDS patients treated with DFX (MDS-15, 16, 18, 19).

References

1. Richardson, C.; Yan, S.; Vestal, C. Oxidative Stress, Bone Marrow Failure, and Genome Instability in Hematopoietic Stem Cells. *Int. J. Mol. Sci.* **2015**, *16*, 2366–2385. [CrossRef] [PubMed]

2. Gangat, N.; Patnaik, M.M.; Tefferi, A. Myelodysplastic syndromes: Contemporary review and how we treat: MDS treatment. *Am. J. Hematol.* **2016**, *91*, 76–89. [CrossRef] [PubMed]

3. Chung, Y.J.; Robert, C.; Gough, S.M.; Rassool, F.V.; Aplan, P.D. Oxidative stress leads to increased mutation frequency in a murine model of myelodysplastic syndrome. *Leuk. Res.* **2014**, *38*, 95–102. [CrossRef] [PubMed]

4. Peddie, C.M.; Wolf, C.R.; Mclellan, L.I.; Collins, A.R.; Bowen, D.T. Oxidative DNA damage in CD34+ myelodysplastic cells is associated with intracellular redox changes and elevated plasma tumour necrosis factor-α concentration. *Br. J. Haematol.* **1997**, *99*, 625–631. [CrossRef] [PubMed]

5. Ivars, D.; Orero, M.T.; Javier, K.; Díaz-Vico, L.; García-Giménez, J.L.; Mena, S.; Tormos, C.; Egea, M.; Pérez, P.L.; Arrizabalaga, B.; et al. Oxidative imbalance in low/intermediate-1-risk myelodysplastic syndrome patients: The influence of iron overload. *Clin. Biochem.* **2017**, *50*, 911–917. [CrossRef]

6. Jiménez-Solas, T.; López-Cadenas, F.; Aires-Mejía, I.; Caballero-Berrocal, J.C.; Ortega, R.; Redondo, A.M.; Sánchez-Guijo, F.; Muntión, S.; García-Martín, L.; Albarrán, B.; et al. Deferasirox reduces oxidative DNA damage in bone marrow cells from myelodysplastic patients and improves their differentiation capacity. *Br. J Haematol.* **2019**, *187*, 93–104. [CrossRef]

7. Dalle-Donne, I. Biomarkers of Oxidative Damage in Human Disease. *Clinical Chemistry* **2006**, *52*, 601–623. [CrossRef]

8. Evans, J.L.; Maddux, B.A.; Goldfine, I.D. The Molecular Basis for Oxidative Stress-Induced Insulin Resistance. *Antioxid. Redox Signal.* **2005**, *7*, 1040–1052. [CrossRef]

9. De Souza, G.F.; Ribeiro, H.L.; De Sousa, J.C.; Heredia, F.F.; De Freitas, R.M.; Martins, M.R.A.; Gonçalves, R.P.; Pinheiro, R.F.; Magalhaes, S.M.M. HFE gene mutation and oxidative damage biomarkers in patients with myelodysplastic syndromes and its relation to transfusional iron overload: An observational cross-sectional study. *BMJ Open* **2015**, *5*, e006048. [CrossRef]

10. Gonçalves, A.C.; Cortesão, E.; Oliveiros, B.; Alves, V.; Espadana, A.I.; Rito, L.; Magalhães, E.; Lobão, M.J.; Pereira, A.; Nascimento Costa, J.M.; et al. Oxidative stress and mitochondrial dysfunction play a role in myelodysplastic syndrome development, diagnosis, and prognosis: A pilot study. *Free Radic. Res.* **2015**, *49*, 1081–1094. [CrossRef]

11. Gonçalves, A.C.; Cortesão, E.; Oliveiros, B.; Alves, V.; Espadana, A.I.; Rito, L.; Magalhães, E.; Pereira, S.; Pereira, A.; Costa, J.M.N.; et al. Oxidative stress levels are correlated with P15 and P16 gene promoter methylation in myelodysplastic syndrome patients. *Clin. Exp. Med.* **2016**, *16*, 333–343. [CrossRef] [PubMed]

12. Ghoti, H.; Amer, J.; Winder, A.; Rachmilewitz, E.; Fibach, E. Oxidative stress in red blood cells, platelets and polymorphonuclear leukocytes from patients with myelodysplastic syndrome. *Eur. J. Haematol.* **2007**, *79*, 463–467. [CrossRef] [PubMed]

13. Cassarino, D.S.; Fall, C.P.; Swerdlow, R.H.; Smith, T.S.; Halvorsen, E.M.; Miller, S.W.; Parks, J.P.; Parker, W.D.; Bennett, J.P. Elevated reactive oxygen species and antioxidant enzyme activities in animal and cellular models of Parkinson's disease. *Biochim. Biophys. Acta (BBA) Mol. Basis Dis.* **1997**, *1362*, 77–86. [CrossRef]

14. Cheresh, P.; Kim, S.-J.; Tulasiram, S.; Kamp, D.W. Oxidative stress and pulmonary fibrosis. *Biochim. Biophys. Acta (BBA) Mol. Basis Dis.* **2013**, *1832*, 1028–1040. [CrossRef] [PubMed]

15. Niu, Y.; DesMarais, T.L.; Tong, Z.; Yao, Y.; Costa, M. Oxidative stress alters global histone modification and DNA methylation. *Free Radic. Biol. Med.* **2015**, *82*, 22–28. [CrossRef]

16. Bhatti, J.S.; Bhatti, G.K.; Reddy, P.H. Mitochondrial dysfunction and oxidative stress in metabolic disorders — A step towards mitochondria based therapeutic strategies. *Biochim. Biophys. Acta (BBA) Mol. Basis Dis.* **2017**, *1863*, 1066–1077. [CrossRef]

17. Afiuni-Zadeh, S.; Rogers, J.C.; Snovida, S.I.; Bomgarden, R.D.; Griffin, T.J. AminoxyTMT: A novel multi-functional reagent for characterization of protein carbonylation. *BioTechniques* **2016**, *60*, 186–196. [CrossRef]

18. Linares, M.; Marín-García, P.; Méndez, D.; Puyet, A.; Diez, A.; Bautista, J.M. Proteomic Approaches to Identifying Carbonylated Proteins in Brain Tissue. *J. Proteome Res.* **2011**, *10*, 1719–1727. [CrossRef]

19. Frank, J.; Pompella, A.; Biesalski, H.K. Immunohistochemical Detection of Protein Oxidation. In *Oxidants and Antioxidants. Methods in Molecular Biology*$^{TM.}$; Armstrong, D., Ed.; Humana Press: Totowa, NJ, USA, 2002; Volume 196, pp. 35–40. ISBN 978-0-89603-851-6.

20. Barrera, G.; Pizzimenti, S.; Daga, M.; Dianzani, C.; Arcaro, A.; Cetrangolo, G.P.; Giordano, G.; Cucci, M.A.; Graf, M.; Gentile, F. Lipid Peroxidation-Derived Aldehydes, 4-Hydroxynonenal and Malondialdehyde in Aging-Related Disorders. *Antioxidants* **2018**, *7*, 102. [CrossRef]

21. Schneider, R.K.; Schenone, M.; Ferreira, M.V.; Kramann, R.; Joyce, C.E.; Hartigan, C.; Beier, F.; Brümmendorf, T.H.; Germing, U.; Platzbecker, U.; et al. Rps14 haploinsufficiency causes a block in erythroid differentiation mediated by S100A8 and S100A9. *Nat. Med.* **2016**, *22*, 288–297. [CrossRef]

22. Liggins, J.; Furth, A.J. Role of protein-bound carbonyl groups in the formation of advanced glycation endproducts. *Biochim. Biophys. Acta (BBA) Mol. Basis Dis.* **1997**, *1361*, 8. [CrossRef]

23. Hole, P.S.; Darley, R.L.; Tonks, A. Do reactive oxygen species play a role in myeloid leukemias? *Blood* **2011**, *117*, 5816–5826. [CrossRef] [PubMed]

24. Fei, C.; Zhao, Y.; Guo, J.; Gu, S.; Li, X.; Chang, C. Senescence of bone marrow mesenchymal stromal cells is accompanied by activation of p53/p21 pathway in myelodysplastic syndromes. *Eur. J. Haematol.* **2014**, *93*, 476–486. [CrossRef] [PubMed]

25. Deville, L.; Hillion, J.; Ségal-Bendirdjian, E. Telomerase regulation in hematological cancers: A matter of stemness? *Biochim. Biophys. Acta (BBA) Mol. Basis Dis.* **2009**, *1792*, 229–239. [CrossRef] [PubMed]

26. Pilo, F.; Angelucci, E. A storm in the niche: Iron, oxidative stress and haemopoiesis. *Blood Rev.* **2018**, *32*, 29–35. [CrossRef] [PubMed]

27. Zhao, Y.; Guo, J.; Zhang, X.; Zhang, Z.; Gu, S.; Fei, C.; Li, X.; Chang, C. Downregulation of p21 in Myelodysplastic Syndrome Is Associated With p73 Promoter Hypermethylation and Indicates Poor Prognosis. *Am. J. Clin. Pathol.* **2013**, *140*, 819–827. [CrossRef] [PubMed]

28. Chai, X.; Li, D.; Cao, X.; Zhang, Y.; Mu, J.; Lu, W.; Xiao, X.; Li, C.; Meng, J.; Chen, J.; et al. ROS-mediated iron overload injures the hematopoiesis of bone marrow by damaging hematopoietic stem/progenitor cells in mice. *Sci. Rep.* **2015**, *5*, 10181. [CrossRef]

29. Silva, B.; Faustino, P. An overview of molecular basis of iron metabolism regulation and the associated pathologies. *Biochim. Biophys. Acta (BBA) Mol. Basis Dis.* **2015**, *1852*, 1347–1359. [CrossRef]

30. Leitch, H.A. Improving clinical outcome in patients with myelodysplastic syndrome and iron overload using iron chelation therapy. *Leuk. Res.* **2007**, *31*, S7–S9. [CrossRef]

31. Rose, C.; Brechignac, S.; Vassilief, D.; Pascal, L.; Stamatoullas, A.; Guerci, A.; Larbaa, D.; Dreyfus, F.; Beyne-Rauzy, O.; Chaury, M.P.; et al. Does iron chelation therapy improve survival in regularly transfused lower risk MDS patients? A multicenter study by the GFM. *Leuk. Res.* **2010**, *34*, 864–870. [CrossRef]

32. Neukirchen, J.; Fox, F.; Kündgen, A.; Nachtkamp, K.; Strupp, C.; Haas, R.; Germing, U.; Gattermann, N. Improved survival in MDS patients receiving iron chelation therapy—A matched pair analysis of 188 patients from the Düsseldorf MDS registry. *Leuk. Res.* **2012**, *36*, 1067–1070. [CrossRef] [PubMed]

33. Pullarkat, V.; Sehgal, A.; Li, L.; Meng, Z.; Lin, A.; Forman, S.; Bhatia, R. Deferasirox exposure induces reactive oxygen species and reduces growth and viability of myelodysplastic hematopoietic progenitors. *Leuk. Res.* **2012**, *36*, 966–973. [CrossRef] [PubMed]

34. Tataranni, T.; Agriesti, F.; Mazzoccoli, C.; Ruggieri, V.; Scrima, R.; Laurenzana, I.; D'Auria, F.; Falzetti, F.; Di Ianni, M.; Musto, P.; et al. The iron chelator deferasirox affects redox signalling in haematopoietic stem/progenitor cells. *Br. J. Haematol.* **2015**, *170*, 236–246. [CrossRef] [PubMed]

35. Gattermann, N.; Finelli, C.; Porta, M.D.; Fenaux, P.; Ganser, A.; Guerci-Bresler, A.; Schmid, M.; Taylor, K.; Vassilieff, D.; Habr, D.; et al. Deferasirox in iron-overloaded patients with transfusion-dependent myelodysplastic syndromes: Results from the large 1-year EPIC study. *Leuk. Res.* **2010**, *34*, 1143–1150. [CrossRef] [PubMed]

36. Hatcher, H.C.; Singh, R.N.; Torti, F.M.; Torti, S.V. Synthetic and natural iron chelators: Therapeutic potential and clinical use. *Futur. Med. Chem.* **2009**, *1*, 1643–1670. [CrossRef] [PubMed]

37. Vardiman, J.W.; Thiele, J.; Arber, D.A.; Brunning, R.D.; Borowitz, M.J.; Porwit, A.; Harris, N.L.; Le Beau, M.M.; Hellstrom-Lindberg, E.; Tefferi, A.; et al. The 2008 revision of the World Health Organization (WHO) classification of myeloid neoplasms and acute leukemia: Rationale and important changes. *Blood* **2009**, *114*, 937–951. [CrossRef]

38. Caceres, G.; McGraw, K.; Yip, B.H.; Pellagatti, A.; Johnson, J.; Zhang, L.; Liu, K.; Zhang, L.M.; Fulp, W.J.; Lee, J.-H.; et al. TP53 suppression promotes erythropoiesis in del(5q) MDS, suggesting a targeted therapeutic strategy in lenalidomide-resistant patients. *Proc. Natl. Acad. Sci. USA* **2013**, *110*, 16127–16132. [CrossRef]

39. Perez-Riverol, Y.; Csordas, A.; Bai, J.; Bernal-Llinares, M.; Hewapathirana, S.; Kundu, D.J.; Inuganti, A.; Griss, J.; Mayer, G.; Eisenacher, M.; et al. The PRIDE database and related tools and resources in 2019: Improving support for quantification data. *Nucl. Acids Res.* **2019**, *47*, D442–D450. [CrossRef]

40. Livak, K.J.; Schmittgen, T.D. Analysis of Relative Gene Expression Data Using Real-Time Quantitative PCR and the 2−ΔΔCT Method. *Methods* **2001**, *25*, 402–408. [CrossRef]

41. Hlaváčková, A.; Štikarová, J.; Pimková, K.; Chrastinová, L.; Májek, P.; Kotlín, R.; Čermák, J.; Suttnar, J.; Dyr, J.E. Enhanced plasma protein carbonylation in patients with myelodysplastic syndromes. *Free Radic. Biol. Med.* **2017**, *108*, 1–7. [CrossRef]

42. Meunier, M.; Ancelet, S.; Lefebvre, C.; Arnaud, J.; Garrel, C.; Pezet, M.; Wang, Y.; Faure, P.; Szymanski, G.; Duployez, N.; et al. Reactive oxygen species levels control NF-κB activation by low dose deferasirox in erythroid progenitors of low risk myelodysplastic syndromes. *Oncotarget* **2017**, *8*, 105510–105524. [CrossRef] [PubMed]

43. Castro, J.P.; Jung, T.; Grune, T.; Almeida, H. Actin carbonylation: From cell dysfunction to organism disorder. *J. Proteom.* **2013**, *92*, 171–180. [CrossRef] [PubMed]

44. Dalle-Donne, I.; Rossi, R.; Giustarini, D.; Gagliano, N.; Lusini, L.; Milzani, A.; Di Simplicio, P.; Colombo, R. Actin carbonylation: From a simple marker of protein oxidation to relevant signs of severe functional impairment. *Free Radic. Biol. Med.* **2001**, *31*, 1075–1083. [CrossRef]

45. Shonnard, G.C.; Hud, N.V.; Mohrenweiser, H.W. Arginine to tryptophan substitution in the active site of a human lactate dehydrogenase variant-LDHB GUA1: Postulated effects on subunit structure and catalysis. *Biochim. Biophys. Acta (BBA) Mol. Basis Dis.* **1996**, *1315*, 6. [CrossRef]

46. Vander Heiden, M.G.; Cantley, L.C.; Thompson, C.B. Understanding the Warburg Effect: The Metabolic Requirements of Cell Proliferation. *Science* **2009**, *324*, 1029–1033. [CrossRef] [PubMed]

47. Boike, L. An Analysis of Oxidative Damage to Lactate Dehydrogenase in Context of Neurodegeneration and Catechol-Based Phenolic Antioxidant Chemistry. Bachelor's Thesis, College of William & Mary, Williamsburg, VA, USA, December 2017.

48. Di Stefano, G.; Manerba, M.; Di Ianni, L.; Fiume, L. Lactate dehydrogenase inhibition: Exploring possible applications beyond cancer treatment. *Futur. Med. Chem.* **2016**, *8*, 713–725. [CrossRef]

49. Fiume, L.; Manerba, M.; Vettraino, M.; Di Stefano, G. Inhibition of lactate dehydrogenase activity as an approach to cancer therapy. *Futur. Med. Chem.* **2014**, *6*, 429–445. [CrossRef]

50. Ros, J. Protein carbonylation: Principles, analysis, and biological implications. In *Wiley Series on Mass Spectrometry*; Wiley: Hoboken, NJ, USA, 2017; pp. 110–123. ISBN 978-1-119-07491-5.

51. Gardino, A.K.; Yaffe, M.B. 14-3-3 proteins as signaling integration points for cell cycle control and apoptosis. *Semin. Cell Dev. Boil.* **2011**, *22*, 688–695. [CrossRef]

52. Yang, X.; Lee, W.H.; Sobott, F.; Papagrigoriou, E.; Robinson, C.V.; Grossmann, J.G.; Sundstrom, M.; Doyle, D.A.; Elkins, J.M. Structural basis for protein-protein interactions in the 14-3-3 protein family. *Proc. Natl. Acad. Sci. USA* **2006**, *103*, 17237–17242. [CrossRef]

53. Jayaraman, T.; Tejero, J.; Chen, B.B.; Blood, A.B.; Frizzell, S.; Shapiro, C.; Tiso, M.; Hood, B.L.; Wang, X.; Zhao, X.; et al. 14-3-3 Binding and Phosphorylation of Neuroglobin during Hypoxia Modulate Six-to-Five Heme Pocket Coordination and Rate of Nitrite Reduction to Nitric Oxide. *J. Biol. Chem.* **2011**, *286*, 42679–42689. [CrossRef]

54. Watanabe, K.; Thandavarayan, R.; Gurusamy, N.; Zhang, S.; Muslin, A.; Suzuki, K.; Tachikawa, H.; Kodama, M.; Aizawa, Y. Role of 14-3-3 protein and oxidative stress in diabetic cardiomyopathy. *Acta Physiol. Hung.* **2009**, *96*, 277–287. [CrossRef] [PubMed]

55. Zoila, I. The Role of Nox1 and 14-3-3 in the Regulation of Slingshot Phosphatase in Vascular Smooth Muscle Cells. Bachelor's Thesis, Faculty of Emory Collegue of Arts and Sciences, Atlanta, GA, USA, April 2012.

56. Kim, H.S.; Ullevig, S.L.; Nguyen, H.N.; Vanegas, D.; Asmis, R. Redox Regulation of 14-3-3 Controls Monocyte Migration. *Arterioscler. Thromb. Vasc. Biol.* **2014**, *34*, 1514–1521. [CrossRef] [PubMed]

57. Ravasi, T. Systematic Characterization of the Zinc-Finger-Containing Proteins in the Mouse Transcriptome. *Genome Res.* **2003**, *13*, 1430–1442. [CrossRef] [PubMed]

58. Kröncke, K.-D.; Klotz, L.-O. Zinc Fingers as Biologic Redox Switches? *Antioxid. Redox Signal.* **2009**, *11*, 1015–1027. [CrossRef] [PubMed]

59. Oteiza, P.I. Zinc and the modulation of redox homeostasis. *Free Radic. Biol. Med.* **2012**, *53*, 1748–1759. [CrossRef] [PubMed]

 *antioxidants*

Article

Cancer Chemotherapy and Chemiluminescence Detection of Reactive Oxygen Species in Human Semen

Teppei Takeshima *, Shinnosuke Kuroda and Yasushi Yumura

Department of Urology, Reproduction Center, Yokohama City University Medical Center, Yokohama 232-0024, Japan; shinnosuke_1014@yahoo.co.jp (S.K.); yumura@yokohama-cu.ac.jp (Y.Y.)
* Correspondence: teppeitalia@gmail.com

Received: 16 June 2019; Accepted: 25 September 2019; Published: 1 October 2019

Abstract: Advanced treatments have improved the prognosis of cancer survivors. Anticancer drugs generate large amounts of cellular reactive oxygen species (ROS), but their direct effects on sperm ROS production are unclear. We examined 64 semen samples of men who had received cancer chemotherapy, 467 semen samples of men consulting for idiopathic infertility, and 402 semen samples of partners of female patients as a control group. ROS production was calculated as the integrated chemiluminescence between 0 and 200 seconds after the addition of luminol to unwashed semen. We found that their ROS-positive rate of semen samples in the chemotherapy group was significantly higher than that in the control group. We compared the sperm parameters (concentration and motility) and the ROS production levels between chemotherapy subgroups and one of the remaining subgroups with positive ROS, and we found that only sperm motility was significantly lower in the samples in the postchemotherapy subgroup than in the idiopathic infertility subgroup, and that both sperm parameters were significantly lower in those from postchemotherapy subgroup than in the control subgroup. The ROS production level per million spermatozoa in the postchemotherapy subgroup was significantly higher than that in the control subgroup. Additionally, we compared variables, such as age, sperm features, and the duration from the end of the treatment to the first consultation between ROS-positive and ROS-negative subgroups in samples from men in the postchemotherapy group, but we found no significant differences. Of the men in the postchemotherapy group, three underwent a long-term antioxidant therapy, and all of them had low ROS semen levels after that. In conclusion, the production of ROS in semen detected by chemiluminescence from men who undergo cancer chemotherapy is similar to that of men with idiopathic infertility, and long-term oral antioxidant therapy may reduce the amount of ROS in the semen.

Keywords: reactive oxygen species; oxidative stress; sperm; cancer chemotherapy; antioxidant therapy

---

1. Introduction

Recently, advances in the treatment modalities for cancer have improved the prognosis of cancer survivors. The five-year survival rate for most types of cancer now exceeds 70% in Japan [1], and the disease has become curable or even chronic for some men. The ability to have children after cancer treatment is a major concern, especially for patients with cancer in the adolescent or young adult populations. Saito et al. reported that 70% of young patients with cancer wish to have children after cancer chemotherapy [2]. However, cancer treatments may impair the fertile capacity of these patients, and approximately 15–30% of patients with cancer have been reported to remain permanently sterile after treatment [3].

Reactive oxygen species (ROS) act as second messengers in cell signaling, whose low levels are essential for various biological processes in cells [4–6]. However, oxidative stress, resulting from the imbalances between excessive ROS and low cellular antioxidant levels, leads to deoxyribonucleic acid (DNA) damage and cellular lipid peroxidation.

It has been demonstrated in some studies that anticancer drugs generate elevated cellular levels of ROS via mitochondrial ROS generation and inhibition of the cellular antioxidant system [7,8]. Arsenic trioxide, alkylating agents, vinca alkaloids, topoisomerase inhibitors, platinum compounds, anthracyclines, and bleomycin have all been reported to induce a loss of mitochondrial potential, disruption of the mitochondrial electron transport chain (ETC), electron leakage, and elevated ROS production [9–11]. On the other hand, taxanes and nitrogen mustards inhibit the antioxidant system [8]. Amplification of cellular ROS levels by these mechanisms increases the cancer cell ROS threshold to induce cell death [8]. Thus, these drugs have an anticancer effect, but as far as we know, their effects on sperm ROS have not been reported.

For this study, we measured the seminal ROS levels in patients who had undergone cancer chemotherapy and compared them to the levels in patients with idiopathic infertility. Furthermore, we compared semen parameters and ROS production levels between groups and ROS-positive and -negative samples in the postchemotherapy group.

2. Materials and Methods

2.1. Subjects

We examined 64 semen samples from 64 patients who had received cancer chemotherapy and from 467 patients with idiopathic infertility who had visited our male infertility clinic at the Reproduction Center, Yokohama City University Medical Center, from February 2011 to February 2017, with the chief complaint of low sperm quality. Moreover, we also examined partners of 402 female patients in our center who had normal sperm quality at the first visit and never visited our male infertility clinic as a control group. We formed a "postchemotherapy group", an "idiopathic group", and a "control group" with the semen samples. We excluded patients with azoospermia, varicocele, leukocytospermia, and other organic disorders.

We interviewed all the participants and conducted physical and endocrine examinations (testosterone, luteinizing hormone, follicle stimulating hormone, and estradiol) during their first consultation for ruling out varicoceles, seminal tract obstruction, testicular cancer, and other organic disorders. We also compared the patient characteristics, semen features, and semen ROS status between the samples in the three groups. All patients signed informed consent for participation, and the Institutional Review Board of Yokohama City University Medical Center approved this study design (ethical code: IR2502).

2.2. Semen Collection and Assessment of Semen Features

We instructed the men to deposit their semen samples through masturbation at our hospital after 48–120 h of sexual abstinence. We then conducted semen analyses using the Sperm Motility Analyzing System (SMAS™; DITECT, Tokyo, Japan), a computer-assisted semen analyzer, at 37 °C after 30 min of complete liquefaction. The semen volume (mL), sperm concentration ($\times 10^6$/mL), and sperm motility (%) were measured in accordance with the criteria of the World Health Organization standards of 2010, of which lower reference limits were 1.5 mL for semen volume, 15 million per mL for sperm concentration, and 40% for sperm motility. The reference number of leukocytes was $< 10^6$ per mL.

2.3. ROS Measurement

We used the Monolight 3010 Luminometer™ (BD Biosciences Pharmingen, San Diego, CA, USA) to measure the ROS production levels in unwashed semen while performing the routine semen analysis. We also measured the integrated chemiluminescence between 0 and 200 s before adding luminol to record the chemiluminescence of the samples. Then, we added 40 μL of 100 mM luminol (5-amino-2,3-dihydro-1,4-phthalazinedione) to 500 μL of unwashed semen to obtain the chemiluminescence of the samples expressed as relative light units (RLUs). Then, we calculated the integrated chemiluminescence as the difference between the values, before and after the addition of luminol to the semen samples.

We considered samples as positive for ROS production if the integrated chemiluminescence was over 4332.4 RLUs/200 s [12] (Figure 1). We then divided the data from the semen samples of the postchemotherapy and control groups into ROS-positive and ROS-negative subgroups.

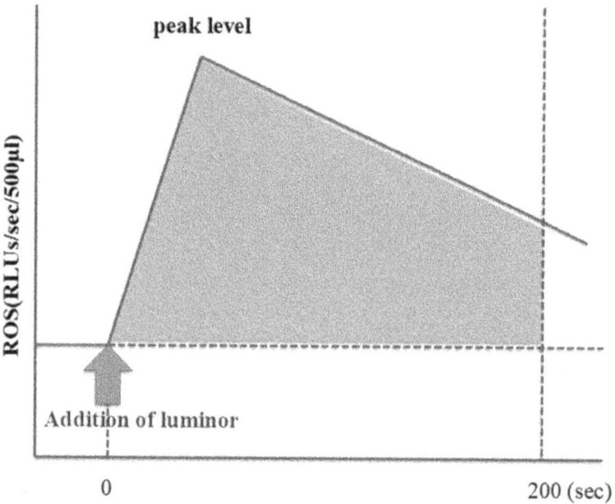

Figure 1. Measurement of reactive oxygen species (ROS) by chemiluminescence method. ROS production in the present study was calculated as the integrated chemiluminescence between 0 and 200 sec after the addition of luminol (5-amino-2,3-dihydro-1,4-phtalazinedione) to unwashed semen after baseline subtraction (expressed as relative light units (RLUs)/s/200 sec).

2.4. Statistical Analysis

All statistical analyses were performed using the JMP® Pro 12 software (SAS Institute, Cary, NC, USA). All data are reported as the means ± standard deviation. Group differences (postchemotherapy versus idiopathic groups, postchemotherapy versus control groups, and ROS-positive versus ROS-negative subgroups in the postchemotherapy group) were evaluated using unpaired t-tests (parametric variant). We compared the ROS-positive rate between the postchemotherapy and idiopathic groups and the postchemotherapy and control groups using the chi-squared test. p values < 0.05 were considered statistically significant in all cases.

3. Results

Table 1 shows the characteristics of patients in the three groups. The age at the first consultation was significantly lower in the postchemotherapy group than in the idiopathic and control groups. According to the sperm parameters at the consultation, both sperm concentration and motility were significantly lower in the postchemotherapy group than in the idiopathic infertility and control group, whereas, the ROS-positive rates between the samples from the postchemotherapy and idiopathic infertility groups were similar (42.2% versus 34.4%, $p = 0.226$), that from the control group was significantly lower than that from the postchemotherapy group (20.4% versus 42.2%, $p < 0.001$). Figure 2 shows the breakdown of original diseases and chemotherapeutic regimens of those positive for ROS. Testicular cancer and malignant lymphoma were equally frequent; and as for the therapeutic regimen, BEP (bleomycin, etoposide, and cisplatin) therapy was the most frequently used regimen.

Table 1. Characteristics of patients and semen features in the three groups.

Characteristic	Postchemotherapy (A)	Idiopathic (B)	Control (C)	*p*	
				A–B	*A–C*
n	64	467	402		
Age (years)	34.9 ± 7.46	37.02 ± 7.00	37.40 ± 6.28	0.015	0.004
Sperm concentration (million /mL)	22.10 ± 31.32	34.75 ± 34.98	44.18 ± 28.93	0.003	<0.001
Sperm motility (%)	23.85 ± 20.63	29.75 ± 21.17	26.90 ± 16.08	0.018	0.03
ROS-positive rate (%)	42.2 (27/64)	34.4 (161/467)	20.4 (82/402)	0.226	<0.001

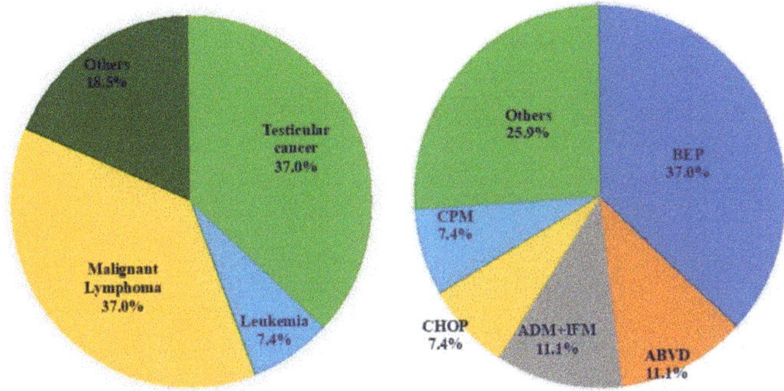

Figure 2. Breakdown of original diseases and chemotherapeutic regimens of those positive for ROS. Testicular cancer and malignant lymphoma were equally frequent. As for the therapeutic regimen, BEP was the most frequent. BEP: Bleomycin, etoposide, and cisplatin, ABVD: Doxorubicin, bleomycin, vinblastine, and dacarbazine, ADM: Doxorubicin, IFM: Ifosfamide, CHOP: Cyclophosphamide, doxorubicin, vincristine, and prednisone, CPM: Cyclophosphamide.

Subsequently, we compared sperm parameters (concentration and motility), total ROS production levels and ROS production levels per one million spermatozoa between postchemotherapy and the other groups positive for ROS (Table 2). Only sperm motility was significantly lower in the samples from the postchemotherapy group than in those from the idiopathic group, but both sperm concentration and motility were significantly lower in those from the postchemotherapy group than in those from the control group positive for ROS. In the control group, sperm parameters were within the normal range at the first visit, but the sperm motility was unexpectedly lower. The total ROS production level in the postchemotherapy group was slightly higher than those in the other groups, but the difference was not statistically significant. As for the ROS production level per one million spermatozoa, no significant difference was seen between the chemotherapy and idiopathic group, but that for the chemotherapy group was significantly higher than for the control group. Additionally, our comparison of the age, sperm features, and the duration from the end of the treatment to the first consultation between ROS-positive and ROS-negative samples in the postchemotherapy group yielded no significant differences between the two groups (Table 3).

Table 2. Sperm parameters and ROS production levels between both groups positive for ROS.

Parameter	Postchemotherapy (A)	Idiopathic (B)	Control (C)	*p*	
				A–B	*A–C*
n	27	161	82		
Sperm concentration (million/mL)	28.66 ± 37.19	28.97 ± 31.55	43.48 ± 24.70	0.482	0.010
Sperm motility (%)	19.36 ± 20.89	26.37 ± 19.89	29.03 ± 14.17	0.047	0.004
ROS level (RLUs)	81032.2 ± 139294.0	76906.7 ± 155068.0	79374.3 ± 273599.9	0.448	0.488
ROS level per million spermatozoa (RLUs)	25937.0 ± 62187.3	21389.6 ± 134787.0	6696.7 ± 26468.2	0.432	0.013

Table 3. Comparison between ROS-positive and ROS-negative samples in the postchemotherapy group.

Parameter	ROS-positive	ROS-negative	*p*
n	27	37	
Age	35.1 ± 7.3	34.9 ± 7.7	0.456
Sperm concentration (million/mL)	28.66 ± 37.19	17.31 ± 25.74	0.077
Sperm motility (%)	19.36 ± 20.89	21.14 ± 20.08	0.069
The period from treatment to the first visit (month)	34.67 ± 52.57	30.07 ± 70.02	0.387

Of the patients in the postchemotherapy group, three underwent a long-term antioxidant therapy (combined vitamin C and E supplementation), and all of their semen samples had diminished ROS levels (from 129,745.5 ± 5,383.8 RLUs at the first visit to 15,144.3 ± 10,473.9 RLUs). The mean period from the end of the chemotherapy treatment to the first visit was 20 months, and the mean follow-up duration was 20 months.

4. Discussion

Excessive ROS generation in the semen has been observed in approximately 30–40% of men with infertility [12,13], and this unbalance is known to cause lipid peroxidation of the membrane polyunsaturated fatty acids (particularly of docosahexaenoic acids) and to impair sperm mitochondrial respiration, leading to single- or double-stranded DNA fragmentation [14,15]. Lipid peroxidation of sperm cellular membranes causes a loss of membrane fluidity and integrity, which are required for sperm–oocyte fusion [16]. In addition, DNA fragmentation of sperm has adverse effects on embryo development, blastulation, implantation, and pregnancy [17,18]. There have been many reports on the negative effect of ROS on sperm motion parameters [13] and fertile capacity [19,20].

It has been demonstrated in various studies that anticancer drugs produce cellular ROS mainly by two mechanisms: Inducing mitochondrial ROS generation and impairing the cellular antioxidant system [8]. Arsenic trioxide, which was used for leukemia, has been reported to cause a loss of mitochondrial membrane potential and to inhibit complexes I and II, leading to the disruption of the mitochondrial ETC, to electron leakage, and to an elevated ROS production [9,10]. Similarly, alkylating agents, vinca alkaloids, topoisomerase inhibitors, platinum compounds, anthracyclines, and bleomycin generate ROS by the same mechanism [8]. On the other hand, imexon binds to thiols, such as glutathione and causes the suppression of cellular glutathione and accumulation of cellular ROS [21]. Similarly, taxanes and nitrogen mustards inhibit the antioxidant system. In all, these mechanisms amplify the cellular ROS levels and increase the cellular concentrations in cancer cells, resulting in death [8]. Thus, we hypothesized that these anticancer effects might also affect the ROS levels in semen. Anticancer drugs can cross the blood–testicular barrier and impair spermatogenesis. As a result, excessive ROS production in immature sperm may lead to DNA damage. In some studies, there have reported on the effect of BEP therapy for testicular germ cell tumors on sperm DNA integrity [22]. BEP therapy causes single- or double-stranded DNA breaks for more than two years after treatment withdrawal [22]. This may be partly explained by the fact that platinum analogs, such as cisplatin, form cross-links with single- or double-stranded DNA in the sperm, which result in sperm DNA fragmentation and apoptosis [23]. Another mechanism of sperm DNA fragmentation may occur by the accumulation of ROS generated by immature spermatozoa as a result of impaired spermatogenesis, due to cancer chemotherapy. On the contrary, Smit et al. reported that DNA fragmentation index was higher in patients treated with radiotherapy compared with those treated with chemotherapy [24]. In our study, BEP therapy was the most frequently used regimen in men of the ROS-positive group. As mentioned, platinum compounds and bleomycin may disrupt the mitochondrial ETC, leading to the amplification of cellular ROS and subsequent cellular DNA fragmentation [8]. Similar events are thought to occur in sperm cells. Recently, electron spin resonance spectroscopy applying a spin trapping procedure that can measure the oxidative stress, lipid peroxidation, and DNA damage and repair simultaneously has been reported [25].

The total ROS levels were almost equivalent between the samples in the postchemotherapy group, those in the idiopathic group, and the control group. However, produced ROS per unit spermatozoa was significantly higher in the chemotherapy group than the control group. The principal sources of endogenous ROS in semen are immature spermatozoa and seminal leukocytes. In this study, samples with leukocytospermia were excluded. Therefore, in the chemotherapy group, ROS productivity of individual sperm was significantly high. This suggests that defects of spermatogenesis, because of the cancer chemotherapy may result in the accumulation of immature spermatozoa in semen, which produce high levels of ROS. In order to minimize the detrimental effect of ROS on the fertile capacity, an oral antioxidant is considered as an option. According to systematic reviews in the Cochrane database, 48 randomized controlled trials (RCTs) have compared single and combined antioxidants with placebo in a population of 4179 men with infertility [26]. The results suggested that the oral antioxidant therapy may increase clinical pregnancy rates (odds ratio [OR]: 3.43, $p < 0.0001$, seven RCTs, 522 men) and live birth rates (OR: 4.21, $p < 0.0001$, four RCTs, 277 men) [26]. Despite the small number of cases, our results also suggest that long-term antioxidant therapy (combined vitamin C and vitamin E) reduces ROS levels regardless of the period after the cancer treatment and that the oxidative stress is reversible.

The main limitation of this study was the inability to compare ROS levels in semen before and after the administration of cancer chemotherapy, because the semen collected before chemotherapy was all cryopreserved for fertility preservation in consideration of the risk of developing azoospermia permanently after the treatment. Therefore, we cannot be sure that the semen features reflect the cancer chemotherapy effects on ROS in semen.

Another limitation has to do with the inability to evaluate ROS levels depending on the types of cancer and regimens used, because the main subjects were patients whose diseases were in remission, due to cancer treatment and who wished to have babies. However, we believe that cancer chemotherapy may generate excessive ROS in sperm cells, depending on the type of anticancer drugs used.

5. Conclusions

In conclusion, our results suggest that the ROS production levels in semen from men who underwent cancer chemotherapy are similar to those of men with idiopathic infertility and that long-term oral antioxidant therapy may reduce ROS levels in the semen.

Author Contributions: Conceptualization, T.T. and Y.Y.; methodology, T.T.; formal analysis, T.T.; investigation, T.T, S.K. and Y.Y.; data curation, T.T.; writing—original draft preparation, T.T.; writing—review and editing, T.T.; supervision, S.K. and Y.Y.

Funding: This research received no external funding.

Conflicts of Interest: The authors declare having no conflicts of interest.

References

1. Ito, Y.; Miyashiro, I.; Ito, H.; Hosono, S.; Chihara, D.; Nakata-Yamada, K.; Nakayama, M.; Matsuzaka, M.; Hattori, M.; Sugiyama, H.; et al. Long-term survival and conditional survival of cancer patients in Japan using population-based cancer registry data. *Cancer Sci.* **2014**, *105*, 1480–1486. [CrossRef] [PubMed]
2. Saito, K.; Suzuki, K.; Iwasaki, A.; Yumura, Y.; Kubota, Y. Sperm cryopreservation before cancer chemotherapy helps in the emotional battle against cancer. *Cancer* **2005**, *104*, 521–524. [CrossRef] [PubMed]
3. Schrader, M.; Müller, M.; Straub, B.; Miller, K. The impact of chemotherapy on male fertility: A survey of the biologic basis and clinical aspects. *Reprod. Toxicol.* **2001**, *15*, 611–617. [CrossRef]
4. Redza-Dutordoir, M.; Averill-Bates, D.A. Activation of apoptosis signalling pathways by reactive oxygen species. *Biochim. Biophys. Acta* **2016**, *1863*, 2977–2992. [CrossRef] [PubMed]
5. Aitken, R.J.; Jones, K.T.; Robertson, S.A. Reactive Oxygen Species and Sperm Function—In Sickness and In Health. *J. Androl.* **2012**, *33*, 1096–1106. [CrossRef] [PubMed]

6. Aitken, R.J.; Smith, T.B.; Jobling, M.S.; Baker, M.A.; De Iuliis, G.N. Oxidative stress and male reproductive health. *Asian J. Androl.* **2014**, *16*, 31–38. [CrossRef] [PubMed]

7. Kardeh, S.; Ashkani-Esfahani, S.; Alizadeh, A.M. Paradoxical action of reactive oxygen species in creation and therapy of cancer. *Eur. J. Pharmacol.* **2014**, *735*, 150–168. [CrossRef]

8. Yang, H.; Villani, R.M.; Wang, H.; Simpson, M.J.; Roberts, M.S.; Tang, M.; Liang, X. The role of cellular reactive oxygen species in cancer chemotherapy. *J. Exp. Clin. Cancer Res.* **2018**, *37*, 266. [CrossRef] [PubMed]

9. Shi, H.; Shi, X.; Liu, K.J. Oxidative mechanism of arsenic toxicity and carcinogenesis. *Mol. Cell. Biochem.* **2004**, *255*, 67–78. [CrossRef] [PubMed]

10. Yen, Y.-P.; Tsai, K.-S.; Chen, Y.-W.; Huang, C.-F.; Yang, R.-S.; Liu, S.-H. Arsenic induces apoptosis in myoblasts through a reactive oxygen species-induced endoplasmic reticulum stress and mitochondrial dysfunction pathway. *Arch. Toxicol.* **2012**, *86*, 923–933. [CrossRef]

11. Marullo, R.; Werner, E.; Degtyareva, N.; Moore, B.; Altavilla, G.; Ramalingam, S.S.; Doetsch, P.W. Cisplatin Induces a Mitochondrial-ROS Response That Contributes to Cytotoxicity Depending on Mitochondrial Redox Status and Bioenergetic Functions. *PLoS ONE* **2013**, *8*, e81162. [CrossRef] [PubMed]

12. Yumura, Y.; Takeshima, T.; Kawahara, T.; Sanjo, H.; Kuroda, S.; Asai, T.; Mori, K.; Kondou, T.; Uemura, H.; Iwasaki, A. Reactive oxygen species measured in the unprocessed semen samples of 715 infertile patients. *Reprod. Med. Biol.* **2017**, *16*, 354–363. [CrossRef] [PubMed]

13. Takeshima, T.; Yumura, Y.; Yasuda, K.; Sanjo, H.; Kuroda, S.; Yamanaka, H.; Iwasaki, A. Inverse correlation between reactive oxygen species in unwashed semen and sperm motion parameters as measured by a computer-assisted semen analyzer. *Asian J. Androl.* **2017**, *19*, 350–354. [CrossRef] [PubMed]

14. Agarwal, A.; Saleh, R.A.; Bedaiwy, M.A. Role of reactive oxygen species in the pathophysiology of human reproduction. *Fertil. Steril.* **2003**, *79*, 829–843. [CrossRef]

15. Ollero, M.; Gil-Guzman, E.; Lopez, M.C.; Sharma, R.K.; Agarwal, A.; Larson, K.; Evenson, D.; Thomas, A.J., Jr.; Alvarez, J.G. Characterization of subsets of human spermatozoa at different stages of maturation: Implications in the diagnosis and treatment of male infertility. *Hum. Reprod.* **2001**, *16*, 1912–1921. [CrossRef] [PubMed]

16. Aitken, R.J. Reactive oxygen species as mediators of sperm capacitation and pathological damage. *Mol. Reprod. Dev.* **2017**, *84*, 1039–1052. [CrossRef]

17. Tesarik, J.; Greco, E.; Mendoza, C. Late, but not early, paternal effect on human embryo development is related to sperm DNA fragmentation. *Hum. Reprod.* **2004**, *19*, 611–615. [CrossRef]

18. Sakkas, D.; Alvarez, J.G. Sperm DNA fragmentation: Mechanisms of origin, impact on reproductive outcome, and analysis. *Fertil. Steril.* **2010**, *93*, 1027–1036. [CrossRef]

19. Yumura, Y.; Iwasaki, A.; Saito, K.; Ogawa, T.; Hirokawa, M. Effect of reactive oxygen species in semen on the pregnancy of infertile couples. *Int. J. Urol.* **2009**, *16*, 202–207. [CrossRef]

20. O'Flaherty, C.; Hales, B.F.; Chan, P.; Robaire, B. Impact of chemotherapeutics and advanced testicular cancer or Hodgkin lymphoma on sperm deoxyribonucleic acid integrity. *Fertil. Steril.* **2010**, *94*, 1374–1379. [CrossRef]

21. Dragovich, T.; Gordon, M.; Mendelson, D.; Wong, L.; Modiano, M.; Chow, H.-H.S.; Samulitis, B.; O'Day, S.; Grenier, K.; Hersh, E.; et al. Phase I Trial of Imexon in Patients with Advanced Malignancy. *J. Clin. Oncol.* **2007**, *25*, 1779–1784. [CrossRef] [PubMed]

22. Ghezzi, M.; Berretta, M.; Bottacin, A.; Palego, P.; Sartini, B.; Cosci, I.; Finos, L.; Selice, R.; Foresta, C.; Garolla, A. Impact of Bep or Carboplatin Chemotherapy on Testicular Function and Sperm Nucleus of Subjects with Testicular Germ Cell Tumor. *Front. Pharmacol.* **2016**, *7*, 673. [CrossRef] [PubMed]

23. Lee, S.J.; Schover, L.R.; Partridge, A.H.; Patrizio, P.; Wallace, W.H.; Hagerty, K.; Beck, L.N.; Brennan, L.V.; Oktay, K. American Society of Clinical Oncology Recommendations on Fertility Preservation in Cancer Patients. *J. Clin. Oncol.* **2006**, *24*, 2917–2931. [CrossRef] [PubMed]

24. Smit, M.; Van Casteren, N.J.; Wildhagen, M.F.; Romijn, J.C.; Dohle, G.R. Sperm DNA integrity in cancer patients before and after cytotoxic treatment. *Hum. Reprod.* **2010**, *25*, 1877–1883. [CrossRef]

25. Guerriero, G.; D'Errico, G.; Di Giaimo, R.; Rabbito, D.; Olanrewaju, O.S.; Ciarcia, G. Reactive oxygen species and glutathione antioxidants in the testis of the soil biosentinel Podarcis sicula (Rafinesque 1810). *Environ. Sci. Pollut. Res. Int.* **2018**, *25*, 18286–18296. [CrossRef]

26.   Showell, M.G.; Mackenzie-Proctor, R.; Brown, J.; Yazdani, A.; Stankiewicz, M.T.; Hart, R.J. Antioxidants for male subfertility. *Cochrane Database Syst. Rev.* **2014**, *12*, CD007411. [CrossRef]

Data Availability: The data used to support the findings of this study are available from the corresponding author upon request.

Article

Cancer Cell Sensitivity to Redox-Cycling Quinones is Influenced by NAD(P)H: Quinone Oxidoreductase 1 Polymorphism

Christophe Glorieux [1,†] and Pedro Buc Calderon [1,2,*]

[1] Metabolism and Nutrition Research Group, Louvain Drug Research Institute, Université catholique de Louvain, 1200 Brussels, Belgium
[2] Facultad de Ciencias de la Salud, Universidad Arturo Prat, 1110939 Iquique, Chile
* Correspondence: pedro.buccalderon@uclouvain.be; Tel.: +32-2-764-73-69
† Current address: Sun Yat-Sen University Cancer Center, State Key Laboratory of Oncology in South China, Collaborative Innovation Center of Cancer Medicine, Guangzhou 510275, China.

Received: 3 August 2019; Accepted: 28 August 2019; Published: 2 September 2019

Abstract: Background: Cancer cell sensitivity to drugs may be associated with disturbed antioxidant enzymes expression. We investigated mechanisms of resistance by using oxidative stress-resistant MCF-7 breast cancer cells (Resox cells). Since nicotinamide adenine dinucleotide phosphate (NAD(P)H): quinone oxidoreductase-1 (NQO1) is modified in tumors and oxidative stress-resistant cells, we studied its role in cells exposed to β-lapachone, menadione, and doxorubicin. Methods: Normal mammary epithelial 250MK, MCF-7, and Resox cells were employed. NQO1 expression and enzyme activity were determined by quantitative polymerase chain reaction (RT-PCR), immunoblotting, and biochemical assays. Dicoumarol and gene silencing (siRNA) were used to modulate NQO1 expression and to assess its potential drug-detoxifying role. MTT (3-(4,5-dimethylthia-zolyl-2)-2,5-diphenyltetrazolium bromide) or clonogenic assays were used to investigate cytotoxicity. NQO1 variants, NQO1*1 (wt), and NQO1*2 (C609T), were obtained by transfecting NQO1-null MDA-MB-231 cell line. Results: Resox cells have higher NQO1 expression than MCF-7 cells. In 250MK cells its expression was low but enzyme activity was higher suggesting a variant NQO1 form in MCF-7 cells. MCF-7 and Resox cells are heterozygous NQO1*1 (wt)/NQO1*2 (C609T). Both NQO1 polymorphism and NQO1 overexpression are main determinants for cell resistance during oxidative stress. NQO1 overexpression increases cell sensitivity to β-lapachone whereas NQO1*2 polymorphism triggers quinone-based chemotherapies-sensitivity. Conclusions: NQO1 influences cancer cells redox metabolism and their sensitivity to drugs. We suggest that determining *NQO1* polymorphism may be important when considering the use of quinone-based chemotherapeutic drugs.

Keywords: NQO1; NQO1*2; polymorphism; quinone; breast cancer; menadione; lapachone; doxorubicin; ascorbate; oxidative stress

---

1. Introduction

It has been generally accepted that nicotinamide adenine dinucleotide phosphate (NAD(P)H): quinone oxidoreductase 1 (NQO1, also known as DT-diaphorase) facilitates detoxification of quinone-based compounds by reducing their quinone nucleus [1–3]. Indeed, NQO1 through a two-electron reduction process, transforms quinone (Q) into a hydroquinone (QH2). In some occasions and depending on its stability, QH2 may be back oxidized to a semiquinone free radical (SQ•) and further to the original quinone while molecular oxygen is reduced to superoxide anion (Figure 1). In this context, since NQO1 is frequently overexpressed in a variety of tumors [4–6], the use of bioactive quinones has been exploited therapeutically because they are activated by NQO1 [7,8].

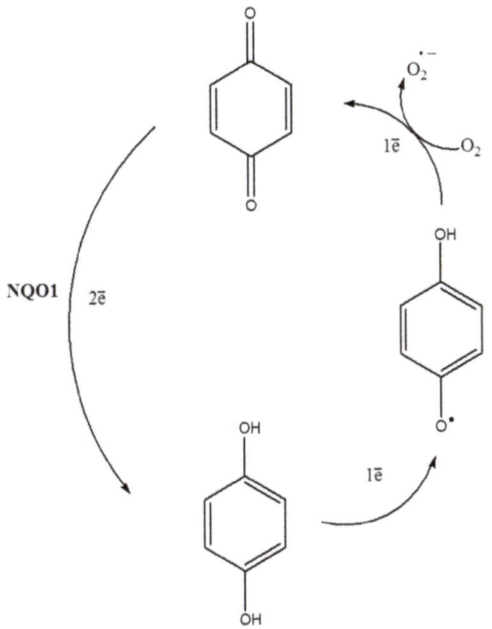

Figure 1. Quinone reduction by nicotinamide adenine dinucleotide phosphate (NAD(P)H): quinone oxidoreductase 1 (NQO1) and quinone redox cycling generating reactive oxygen species (ROS).

A different NQO1 expression pattern, at both protein levels and enzyme activity, has been found in tumors and normal tissues [9]. Recently, we found that some antioxidant enzymes, including NQO1, were overexpressed in MCF-7 breast cancer cells chronically exposed to hydrogen peroxide (H_2O_2) and by consequence becoming resistant against such oxidative stress, the so-called Resox cells [10]. Moreover, we detected a genomic gain of the chromosomal band 16q22 (where *NQO1* gene is located) in Resox cells as compared to parental MCF-7 cells suggesting an amplification of *NQO1* gene [11].

In addition, the existence of a polymorphism has also been noted. Indeed, two single nucleotide mutations have been reported: The C609T polymorphism, corresponding to a Pro187Ser change in the enzyme and described as *NQO1*2*, and C465T polymorphism, corresponding to an Arg139Trp change in the enzyme and described as *NQO1*3*. These polymorphisms are associated with a decreased enzyme activity [12,13]. Here, we paid particular attention to the study of *NQO1* polymorphism, using a model of NQO1-null MDA-MB-231 cells stably transfected with either *NQO1*1*, the wild-type form of *NQO1*, or the *NQO1*2* polymorphism. Depending on both the genotype and the chemotherapeutic drug, the final antitumor outcome can be dramatically influenced by NQO1 activity. The aim of the study was to investigate in such experimental model the role of NQO1 polymorphism on cancer cell sensitivity to quinone-based therapeutic drugs.

2. Materials and Methods

2.1. Cell Lines and Culture Conditions

MCF-7 cells, a human breast derived cell line, was obtained from ATCC (Manassas, VA, USA). By exposing them to chronic oxidative stress, they acquired resistance against a pro-oxidant treatment; therefore, they were named Resox cells [10]. MDA-MB-231 cells (ATCC) were kindly offered by Dr. Akeila Bellahcene (Metastasis Research Laboratory, Giga Cancer, Liège, Belgium). DMEM medium containing 10% fetal calf serum (10%), penicillin (100 U/mL), and streptomycin (100 μg/mL), obtained from Gibco (Grand Island, NY, USA), was used for cell cultures. Dr. Martha Stampfer (Lawrence

Berkeley National Laboratory, Berkeley, CA, USA) kindly provided 250MK cells, a human mammary epithelial cell line. They were maintained in a special medium (M87A + CT + X) and further used between eight and 10 passages [14]. Cell cultures were kept at 37 °C under an atmosphere of 95% air/5% CO_2 and 100% humidity. Dicoumarol, sodium L-ascorbate, menadione sodium bisulfite, β-lapachone, and doxorubicin hydrochloride were purchased from Sigma (St Louis, MO, USA).

2.2. Stable Transfection

pKK233-2 plasmids containing human *NQO1*1* (wild-type) and *NQO1*2* (C609T) cDNA (NCBI Reference Sequence: NM_000903.3) were a kind gift of Dr. David Ross [15]. The following primers were used to amplify by PCR the different cDNAs. Forward 5′-ccgaagcttgccatggtcggcagaagagc-3′ and Reverse 5′-ccgggtacctcattttctagctttgatct-3′ (Sigma, St Louis, MO, USA). HindIII and KpnI (Fermentas, Vilnius, Lithuania) were used as restriction enzymes and insert DNA were then cloned into pcDNA3.1 plasmid from Invitrogen (Grand Island, NY, USA). The transfection of MDA-MB-231 cells were done with different plasmids (1 µg), followed by four week-selection of exposure to 1 mg/mL neomycin (Invivogen, San Diego, CA, USA). Both NQO1 enzyme activity and protein levels were used to characterize stable transfecting clones. Only clones with high NQO1 activity and similar NQO1 protein levels were chosen for further studies.

2.3. Small Interfering RNA Transfection Procedure

The transfection of cells with siRNA against NQO1 (ON-TARGET plus SMART pool siRNA) was done with Dharmafect reagent 1, according to Dharmacon protocols (Lafayette, CO, USA). The transfection technique was conducted for 24 h at 50% cell confluence, using 0.1 µmol/L siRNA solution. Transfected cells were utilized 48 h after such transfection procedure.

2.4. Western Blots Assay

Protein sample preparation, protein quantification, and western blot analyses were done as reported elsewhere [16]. Primary mouse antibodies were: β-actin (ab6276) from Abcam (Cambridge, UK) and NQO1 (sc-32793) from Santa Cruz Biotechnology (Santa Cruz, CA, USA). Protein bands were revealed by chemiluminescence, according to procedures given by the ECL detection kit (Pierce, Thermo Scientific, Rockford, IL, USA). ImageJ software (http://rsb.info.nih.gov/ij/) was used to quantify protein bands.

2.5. Measurement of NQO1 Enzyme Activity

The activity of NQO1 was measured following the reduction of cytochrome C in the presence of NADH (reduced nicotinamide adenine dinucleotide) as reported by Fitzsimmons et al. [17]. Briefly, 2×10^6 cells were seeded in a 100 mm-culture dish containing 7 mL of incubation medium. Afterwards, when cells reached confluence, they were washed twice with ice-cold PBS (phosphate buffer saline) and further lysed with 1% Triton X-100 in PBS (500 µL), containing cocktails to inhibit both proteases (Sigma, St Louis, MO, USA) and phosphatases (Millipore, Merck KGaA, Darmstadt, Germany). Afterwards, samples were cold conserved for 5 min followed by sonication for 5 s at 4 °C using the Labsonic U sonicator (B Braun Biotech International, Melsungen, Germany). Each sample was prepared in duplicate, in the absence or in the presence of 10 µmol/L of dicoumarol (a well-known NQO1 inhibitor). The mixture (1 mL) contains 77 µmol/L of cytochrome C, 200 µmol/L NADH, 10 µmol/L menadione, 0.14% bovine serum albumin, and 50 mmol/L Tris-HCl pH 7.5. This mixture was incubated for 20 min at 37 °C and the tested sample (5 µL) was added. Cytochrome C reduction was read during 2 min at 550 nm and the calculated ΔOD/min, in the absence and in the presence of dicoumarol, was used to determine enzyme activity. The specific NQO1 activity was calculated by using a cytochrome C molar extinction coefficient of 21.1 mM/cm and results were expressed as nmol of cytochrome C reduced/minute/mg of protein. Intracellular protein levels were measured using a BCA (bicinchoninic acid) protein kit (Thermo Scientific, Rockford, IL, USA). All reagents were from Sigma (St Louis, MO, USA).

2.6. MTT Reduction Assay

The metabolic status of cells was evaluated by recording the formation of blue formazan crystals due to the reduction of MTT (3-(4,5-dimethylthia-zolyl-2)-2,5-diphenyltetrazolium bromide) by cellular dehydrogenases. In brief, 1×10^4 cells/well were plated onto 96-well plates and, when confluence was achieved, cells were further exposed to the indicated treatments. Afterwards, cells were washed two-times using PBS and further incubated with MTT (0.5 mg/mL) for 2 h. The blue formazan crystals were solubilized with DMSO (100 µL/well) and the absorbance of blue-dyed solutions were read at a wavelength of 550 nm. The absorbance obtained by measuring untreated control cells was taken as 100%.

2.7. Proliferation Assay

The clonogenic assay was done as follows: about 500 cells were seeded in six-well plates at a single-cell density. After overnight incubation, cells were exposed for 24 h to the respective quinones. Furthermore, they were washed with warm PBS, a new fresh medium was added, and cells were allowed to proliferate for 10 days. Clonogenic survival was measured by fixing and staining colonies using crystal violet and further counting their number. The number of colonies calculated under control conditions was set as 100%.

2.8. Quantitative Polymerase Chain Reaction

The extraction of total cellular RNA was performed using TriPure reagent (Roche Applied Science Diagnostics, Mannheim, Germany). Reverse transcription was done using SuperScript II RNase H-reverse transcriptase and random hexamer primers were acquired from Invitrogen (Grand Island, NY, USA). For real-time PCR, the Sybr Green Supermix (Bio-Rad, Hercules, CA, USA) was used. Primer sequences were 5′-caaatcctggaaggatggaa-3′ (forward) and 5′-aagtgatggcccacagaaag-3′ (reverse) for human NQO1 (NCBI Reference Sequence: NM_000903.3); 5′-cttcactgctcaggtgat-3′ (forward) and 5′-gccgtgtggcaatccaat-3′ (reverse) for human EF1 (Elongation factor 1, as reference gene; NCBI Reference Sequence: NM_001402.6). Primers were designed by Sigma (St Louis, MO, USA). The incubation of samples was conducted for 5 min at 95 °C, and subsequently for 40 cycles of 10 s at 95 °C and 30 s at 60 °C followed by a melting curve. The fluorescence of samples was assessed after each cycle in a Bio-Rad IQ5 thermocycler. Results were calculated by means of the following equation: $2^{-(Ct\,NQO1\,-\,Ct\,EF1)}$ and then compared to values obtained with untreated control cells.

2.9. Statistical Data Analyses

All experiments were done by at least three times. The experimental data were examined using either a one-way ANOVA or an unpaired two-tailed *t*-test, using GraphPad Prism software (GraphPad Software, San Diego, CA, USA). A value of $p < 0.05$ was set as level of significance.

3. Results

3.1. NQO1 Expression and Activity in Non-Cancerous and Cancerous Breast Cell Lines.

We first measured NQO1 expression in three cell lines: The normal human mammary epithelial 250MK, MCF-7 breast cancer, and Resox (MCF-7 resistant to an oxidative stress) cells. NQO1 was highly expressed in the breast cancer MCF-7 cells compared to 250MK cells (Figure 2A–C). Compared to parental MCF-7 cells, Resox cells have increased NQO1 mRNA (Figure 2A) and protein levels (Figure 2B,C). We then measured NQO1 enzyme activities and found those are also augmented in Resox compared to MCF-7 cells (Figure 2D). Surprisingly, NQO1 activity is enhanced in 250MK despite a low protein expression (Figure 2C,D). One explanation could be MCF-7 cells harbor a variant form of

NQO1 and sequencing of the complete open reading frame (ORF) of the human *NQO1* gene in the three cell lines confirmed that 250MK cells were homozygous *NQO1*1* (wild-type NQO1), whereas MCF-7 and Resox cells were heterozygous *NQO1*1/NQO1*2* (data not shown).

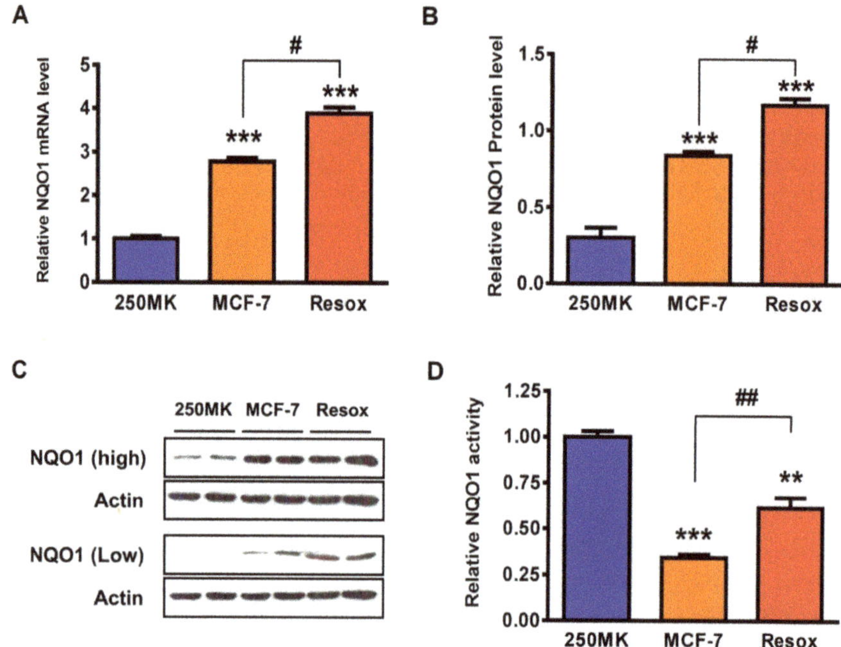

Figure 2. Basal levels of NQO1 expression in 250MK, MCF-7 and Resox cells. (**A**) NQO1 mRNA level was measured by real-time PCR. Data are means ± SEM ($n = 3$). (**B**) NQO1 protein levels normalized to β-actin were measured by immunoblotting. Data are means ± SEM ($n = 3$). (**C**) NQO1 expression detected by immunoblotting (low and high exposure times). (**D**) NQO1 enzyme activity. Data are means ± SEM ($n = 3$). Statistics: One-way ANOVA for (**A,B,D**). ** $p < 0.01$, *** $p < 0.001$ versus 250MK. # $p < 0.05$, ## $p < 0.01$ versus MCF-7. Abbreviations: NQO1: NAD(P)H: quinone oxidoreductase 1.

3.2. NQO1 Activity Correlates with Cancer Cell Sensitivity to Pro-oxidant Treatment.

Figure 3 shows that 250MK cells were resistant to the pro-oxidant treatment, in contrast to MCF-7 and Resox cells. Such oxidant treatment was induced by exposing cells to a H_2O_2-generating system (ascorbate and menadione), a mixture widely used in our laboratory to induce an oxidative stress [18–22]. Pharmacological inhibition of NQO1, by using dicoumarol, considerably increased the cytotoxicity in the three cell lines (Figure 3A). These results were then confirmed by genetic inactivation using a specific siRNA against NQO1 (Figure 3B,C). Unless ascorbate (Asc)/menadione (Men)-mediated cytotoxicity was not modified by dicoumarol in MDA-MB-231 cells (Figure 3D).

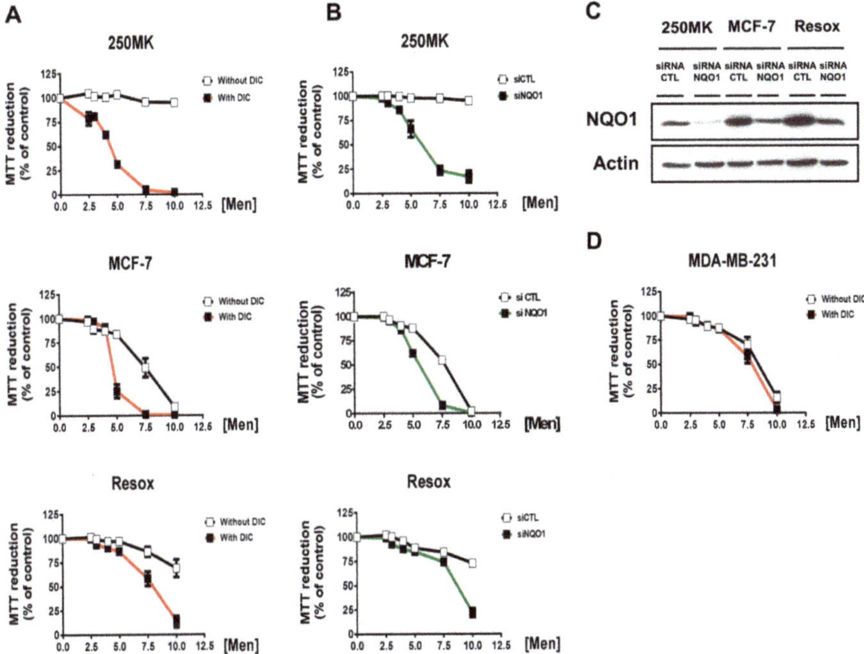

Figure 3. Sensitivity of mammary cells towards a pro-oxidant treatment. Cells were incubated for 24 h with various concentrations of ascorbate (Asc) ranging from 0 to 1 mmol/L, associated with menadione (Men) ranging from 0 to 10 µmol/L (ratio Asc/Men 100:1). (**A**) The specific NQO1 inhibitor dicoumarol (DIC) was tested at 25 µmol/L. Data are means (% of control) ± SEM (*n* = 3). (**B**) For genetic invalidation cells were transfected for 48 h with scrambled siRNA (siCTL) or specific siRNA against NQO1 mRNA (siNQO1). Cytotoxicity was evaluated using MTT assays. Data are means (% of control) ± SEM (*n* = 3). (**C**) NQO1 expression detected by immunoblotting. (**D**) MDA-MB-231 (pcDNA3.1) cells were treated with various concentrations of Asc/Men in presence or absence of 25 µmol/L dicoumarol. Cell viability was quantified by MTT assay. Abbreviations: DIC: dicoumarol; Men: menadione; NQO1: NAD(P)H: quinone oxidoreductase 1; si(RNA)CTL: control siRNA; si(RNA)NQO1: siRNA against NQO1.

3.3. NQO1 Polymorphism Influences Cancer Cell Sensitivity to Redox-Cycling Quinones.

Because MCF-7 cells are heterozygous *NQO1*1/NQO1*2* [12], we decided to study the importance of the *NQO1* polymorphism on the sensitivity of breast cancer cells to quinone-containing drugs. To this end, the wild-type form of *NQO1* (*NQO1*1*), or the variant form of the enzyme *NQO1*2*, were overexpressed in NQO1-null MDA-MB-231 cells.

As shown in Figure 4A, NQO1*1-overexpressing cells had about a 10-fold greater NQO1 activity than NQO1*2-overexpressing cells, despite the fact that both cells have similar NQO1 protein levels (Figure 4B,C). This result confirms that the presence of the polymorphism *NQO1*2* is associated with a decrease in NQO1 activity [13].

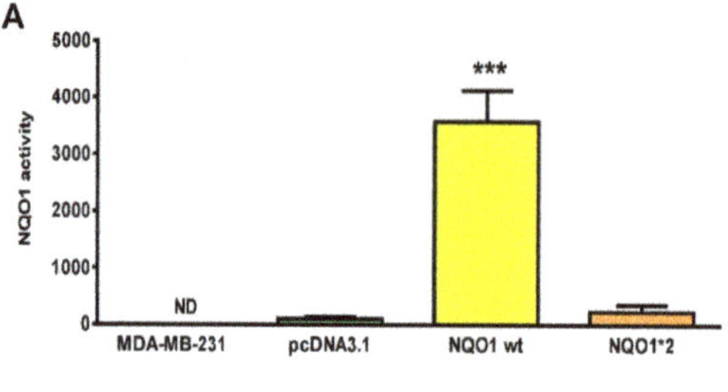

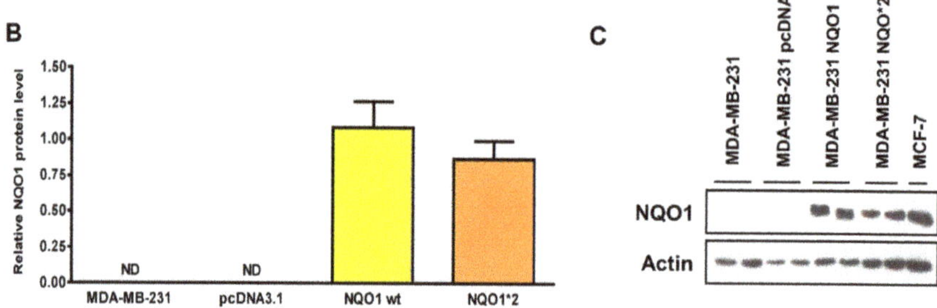

Figure 4. MDA-MB-231 cells overexpressing WT or C609T mutant forms of NQO1. (**A**) The specific NQO1 enzyme activity was calculated using a kinetic spectrophotometric method as explained in the respective experimental section. Results are expressed as nmol of cytochrome C reduced/minute/mg of protein. Data are means ± SEM (n = 3). *** p < 0.001 versus pcDNA3.1 (unpaired t-test) (**B**) NQO1 protein levels were quantified and normalized to β-actin. Data are means ± SEM (n = 3). (**C**) NQO1 expression was detected by immunoblotting. Abbreviations: NQO1: NAD(P)H: quinone oxidoreductase 1.

These newly generated cell lines were further exposed to menadione, doxorubicin, and β-lapachone. Using clonogenic survival assays (Table 1), we demonstrated that the expression of NQO1*1, the wild-type form with normal activity, was a main factor of cancer cell resistance vis-à-vis menadione and doxorubicin. In addition, the NQO1*1 expression was essentially linked to cell sensitivity toward β-lapachone. In contrast, expression of the NQO1*2 variant, which presented virtually no NQO1 activity, had no significant influence on cell survival, compared to MDA-MB-231 cells expressing the empty vector.

Table 1. Sensitivity of MDA-MB-231 cells overexpressing WT or C609T (NQO1*2) mutant form of NQO1 to quinones (clonogenic assay).

Parameter	Cell Type	Menadione (60 µmol/L)	Doxorubicin (0.05 µmol/L)	β-Lapachone (1.5 µmol/L)
Clonogenic Assay (Survival fraction)	MDA-MB-231 (pcDNA3.1)	3.7 ± 3.2	34.3 ± 8.5	87.1 ± 21.8
	MDA-MB-231 (NQO1 WT)	23.8 ± 5.1 **	58.6 ± 0.9 **	41.0 ± 28.9
	MDA-MB-231 (NQO1*2)	5.5 ± 4.6	40.4 ± 0.6	98.8 ± 18.0

Clonogenic survival was determined as described in Section 2. Cells were transfected with either the empty vector pcDNA3.1, NQO1 wild-type or NQO1*2 and incubated with quinones for 24h. Data are means ± SEM from three separate experiments. ** p < 0.01 versus MDA-MB-231 pcDNA3.1 (one-way ANOVA). Abbreviations: NQO1: NAD(P)H: quinone oxidoreductase 1.

Similar findings were observed with MTT assay whereas inhibition of NQO1 with dicoumarol abolished the effect of the quinones (menadione and β-lapachone) on the cell viability (Table 2). Altogether these data suggest that, beyond the expression of NQO1 itself, the *NQO1* polymorphism has a major influence on the sensitivity of cancer cells to some chemotherapeutic drugs.

Table 2. Sensitivity of MDA-MB-231 cells overexpressing WT or C609T (NQO1*2) mutant form of NQO1 to quinones and dicoumarol (MTT assay).

Parameter	Cell Type	Menadione (60 μmol/L)	Menadione + DIC	β-Lapachone (1.5 μmol/L)	β-Lapachone + DIC
MTT Assay (% of control)	MDA-MB-231 (pcDNA3.1)	4.6 ± 3.7	0.6 ± 0.1	44.3 ± 3.5	43.9 ± 4.9
	MDA-MB-231 (NQO1 WT)	30.0 ± 8.6 **	0.5 ± 0.2	21.4 ± 9.4 *	53.3 ± 3.1
	MDA-MB-231 (NQO1*2)	7.3 ± 3.8	1.4 ± 0.3 **	46.6 ± 8.3	38.7 ± 5.5

Cells were transfected with either the empty vector pcDNA3.1, NQO1 wild-type or NQO1*2 and incubated with quinones (menadione or β-lapachone) and/or dicoumarol (25 μmol/L) for 24h. Cell viability was determined by MTT assay. Data are means ± SEM from three separate experiments. * $p < 0.05$; ** $p < 0.01$ versus MDA-MB-231 pcDNA3.1 (one-way ANOVA). Abbreviations: NQO1: NAD(P)H: quinone oxidoreductase 1; DIC: dicoumarol.

4. Discussion

To explore how NQO1 may affect both the activity and the detoxification of quinone-bearing compounds, we selected three different types of quinones displaying different physical-chemistry properties and molecular descriptors such as lipophilia and redox potential. Indeed, β-lapachone is an ortho-naphthoquinone derivative and menadione is bearing an 1,4-naphthoquinone scaffold while doxorubicin belongs to the anthraquinone antitumor class. Since NQO1 protein levels are often increased in tumors as compared to healthy tissues [4–6], this enzyme emerges as an attractive target for cancer therapy, because NQO1 bioactivates compounds like β-lapachone leading to a selective cancer cells toxicity [8,23–25].

A polymorphism has been described for the *NQO1* gene [12,13], resulting in the production of three variants: *NQO1*1*, the wild-type form; *NQO1*2*, a variant with a C609T substitution in exon 6; and *NQO1*3*, a variant with a C465T substitution in exon 4. Tumors with the *NQO1*2* polymorphism usually have low NQO1 protein levels: For example, MDA-MB-231 cells, which are homozygous *NQO1*2/NQO1*2*, have very low NQO1 proteins and are nearly undetectable by immunoblotting; thereby they are considered as NQO1-null [12,26]. In contrast, MCF-7 cells, which are heterozygous *NQO1*1/NQO1*2*, have high NQO1 expression. Upon NQO1 inhibition, either by using a pharmacological inhibitor or by genetic inactivation, 250MK cells which were naturally resistant to a pro-oxidant treatment (Asc/Men), were dramatically sensitized against this oxidant insult most likely explained because they are *NQO1*1* homozygous. In contrast, dicoumarol has no major impact on Asc/Men-mediated cytotoxicity in MDA-MB-231 cells.

Due to the critical importance of *NQO1* polymorphism, both NQO1*1 and NQO1*2 isoforms were stably overexpressed in NQO1-null MDA-MB-231 cells. The overexpression of wild-type NQO1 made cells more resistant to menadione than cells transfected with pcDNA3.1, the empty vector. In contrast, the overexpression of NQO1*2 did not increase cancer cell resistance against menadione, most probably due to the strongly decreased NQO1 enzyme activity showed by this variant isoform. Doxorubicin follows a similar pattern as menadione, although its mechanism of cytotoxicity is rather more complex. Indeed, doxorubicin toxicity is partially mediated through a redox-cycling reactive oxygen species (ROS) formation. However, a major mechanism is associated with DNA-intercalating effects [27]. Furthermore, we analyzed the impact of *NQO1* polymorphism on β-lapachone, which is bioactivated by the enzyme. MDA-MB-231 overexpressing NQO1 cells, with high NQO1 activity, were more sensitive to β-lapachone treatment and less sensitive to doxorubicin and menadione than empty vector or NQO1*2 overexpressing cells.

5. Conclusions

As conclusion, we suggest that determining *NQO1* polymorphism may be important when considering the use of quinone-based chemotherapeutic drugs. Compounds such as β-lapachone will not be useful, for example, when cancer cells are homozygous for the NQO1*2 isoform, because these types of compounds cannot be bioactivated by the NQO1*2 variant. However, compounds such as doxorubicin (and menadione) may be used in these circumstances because these types of compounds are not detoxified by the NQO1*2 variant.

Author Contributions: C.G. performed experiments; C.G. and P.B.C. analyzed, interpreted the data and wrote the paper.

Funding: This research was funded by FNRS-Télévie, grant number 7.4575.12F.

Conflicts of Interest: The authors declare that they have no competing interests.

References

1. Chesis, P.L.; Levin, D.E.; Smith, M.T.; Ernster, L.; Ames, B.N. Mutagenicity of quinones: Pathways of metabolic activation and detoxification. *Proc. Natl. Acad. Sci. USA* **1984**, *81*, 1696–1700. [CrossRef] [PubMed]
2. Lind, C.; Hochstein, P.; Ernster, L. DT-diaphorase as a quinone reductase: A cellular control device against semiquinone and superoxide radical formation. *Arch. Biochem. Biophys.* **1982**, *216*, 178–185. [CrossRef]
3. Morrison, H.; Jernstrom, B.; Nordenskjold, M.; Thor, H.; Orrenius, S. Induction of DNA damage by menadione (2-methyl-1,4-naphthoquinone) in primary cultures of rat hepatocytes. *Biochem. Pharmacol.* **1984**, *33*, 1763–1769. [CrossRef]
4. Cresteil, T.; Jaiswal, A.K. High levels of expression of the NAD(P)H:quinone oxidoreductase (NQO1) gene in tumor cells compared to normal cells of the same origin. *Biochem. Pharmacol.* **1991**, *42*, 1021–1027. [CrossRef]
5. Schlager, J.J.; Powis, G. Cytosolic NAD(P)H:(quinone-acceptor)oxidoreductase in human normal and tumor tissue: Effects of cigarette smoking and alcohol. *Int. J. Cancer* **1990**, *45*, 403–409. [CrossRef] [PubMed]
6. Siegel, D.; Franklin, W.A.; Ross, D. Immunohistochemical detection of NAD(P)H:quinone oxidoreductase in human lung and lung tumors. *Clin. Cancer Res.* **1998**, *4*, 2065–2070. [PubMed]
7. Fleming, R.A.; Drees, J.; Loggie, B.W.; Russell, G.B.; Geisinger, K.R.; Morris, R.T.; Sachs, D.; McQuellon, R.P. Clinical significance of a NAD(P)H: Quinone oxidoreductase 1 polymorphism in patients with disseminated peritoneal cancer receiving intraperitoneal hyperthermic chemotherapy with mitomycin C. *Pharmacogenetics* **2002**, *12*, 31–37. [CrossRef]
8. Ough, M.; Lewis, A.; Bey, E.A.; Gao, J.; Ritchie, J.M.; Bornmann, W.; Boothman, D.A.; Oberley, L.W.; Cullen, J.J. Efficacy of beta-lapachone in pancreatic cancer treatment: Exploiting the novel, therapeutic target NQO1. *Cancer Biol. Ther.* **2005**, *4*, 95–102. [CrossRef]
9. Hungermann, D.; Schmidt, H.; Natrajan, R.; Tidow, N.; Poos, K.; Reis-Filho, J.S.; Brandt, B.; Buerger, H.; Korsching, E. Influence of whole arm loss of chromosome 16q on gene expression patterns in oestrogen receptor-positive, invasive breast cancer. *J. Pathol.* **2011**, *224*, 517–528. [CrossRef]
10. Dejeans, N.; Glorieux, C.; Guenin, S.; Beck, R.; Sid, B.; Rousseau, R.; Bisig, B.; Delvenne, P.; Buc Calderon, P.; Verrax, J. Overexpression of GRP94 in breast cancer cells resistant to oxidative stress promotes high levels of cancer cell proliferation and migration: Implications for tumor recurrence. *Free Radic. Biol. Med.* **2012**, *52*, 993–1002. [CrossRef]
11. Glorieux, C.; Sandoval, J.M.; Dejeans, N.; Ameye, G.; Poirel, H.A.; Verrax, J.; Calderon, P.B. Overexpression of NAD(P)H:quinone oxidoreductase 1 (*NQO1*) and genomic gain of the NQO1 locus modulates breast cancer cell sensitivity to quinones. *Life Sci.* **2016**, *145*, 57–65. [CrossRef]
12. Pan, S.S.; Han, Y.; Farabaugh, P.; Xia, H. Implication of alternative splicing for expression of a variant NAD(P)H:quinone oxidoreductase-1 with a single nucleotide polymorphism at 465C>T. *Pharmacogenetics* **2002**, *12*, 479–488. [CrossRef] [PubMed]
13. Siegel, D.; McGuinness, S.M.; Winski, S.L.; Ross, D. Genotype-phenotype relationships in studies of a polymorphism in NAD(P)H:quinone oxidoreductase 1. *Pharmacogenetics* **1999**, *9*, 113–121. [CrossRef] [PubMed]

14. Garbe, J.C.; Bhattacharya, S.; Merchant, B.; Bassett, E.; Swisshelm, K.; Feiler, H.S.; Wyrobek, A.J.; Stampfer, M.R. Molecular distinctions between stasis and telomere attrition senescence barriers shown by long-term culture of normal human mammary epithelial cells. *Cancer Res.* **2009**, *69*, 7557–7568. [CrossRef] [PubMed]

15. Siegel, D.; Yan, C.; Ross, D. NAD(P)H:quinone oxidoreductase 1 (NQO1) in the sensitivity and resistance to antitumor quinones. *Biochem. Pharmacol.* **2012**, *83*, 1033–1040. [CrossRef]

16. Glorieux, C.; Auquier, J.; Dejeans, N.; Sid, B.; Demoulin, J.B.; Bertrand, L.; Verrax, J.; Calderon, P.B. Catalase expression in MCF-7 breast cancer cells is mainly controlled by PI3K/Akt/mTor signaling pathway. *Biochem. Pharmacol.* **2014**, *89*, 217–223. [CrossRef] [PubMed]

17. Fitzsimmons, S.A.; Workman, P.; Grever, M.; Paull, K.; Camalier, R.; Lewis, A.D. Reductase enzyme expression across the National Cancer Institute Tumor cell line panel: Correlation with sensitivity to mitomycin C and EO9. *J. Natl. Cancer Inst.* **1996**, *88*, 259–269. [CrossRef] [PubMed]

18. Beck, R.; Pedrosa, R.C.; Dejeans, N.; Glorieux, C.; Leveque, P.; Gallez, B.; Taper, H.; Eeckhoudt, S.; Knoops, L.; Calderon, P.B.; et al. Ascorbate/menadione-induced oxidative stress kills cancer cells that express normal or mutated forms of the oncogenic protein Bcr-Abl. An in vitro and in vivo mechanistic study. *Investig. New Drugs* **2011**, *29*, 891–900. [CrossRef]

19. Dejeans, N.; Tajeddine, N.; Beck, R.; Verrax, J.; Taper, H.; Gailly, P.; Calderon, P.B. Endoplasmic reticulum calcium release potentiates the ER stress and cell death caused by an oxidative stress in MCF-7 cells. *Biochem. Pharmacol.* **2010**, *79*, 1221–1230. [CrossRef]

20. Verrax, J.; Cadrobbi, J.; Marques, C.; Taper, H.; Habraken, Y.; Piette, J.; Calderon, P.B. Ascorbate potentiates the cytotoxicity of menadione leading to an oxidative stress that kills cancer cells by a non-apoptotic caspase-3 independent form of cell death. *Apoptosis* **2004**, *9*, 223–233. [CrossRef]

21. Verrax, J.; Stockis, J.; Tison, A.; Taper, H.S.; Calderon, P.B. Oxidative stress by ascorbate/menadione association kills K562 human chronic myelogenous leukaemia cells and inhibits its tumour growth in nude mice. *Biochem. Pharmacol.* **2006**, *72*, 671–680. [CrossRef] [PubMed]

22. Verrax, J.; Vanbever, S.; Stockis, J.; Taper, H.; Calderon, P.B. Role of glycolysis inhibition and poly(ADP-ribose) polymerase activation in necrotic-like cell death caused by ascorbate/menadione-induced oxidative stress in K562 human chronic myelogenous leukemic cells. *Int. J. Cancer* **2007**, *120*, 1192–1197. [CrossRef] [PubMed]

23. Bentle, M.S.; Reinicke, K.E.; Dong, Y.; Bey, E.A.; Boothman, D.A. Nonhomologous end joining is essential for cellular resistance to the novel antitumor agent, beta-lapachone. *Cancer Res.* **2007**, *67*, 6936–6945. [CrossRef] [PubMed]

24. Li, L.S.; Bey, E.A.; Dong, Y.; Meng, J.; Patra, B.; Yan, J.; Xie, X.J.; Brekken, R.A.; Barnett, C.C.; Bornmann, W.G.; et al. Modulating endogenous NQO1 levels identifies key regulatory mechanisms of action of beta-lapachone for pancreatic cancer therapy. *Clin. Cancer Res.* **2011**, *17*, 275–285. [CrossRef] [PubMed]

25. Reinicke, K.E.; Bey, E.A.; Bentle, M.S.; Pink, J.J.; Ingalls, S.T.; Hoppel, C.L.; Misico, R.I.; Arzac, G.M.; Burton, G.; Bornmann, W.G.; et al. Development of beta-lapachone prodrugs for therapy against human cancer cells with elevated NAD(P)H:quinone oxidoreductase 1 levels. *Clin. Cancer Res.* **2005**, *11*, 3055–3064. [CrossRef] [PubMed]

26. Blanco, E.; Bey, E.A.; Dong, Y.; Weinberg, B.D.; Sutton, D.M.; Boothman, D.A.; Gao, J. Beta-lapachone-containing PEG-PLA polymer micelles as novel nanotherapeutics against NQO1-overexpressing tumor cells. *J. Control. Release* **2007**, *122*, 365–374. [CrossRef] [PubMed]

27. Fornari, F.A.; Randolph, J.K.; Yalowich, J.C.; Ritke, M.K.; Gewirtz, D.A. Interference by doxorubicin with DNA unwinding in MCF-7 breast tumor cells. *Mol. Pharmacol.* **1994**, *45*, 649–656. [PubMed]

 *antioxidants*

Review

Metabolic Remodelling: An Accomplice for New Therapeutic Strategies to Fight Lung Cancer

Cindy Mendes [1,2] and Jacinta Serpa [1,2,*]

1   CEDOC, Chronic Diseases Research Centre, NOVA Medical School, Faculdade de Ciências Médicas, Universidade NOVA de Lisboa, Campo dos Mártires da Pátria, 130, 1169-056 Lisboa, Portugal; cindymendes8@gmail.com
2   Instituto Português de Oncologia de Lisboa Francisco Gentil (IPOLFG), Rua Prof Lima Basto, 1099-023 Lisboa, Portugal
*   Correspondence: jacinta.serpa@nms.unl.pt

Received: 30 October 2019; Accepted: 27 November 2019; Published: 29 November 2019

Abstract: Metabolic remodelling is a hallmark of cancer, however little has been unravelled in its role in chemoresistance, which is a major hurdle to cancer control. Lung cancer is a leading cause of death by cancer, mainly due to the diagnosis at an advanced stage and to the development of resistance to therapy. Targeted therapeutic agents combined with comprehensive drugs are commonly used to treat lung cancer. However, resistance mechanisms are difficult to avoid. In this review, we will address some of those therapeutic regimens, resistance mechanisms that are eventually developed by lung cancer cells, metabolic alterations that have already been described in lung cancer and putative new therapeutic strategies, and the integration of conventional drugs and genetic and metabolic-targeted therapies. The oxidative stress is pivotal in this whole network. A better understanding of cancer cell metabolism and molecular adaptations underlying resistance mechanisms will provide clues to design new therapeutic strategies, including the combination of chemotherapeutic and targeted agents, considering metabolic intervenients. As cancer cells undergo a constant metabolic adaptive drift, therapeutic regimens must constantly adapt.

Keywords: lung cancer; cancer metabolism; reactive oxygen species (ROS); therapy resistance; new therapeutic strategies

1. Cancer Metabolism

Although cancer metabolism is one of the oldest areas of research in cancer biology, the study of metabolic alterations in tumours has grown exponentially in the past decade [1,2]. The link between cancer and metabolism was first made by Otto Warburg in 1923 with the observation that most cancer cells predominantly produce energy through a high rate of glycolysis followed by lactic acid fermentation, rather than through oxidative phosphorylation (OXPHOS) in the mitochondria [3]. Thus, tumours exhibit an increased rate of glucose uptake with lactate production, even in the presence of oxygen, through aerobic glycolysis with the production of carbon skeletons, NADPH and ATP [4]. This glycolytic switch, known as the "Warburg effect", was first described as a compensation mechanism for mitochondria dysfunction in tumours. However, this view was challenged as several studies found that mitochondrial OXPHOS is active in most types of cancer [5–8]. To fulfil the biosynthetic demands associated with proliferation, mitochondria occupy a core status as providers of ATP and intermediate metabolites, such as citrate to supply anabolic reactions [9]. Moreover, in nontransformed cells, the Warburg effect is a reversible phenomenon linked to proliferation, showing that it reflects proliferation associated changes in metabolism instead of a unique feature of malignancy [10]. Glucose metabolism rewiring is more likely to be driven by an elevated demand of reducing equivalents and molecular precursors of proteins, nucleotides and lipids, which are the building blocks required to maintain cancer cell growth and proliferation [11]. Besides glucose, glutamine also contributes to core metabolic functions of cancer cells since it fuels the tricarboxylic acid (TCA) cycle, nucleotide and fatty acid biosynthesis and redox balance [10,12], replacing completely glucose as a core organic compound.

Despite the fact that energetic metabolism is one of the cancer hallmarks, little has been shown in its role in chemoresistance [13]. Besides producing ATP, OXPHOS is the major source of reactive oxygen species (ROS) within mitochondria and in the entire cell [14]. During OXPHOS, the acetyl-CoA from glucose and fatty acids degradation feeds the TCA cycle, which generates reduced compounds [10,12] that will transfer electrons to the OXPHOS system through a chain of redox reactions with the final reduction of oxygen to water. In some of these reactions, ROS is a by-product [15], changing the mitochondrial membrane potential (MPP) and inducing damages in the respiratory chain, pushing cells towards apoptosis [14]. In response to augmented ROS, many tumours trigger protective antioxidant pathways [16]. Glutathione (GSH), peroxiredoxins, thioredoxin and superoxide dismutase (SOD2) belong to the major antioxidant system, maintaining the redox homeostasis [17–19]. Thus, the activation of those ROS scavenging processes constitutes a mechanism of resistance to chemotherapy exhibited by cancer cells, upon the exposure to alkylating or ROS generator drugs [14]. Changes in endogenous metabolism influence the metabolic course of xenobiotic compounds, which includes drugs used in cancer therapy. This fact reinforces the role of cancer metabolic rewiring as crucial in cancer cells survival and response to therapy. Here, we will discuss the main features of cancer metabolism and its influence on therapy resistance in lung cancer.

2. Metabolic Remodelling in Lung Cancer

Lung cancer is a prevalent cause of cancer-related death [20], being grouped in two principal histological subtypes: small cell lung carcinoma (SCLC; 15% of total) and non-small-cell lung carcinoma (NSCLC; 85% of total) [21]. NSCLC is classified into three histotypes: squamous-cell carcinoma, adenocarcinoma, and large-cell carcinoma [22]. Metabolism was shown to be differently reprogrammed in the major subtypes of non-small cell lung cancer [23]. Metabolic remodelling is one of the emerging hallmarks of cancer [24] and it is well recognised that cancer cells have a high metabolic plasticity to support continuous cell growth and proliferation, meeting their energetic and biomass demands [25]. Recent studies reported metabolic alterations in glucose, lipids, amino acids and nucleic acids metabolism in NSCLC cells (Figure 1).

In the past decade, stable-isotope tracing with ^{13}C-glucose became an important tool for the analysis of metabolic pathways that are differentially activated in tumour cells in vivo, both in cancer mouse models and humans [26–30]. Uniformly labelled ^{13}C-glucose is administered as a bolus by an intraoperative infusion before surgical tumour resection and the distribution of labelled carbons in the various intermediates is analysed by ^{13}C NMR spectroscopy [31,32]. A study using fresh surgical resections from NSCLC patients with mixed histology, after a labelled ^{13}C-glucose infusion, showed contrasting glucose metabolism results; tumour samples displayed high levels of lactate, demonstrating an upregulation in glycolysis, but also increased levels of glucose-derived TCA cycle intermediates, in tumour samples compared with normal tissue [30]. These observations reinforce the fact that glycolysis and OXPHOS can function in simultaneous if not in the same cancer cell at least in the same tumour, in which metabolic symbiosis can be established. Hensley and colleagues combined multimodal imaging analysis (FDG-PET and multiparametric MRI) and ^{13}C-glucose flux profiling of NSCLC in situ to provide quantitative information about glucose metabolism and the tumour microenvironment in NSCLC untreated patients [28]. The activity of PC (pyruvate carboxylase), the enzyme responsible for the conversion of pyruvate into oxaloacetate, was elevated in NSCLC tumours [28,33], and its silencing significantly decreased the proliferative and colony-forming capacity of NSCLC cell lineages and reduced tumour growth in murine xenograft models, suggesting a dependence on PC-mediated and TCA cycle-based anaplerosis [33]. Moreover, it was found that glycolysis and glucose oxidation via PDH (pyruvate dehydrogenase) and the TCA cycle were higher in NSCLC compared to the adjacent normal lung [28]. Glucose-derived metabolic intermediates can be synthesized directly from glucose or indirectly from glucose-derived lactate. This fact was demonstrated by Faubert et al. [29]; lactate is the main carbon source for the TCA cycle in tumours from NSCLC patients and tumour xenografts.

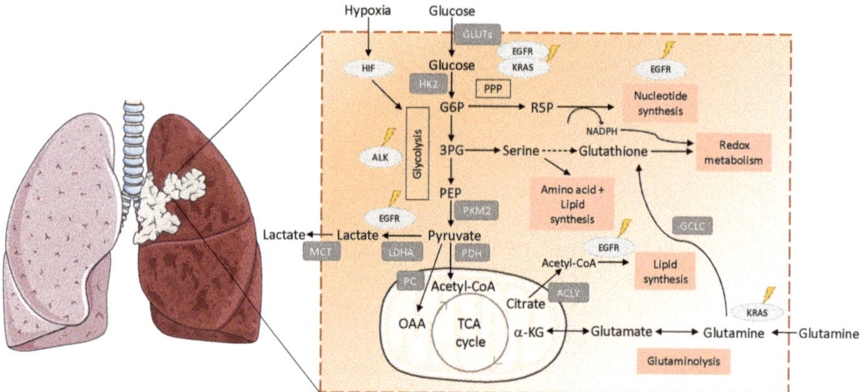

Figure 1. Metabolic remodelling in lung cancer. Metabolic pathways are involved in the synthesis of building blocks for macromolecules and redox homeostasis, needed for cell proliferation are presented. The import of glucose is mediated by the GLUT (glucose transporter) family of membrane transport proteins, which are known to be deregulated in cancer [34]. Hexokinase 2 (HK-2) is the first rate-limiting enzyme in the glycolytic pathway and overexpression of HK-2 has wildly been observed in cancer cells, correlating with poor overall survival in cancer patients [35–37]. Mutant EGFR (epidermal growth factor receptor 1) promotes metabolic remodelling in non-small-cell lung carcinoma (NSCLC) with increased aerobic glycolysis and PPP (pentose phosphate pathway), altered pyrimidine biosynthesis and upregulation of monounsaturated fatty acids [38–40]. *KRAS* (Kirsten rat sarcoma viral oncogene homolog GTPase)-mutated NSCLC cells express higher levels of enzymes involved in glycolysis, such as pyruvate kinase isozyme M2 (PKM2) and lactate dehydrogenase A (LDHA) compared with nonmalignant cells, indicating alterations in glucose metabolism and PPP (glicose-6-fosfato dehydrogenase (G6PD), transketolase (TKT) and 6-phosphogluconate dehydrogenase (6PGD)) [41]. Pyruvate is decarboxylated into acetyl-CoA to be further transported into mitochondria to enter the TCA cycle [42]. *ALK* (anaplastic lymphoma kinase) rearrangements were associated with upregulated glucose metabolism in highly metastatic phenotypes of adenocarcinoma [43]. The expression and activity of PC (pyruvate carboxylase), the enzyme responsible for the conversion of pyruvate into oxaloacetate, was found to be elevated in NSCLC tumours [28,33]. Glycolysis and glucose oxidation via PDH (pyruvate dehydrogenase) and the TCA cycle were enhanced in NSCLC comparing to adjacent benign lung [28]. Cancer cells also show higher levels of monocarboxylate transporters (MCT), which are responsible for lactate export and helps both in maintaining intracellular pH and in continuing glycolysis [44]. Hif-1 (hypoxia inducible factor 1) regulates the transcription of glycolytic enzymes such as, HK-2, LDH-A and PKM2, which upregulate glycolysis [45,46]. The expression of ATP citrate lyase (ACLY), a key enzyme in fatty acid synthesis was upregulated in NSCLC, being associated with poor prognosis [30]. Glutathione cysteine ligase (GCLC), which converts glutamate to Glutathione (GSH), is also highly expressed in several cancers, including lung cancer, and high mRNA expression of GCLC-promoted cisplatin resistance in lung adenocarcinoma cell lines [47]. G6P: glucose 6-phosphate; 3PG: 3-phosphoglyceric acid; PEP: phosphoenolpyruvate; R5P: ribose 5-phosphate; MCT: monocarboxylate transporters; OAA: oxaloacetate; α-KG: alpha ketoglutarate.

Several enzymes and transporters, crucial for carbon and energy metabolism, have been described as differently expressed in a normal lung and in cancer. The expression of ATP citrate lyase (ACLY), a key enzyme in fatty acid synthesis involved in the synthesis of acetyl-CoA and oxaloacetate, was upregulated in NSCLC, being associated with poor prognosis [48]. Glycine decarboxylase (GLDC), which couples decarboxylation of glycine to the biosynthesis of serine and one-carbon metabolism, takes part in pyrimidine metabolism and is upregulated in NSCLC cells [49]. xCT (SLC7A11), a cystine/glutamate antiporter, is overexpressed in the plasma membrane in NSCLC, correlating with patients' worse survival [50]. Interestingly, it was described that cancer cells expressing high levels of xCT relied on glutamine and glutaminolysis dependency for OXPHOS [50]. This glutamine consumer phenotype may also be linked to cyst(e)ine dependency, as xCT concomitantly exports glutamate and imports cyst(e)ine. Glutamate is a direct product of glutamine degradation and maintaining the glutamine import sustains the import of cyst(e)ine, which has been related to increased therapy resistance in different cancer models [51], mostly due to its role in glutathione synthesis [52–55], for which glutamine-derived glutamate is also needed.

Despite similarities in metabolic reprogramming, the metabolic alterations in NSCLC cells or tumours are highly heterogeneous [28,39,56]. Hensley et al., identified metabolically heterogenous regions both within and between tumours, in which reduced perfusion tumour areas preferentially used glucose whereas well perfused regions relied more on non-glucose nutrients indicating that tumour metabolic remodelling is regulated by the microenvironment [28]. The generation of a hypoxic environment (oxygen deprivation) is a common feature of solid cancers as a consequence of the development of a disordered vasculature, which is not able to properly supply oxygen to a rapidly growing tumour [57]. Hypoxia is known to activate HIF signalling in cancer cells and plays an important role in the pathogenesis and prognosis of lung cancer [58]. Induction of HIF activity targets genes involved in glucose metabolism, angiogenesis, cell proliferation and resistance to therapies [59]. Furthermore, both Hif-1α and Hif-2α are frequently overexpressed in NSCLC correlating, in some cases, with poor prognosis [60,61] and Hif-1α expression is associated with resistance to cancer therapy, including EGFR (epidermal growth factor receptor 1) inhibitors in NSCLC [46].

Thus, understanding the influence of cellular or microenvironmental factors, such as oncogene-induced metabolic switches, on cancer cell metabolism is crucial for the development of better-adjusted therapeutic approaches targeting metabolic remodelling in cancer cells.

Role of Oncogenic Mutations (EGFR, ALK, KRAS) in Metabolic Remodelling

Certain genetic alterations have been shown as crucial in lung carcinogenesis. Mutations in *EGFR* (epidermal growth factor receptor 1) and *KRAS* (Kirsten rat sarcoma viral oncogene homolog GTPase) and *ALK* (anaplastic lymphoma kinase) rearrangements are mostly found in lung adenocarcinoma, accounting for 30–40% of NSCLCs [62]. Mutations in these oncogenes have been shown to play a role in metabolic reprogramming of cancer cells to support their high proliferative rate and energetic demands [39,63].

Mutant EGFR promotes metabolic remodelling in NSCLC with increased aerobic glycolysis and PPP (pentose phosphate pathway), altered pyrimidine biosynthesis and redox metabolism [38,39]. Treatment with erlotinib (EGFR inhibitor) and a glutaminase inhibitor (CB-839) generates a metabolic crisis in *EGFR* mutant NSCLC cells, resulting in cell death and in rapid tumour regression in mouse NSCLC xenografts [64]. These facts indicate the need of glutamine as a source for bioenergetics and biosynthesis in *EGFR*-mutated NSCLCs, as glucose is mainly used to sustain PPP and consequently cancer cells proliferation. Another study, showed the role of EGFR in the stabilization by phosphorylation of stearoyl-CoA desaturase-1 (SCD1), augmenting monounsaturated fatty acid synthesis and sustaining cell proliferation [40]. In addition, phosphorylated SCD1 levels were reported as an independent prognostic factor for poor survival in NSCLC [40].

The influence of *ALK* rearrangements on metabolism has not been well described in lung adenocarcinoma. However, a recent study observed the presence of upregulated glucose metabolism in highly metastatic phenotypes in this subset of lung cancer [43]. Again, increased consumption of glucose is linked to more aggressive cancer phenotypes.

Numerous studies showed the involvement of mutant KRAS in the metabolic rewiring of different types of cancer [63,65,66] with an upregulation of glucose uptake and aerobic glycolysis and increased glutamine utilization [67–69]. Proteomic profiles related to metabolism of intrinsic *KRAS* mutant NSCLC cell lines were investigated and compared with those of normal bronchial epithelial cells [41]. *KRAS*-mutated NSCLC cells expressed higher levels of enzymes involved in glycolysis (glyceraldehyde 3-phosphate dehydrogenase—GAPDH, pyruvate kinase isozyme M2—PKM2, lactate dehydrogenase A—LDHA and lactate dehydrogenase B—LDHB) and PPP (glicose-6-fosfato dehydrogenase—G6PD, transketolase—TKT and 6-phosphogluconate dehydrogenase—6PGD) compared with nonmalignant cells [41]. In another study, NSCLC cells carrying *KRAS* mutations showed metabolic remodelling with alterations in redox buffering systems and glutamine dependency [70]. Moreover, an upregulation of lactate production was observed in a mutant *KRAS* lung tumour mouse model [27], however, these mouse models minimally used glutamine as a carbon source for TCA cycle entry, while in the in vitro models, there was a dependence on glutamine. Thus, glutamine dependency can be related to the homeostatic cellular systems, such as GSH production, and not related to the direct and extensive used of glutamine as a carbon and energy source. In agreement with this, oxidative glucose metabolic enzymes, such as PC and PDH (pyruvate dehydrogenase), were shown to be necessary for tumour formation and growth in these mouse models [27]. Homozygous mutant KRAS cells have been shown to have an increased antioxidant capacity that accounts for their selective growth during lung tumour progression [63,71]. Varying oxygen levels within the growing tumour can contribute to this selection, since increased expression of Hif-2α promotes tumour growth and malignant progression of KRAS^{G12D} lung tumours [72]. These results correlated with existing human clinical data, implicating Hif-2α as a negative prognostic factor in human NSCLC [72].

3. Lung Cancer Therapy

Over the last 20 years, lung cancer treatment has evolved from the empiric use of cytotoxic therapies to effective and better tolerated targeted therapies. Platinum-based doublet therapy (combining platinum-based drugs with another cytotoxic/cytostatic agent) has been the standard therapy for both primary and palliative care of patients with advanced stage lung cancer [73,74]. Genotyping studies revealing genetic alterations in the various subtypes of lung cancer accounting for tumorigenesis led to the development of targeted therapies, namely directed to EGFR, ALK, and KRAS mutated variants [75].

In fact, the targeted therapy of patients with *EGFR*-mutated tumours is more effective than conventional therapy; and the efficacy of EGFR tyrosine kinase inhibitors (TKIs) is increasing over generations of drugs. First-generation EGFR inhibitors (e.g., gefitinib and erlotinib) have shown increased objective response (ORRs) and progression-free survival (PFS) compared to conventional cytotoxic treatment of patients suffering from EGFR-mutated tumours [73,76]. Second-generation inhibitors, such as afatinib and dacomitinib, are irreversible inhibitors that additionally target the receptors HER2 and HER4 (epidermal growth factor 2 and 4) and were reported to show higher PFS compared to first-generation EGFR inhibitors, such as gefitinib [77].

The fusion between echinoderm microtubule-associated protein-like 4 (*EML4*) gene and ALK (*EML4-ALK*) was the first fusion oncogene detected in lung cancer [78]. Fusion genes, involving ALK, are usually mutually exclusive with other oncogenic drivers such as EGFR and KRAS [79]. Crizotinib, a first-generation competitive ATP inhibitor of ALK tyrosine kinases with activity against ALK-fusion-positive NSCLC [80], is associated with higher ORRs and PFS in comparison to cytotoxic therapy in both conventionally treated and untreated patients [81]. In preclinical studies, ceritinib, a second-generation ALK inhibitor, has shown greater antitumour activity than first-generation inhibitors, as crizotinib [82]. Another fusion oncogene, encoding a constitutively activated tyrosine kinase, can result from fusion of ROS1 tyrosine kinase domain with CD74 [83]. Because it has a high homology with the kinase domain of ALK, ALK specific inhibitors, including crizotinib [84] and ceritinib [82], revealed marked activity in ROS1-positive tumours [85]. The same efficacy has been observed for crizotinib in NSCLC patients with tumours baring MET tyrosine kinase receptor amplifications [86,87].

Regarding *KRAS* oncogene, despite the fact that *KRAS*-MAPK pathway is downstream of EGFR signalling, *KRAS*-mutation-driven lung cancers, which are mostly adenocarcinomas, do not respond to EGFR TKIs [88] because the mutations in *KRAS* activate and release mutant KRAS from the upstream regulation. *KRAS* is the most frequently mutated oncogene in NSCLC patients, but effective therapies targeting mutant KRAS have yet to be developed. However, different therapies directed to KRAS downstream targets, such as MEK (MAPK/ERK kinase), are currently being tested. In a phase III study, NSCLC patients with *KRAS* mutated tumours treated with MEK inhibitor selumetinib plus docetaxel (taxane) did not show improved PFS compared to taxanes monotherapy [89]. Another MEK inhibitor, trametinib, was also evaluated alone or in combination with taxanes and revealed that combination with chemotherapy increased tolerability and clinical activity in both *KRAS*-mutant and *KRAS*-WT NSCLC patients [90].

Antiangiogenic therapy has also been tested in lung cancer patients. Bevacizumab, a monoclonal antibody against vascular endothelial growth factor (VEGF), in combination with paclitaxel (taxane) or carboplatin, significantly improved the median overall survival (OS) and PFS of NSCLC patients with tumours of nonsquamous cell histology [91].

Recently, immunotherapy has emerged as a potential treatment option against lung cancer, taking advantage of the native antitumour immune response [92]. Immune checkpoint blocker (ICB), generated upon T-cell activation, such as monoclonal antibodies that target cytotoxic T-lymphocyte antigen-4 (CTLA-4) and antibodies against PD-1 or PD-L1, are currently the most relevant targets for immunotherapy [73]. During tumorigenesis, PD-1 signalling inactivates T cells that recognize tumour-specific antigens, permitting tumour progression and metastasis [93]. ICBs, currently used or in development for NSCLC treatment, include the anti-PD-1 antibodies nivolumab (human IgG4) and pembrolizumab (humanized IgG4), along with the anti-PD-L1 antibodies atezolizumab (human IgG1, with the Fc domain engineered to prevent antibody-directed cell cytotoxicity), durvalumab (human IgG1 engineered), and avelumab (human IgG1 showing preclinical antibody-directed cell cytotoxicity activity) [74]. ICBs have been mostly used in patients with advanced NSCLC, whose tumours progress upon first-line cytotoxic therapy. Nivolumab treatment was associated with significantly longer median OS compared to treatment with docetaxel in patients with metastatic NSCLC, who had disease progression during or after platinum-based therapy [94,95]. The combination of anti-PD-(L)1 and anti-CTLA-4 monoclonal antibodies can result in higher and longer responses in NSCLC, as observed in experimental models and clinical studies [96,97]. Most patients, who achieve an initial benefit from an ICB eventually develop resistance, thus, the challenge is to develop rational combinations that will increase responses or delay the onset of resistance [98].

More recently, multiple trials have been investigating combinations of antiangiogenic agents and immunotherapy in NSCLC [99]. In particular, a clinical trial studied the efficiency of bevacizumab plus nivolumab in III/IV NSCLC patients and it was observed that the combinatory treatment improved PFS and decreased associated toxicity [100].

3.1. Mechanisms of Resistance to Conventional Therapy

Cancer chemotherapy resistance is one of the major dilemmas in cancer therapy, resulting in therapeutic failure and increased mortality. NSCLC cells are intrinsically resistant to various anticancer drugs, while SCLC cells can acquire resistance upon cyclic administration of a drug [101]. To address this issue, research has been focusing on how cancer cells modulate their genomes and metabolism to prevent drug influx, to facilitate efflux drugs, to inactivate drugs and/or to repair drug-induced damage [102]. More specifically, mechanisms of drug resistance identified so far include augmented drug eflux, drug inactivation and/or sequestration by enzymes, DNA repair, target modifications and apoptosis defects [101,103,104]. Ineffective drug delivery to the tumour, increased metabolism, lack of drug specificity to the tumour and tumour vasculature are additional contributing factors [105,106].

Additionally, patients treated with chemotherapy develop cumulative genetic mutations which may result in either activation of proto-oncogenes or inactivation of tumour-suppressor genes [101].

3.1.1. Drug Transporters Increase the Efflux of Chemotherapeutic Drugs

The overexpression of ATP-binding cassette (ABC) membrane transporters leads to enhanced cytotoxic drug efflux and diminished intracellular accumulation, granting resistance to drugs such as cisplatin, methotrexate, taxanes, anthracyclines, and vinca alkaloids [101,107]. P-glycoprotein (P-gp) is codified by the multidrug-resistance (MDR-1) gene, belongs to the ABC superfamily and functions as an energy-dependent efflux pump of metabolites [108]. Studies have shown an increased expression of P-gp in lung tumours, with notably higher rates in NSCLCs than in SCLCs [101]. In addition, major vault protein (MVP), also known as the lung resistance-related protein (LRP), has been pointed out to be involved in lung cancer drug resistance [109].

3.1.2. Drug Inactivation by Sulphur-Containing Molecules and Role of Antioxidants as a Cause of Drug Resistance

Another mechanism of resistance is by conjugation of the drug with sulphur-containing macromolecules such as metallothioneins (MTs) and GSH [101,110]. MTs are intracellular proteins rich in cysteine content (30%) that bind to cytotoxic agents such as platinum compounds and alkylating agents [111]. High MTs levels have been observed in tumour cells with acquired resistance to alkylating agents [110,112]. Moreover, augmented expression of MTs was described in NSCLC with squamous cell lung carcinoma and adenocarcinoma histotypes, but it was not demonstrated in SCLC [113]. Studies also show that MTs play a relevant role in the cellular protection against oxidative stress [114]. A strong correlation between MTs expression and cisplatin and doxorubicin resistance was observed in different cell lines of SCLC [115,116]. The GSH S-transferases (GSTs) protect cancer cells from reactive endogenous and exogenous electrophiles, such as prostaglandins, aromatic hydrocarbons and chemotherapeutic agents, through the conjugation with GSH (the most abundant cellular thiol), and scavenges them [101,110,117]. In tumour cells, expression levels of GSTs are increased in comparison to normal cells which may contribute to elevated detoxification of anticancer drugs [118,119]. GST isoenzymes have been reported in lung tumours in higher levels than in normal bronchioles and alveoli [120,121].

More than one mechanism of resistance can act on the same cancer cell/tumour. GS-conjugates are transported out of the cells by efflux transporters, such as multidrug resistance protein 1 (MRP1) and P-gp, thus conferring increased levels of resistance to the cytotoxicity of antineoplastic drugs [122,123]. Multiple studies involving NSCLC and SCLC cell lines suggested that high levels of GSH were associated with decreased platinum-DNA binding and intracellular platinum accumulation, increasing cisplatin resistance [124–126]. Conversely, factors inducing the reduction of cellular GSH content sensitise cancer cells to cisplatin [127]. Nuclear factor erythroid-like 2 (NRF2), a regulator of redox homeostasis upon oxidative stress, is activated by cells in order to upregulate gene networks involved in cytoprotective activities [128]. This transcription factor has been shown to be upregulated in various types of cancer, including skin, breast, prostate, lung and pancreas [129] and has also been associated with chemoresistance [130]. xCT, a downstream target gene of NRF2, responsible for the import of cysteine to support GSH synthesis has been implicated in multidrug resistance of lung cancer [131]. In particular, the x_c^- amino acid transport system maintained intracellular GSH and consequently resulted in cisplatin resistance in ovarian cancer cells [132].

Furthermore, cancer cells may develop resistance by overexpressing antioxidants, which protect cells from chemotherapy-induced oxidative stress and cell death, consequently a new redox balance with higher ROS levels is established, a process called "redox resetting" [101,133]. Epirubicin is known to cause oxidative stress by the generation of superoxide and hydrogen peroxide (H_2O_2) moieties, which drive cancer cell death [134]. Nevertheless, overexpression of antioxidants (e.g., SOD or GSH) neutralizes the oxidative stress leading to drug resistance. Accordingly, high levels of manganese SOD

(MnSOD) seems to protect lung epithelial cells against oxidative injury [135]. Moreover, malignant mesothelioma cells were reported to have higher levels of SOD mRNA and activities compared with nonmalignant mesothelial cells, but also had elevated catalase and GSH levels, being more resistant to H_2O_2 and epirubicin [136]. Accordingly, platinum drugs that generate very high ROS levels can be inactivated by GSH [137].

Alterations in drug metabolism are also associated to resistance since they can lead to drug inactivation or deficient drug activation. Antioxidant systems are able to directly inhibit the antitumour activity of several anticancer agents, such as paclitaxel [138], bortezomib [139] and radiation therapy [140]. Buthionine sulphoximine (BSO) significantly increases paclitaxel cytotoxicity through ROS accumulation [138]. The cellular redox state is associated with enzymatic expression required for the conversion of antimetabolites including 5-fluorouracil (5-FU) and methotrexate to their most active forms [133,141]. Capecitabine is an anticancer agent that is converted into 5-FU by thymidine phosphorylase [142], which is encoded by the *TYMP* gene that can be inactivated by DNA methylation causing capecitabine resistance [143]. These epigenetic alterations can be induced by H_2O_2 as DNA methyltransferase 1 (DNMT1) binds more strongly to chromatin after H_2O_2 exposure altering the methylation status of CpG regions [144]. The inactivation by UDP glucuronosyl transferase 1 (UGT1A1) by the topoisomerase inhibitor, irinotecan, is induced by the redox-sensing NRF2-KEAP1 pathway [145]. On another hand, the expression of UGT1A1 is decreased by promoter DNA methylation, promoting irinotecan activity [146].

3.1.3. DNA-Repair Pathways Inducing Resistance to Chemotherapy

As DNA damage is the main objective of most chemotherapeutic agents, increased capacity of DNA damage repair is one possible mechanism of resistance to the cytotoxic effects of anticancer drugs. Cisplatin for instance induces apoptosis by forming DNA-platinum adducts and by generating ROS, which causes oxidative DNA damage [147]. The nucleotide excision-repair (NER) pathway is one DNA repair pathway involved in the acquisition of platinum-based drug resistance [101,148]. Excision-repair cross-complementation group 1 (ERCC1) protein is a molecular indicator of resistance to platinum salts and forms the molecular complex of the NER pathway along with other proteins that are able to correct nucleotides modified by DNA-platinum adducts [149]. An association was found in in vitro studies between the expression of ERCC1 mRNA in NSCLC and resistance to platinum drugs, showing that a low expression of ERCC1 correlated with prolonged survival of NSCLC patients, who were treated with cisplatin plus gemcitabine [150,151]. Mismatch-repair (MMR) pathway repairs base-base and insertion-deletion mismatches during DNA replication [152]. This pathway can repair DNA-platinum adducts, which often results in mitotic stress and cell death [153]. However, this repair pathway is not considered relevant as a mechanism of chemotherapy resistance in lung cancer [101]. In contrast, base-excision-repair (BER) pathway was correlated with chemoresistance, proved by the fact that N-methylpurine DNA Glycosylase (MPG) and Apurinic/Apyrimidinic Endonuclease (APE) inhibition or elimination lead to increased resistance to alkylating agents, such as platinum-based drugs [154].

3.1.4. Loss of Intracellular Commands of Cell Death as a Cause of Drug Resistance

Cell death inhibition is another way of contributing to drug resistance. Failure of the intracellular death signalling pathways lead to the alteration of various apoptotic and antiapoptotic intracellular proteins, including Bcl-2, Bax and SAPK/JNK in several types of cancer [101,155]. Indeed, in vitro and in vivo evidence have shown that overexpression of Bcl-2 in SCLC contributes to apoptosis resistance [156]. Cancer cells with defective caspases are resistant to drugs, whose main mechanism of action is the induction of apoptosis. In this sense, NSCLC cells showing low expression of caspases 3 and 9 are resistant to cisplatin chemotherapy [157]. Alterations in antiapoptotic proteins such as IAP (inhibitor of apoptosis protein), particularly XIAP (X-linked inhibitor of apoptosis protein), and survivin have been observed in NSCLC patients with implications in resistance. It was shown that

XIAP was overexpressed in human H460 NSCLC cell line, leading to the inhibition of the apoptosome formation, pivotal in caspase-dependent cell death [158]. Low expression of survivin in patients was associated with a significantly better OS in comparison to patients with tumours displaying high expression of this protein [159].

Drugs inducing cell death by affecting microtubule stability, such as paclitaxel, kill cells in a Fas/Fas ligand (FasL)-dependent manner [160]. Blockade of Fas/FasL and inhibition of FasL expression by Bcl-2 overexpression are resistance mechanisms to paclitaxel [160]. Conversely, phosphorylation of Bcl-2 induces the expression of FasL, mediated by the nuclear action of NFAT (nuclear factor of activated T lymphocytes), which is responsive to microtubule damage, thereby restoring paclitaxel sensitivity [161].

3.2. Mechanisms of Resistance to Targeted Therapy

Although targeted therapies are revolutionizing the treatment of advanced NSCLC, resistance appears in most patients sooner or later [162]. In this section, we review the mechanisms of resistance that have been discovered in the past few years, in particular to TKIs directed to EGFR, ALK and ROS1, and later we will discuss strategies to overcome drug resistance. The first-generation erlotinib and gefitinib, and second-generation afatinib are now recognized as the standard first-line therapy in NSCLC patients with activating EGFR mutations [76,163,164]. However, some patients do not respond to EGFR-TKIs (intrinsic resistance). The most common mechanism for the acquired resistance to EGFR-TKIs is the development of the T790M second mutation within the *EGFR* kinase domain, which contributes for 50% of all acquired resistance [165]. The methionine 790 sterically blocks its interaction with TKIs, increasing affinity for ATP and reducing binding of the inhibitor to the kinase domain of EGFR, while keeping the catalytic activity [166]. The *MET* gene amplification is another frequent mechanism of acquired resistance and affects 5–20% EGFR-TKI treated NSCLC patients, irrespective of the T790M mutation status [167–169]. Although *HER2* amplification is a rare event in lung adenocarcinoma, it accounts for about 1–2% of total cases and up to 13% of NSCLC with acquired resistance to EGFR-TKIs [170,171]. Mutated EGFR heterodimerizes with HER2 resulting in heterodimers resistant to degradation [171], supporting EGFR-TKIs resistance in presence of both T790M mutation and *HER2* amplification itself as an acquired mechanism of drug exhaustion. *KRAS* and *EGFR* mutations are usually mutually exclusive but when they coexist, mainly in tumours under EGFR-TKIs treatment, *KRAS* mutations can confer resistance to EGFR inhibitors [172]. A study including 60 lung adenocarcinomas, either refractory or sensitive to both gefitinib and erlotinib, indicated that *KRAS* mutations lead to a lack of sensitivity to these drugs [172].

Another important pathway that has been associated to resistance to EGFR therapy is the PTEN-PI3K-AKT pathway. Phosphatase and tensin homolog (PTEN) acts as a tumour suppressor gene that can decrease tumour growth by inhibiting Akt [173], which can promote cell survival by inactivating several apoptosis mediators, such as Bad and caspases-9. Thus, loss of PTEN leads to increased tumours, since PTEN regulates negatively the PI3K-AKT pathway [174]. In EGFR mutant lung cancer, loss of PTEN led to resistance to EGFR inhibitors such as erlotinib [175], because PI3K activation can somehow interconnect PI3K-AKT and MAPK pathways.

Loss of E-cadherin expression and upregulation of mesenchymal proteins, including vimentin, fibronectin and N-cadherin, are the main characteristic of epithelial–mesenchymal transition (EMT). AXL (AXL receptor tyrosine kinase) is associated with EMT in several tumours and its upregulation in the Hedgehog pathway has been recognized as a mechanism of resistance to targeted drugs in EGFR-mutated NSCLC [176]. In a recent study, gefitinib-mediated ROS promoted EMT and mitochondrial dysfunction concomitant with resistance of lung cancer cells [177]. Moreover, gefitinib treatment in the presence of ROS scavenger provided a partial rescue of mitochondrial aberrations, suggesting that antioxidants may alleviate ROS-mediated resistance.

The evolution of a same tumour from adenocarcinoma to squamous cell carcinoma along the administration of anti-EGFR drugs is a mechanism of acquired drug resistance [178]. However, the cause of resistance to anti-EGFR targeted therapy in 18–30% of NSCLC patients still remains unknown [167,170]. The occurrence of tertiary EGFR mutations has been frequently reported in cases with acquired resistance to third-generation TKIs, being demonstrated in in vitro models [179]. Resistance to osimertinib is mainly caused by the EGFR p.Cys797Ser (C797S) mutation in exon 20, consisting of a substitution of a cysteine to a serine in the tyrosine kinase domain, decreases the action of third-generation TKIs, since it reduces their covalent binding to EGFR [180]. The presence of triple mutants (sensitizing mutation, T790M and C797S) supports the resistance to all three generations of EGFR TKIs [181].

NSCLC cells resistant to EGFR TKIs, gefitinib and erlotinib, were shown to exhibit elevated OXPHOS accompanied by elevated glycolysis and activity in TCA cycle [182]. In A549 NSCLC cell line, erlotinib drove ROS-mediated apoptosis via activation of the c-Jun N-terminal kinase (JNK) pathway, leading ultimately to EGFR inhibition [183]. Administration of the ROS scavenger N-acetyl cysteine reversed this phenomenon [184].

Tumours driven by either *ALK* or *ROS1* involving fusion genes exhibit similar mechanisms of resistance to targeted agents. It is well acknowledged in *ALK* mutated lung cancer the occurrence of mutations in the tyrosine kinase domain, for example L1196M and C1156Y, upon treatment with crizotinib [185]. The L1196 and G1269A substitutions are among the most frequently reported single-nucleotide mutations causing crizotinib resistance in NSCLC [186]. The existence of *ALK* or *ROS1* rearrangements together with *KRAS* mutations in NSCLC may explain primary or acquired resistance to crizotinib [187,188]. Accordingly, KRAS and NRAS activation through mutations promotes the exhaustion of first-generation inhibitors activity in ROS-1 positive cellular models [189]. Relatively to patients treated with second-generation TKIs, the most common *ALK* resistance mutation is G1202R, which is associated with in vitro resistance to all currently available ALK inhibitors excluding lorlatinib, a third-generation ALK inhibitor that shows great efficacy in patients with ALK resistance mutations [190], which are more common after treatment with second-generation ALK inhibitors. After ceritinib and alectinib treatment, missense mutations were observed in more than 50% of the samples, compared with the 30% of target alterations responsible for crizotinib exhaustion [186]. Despite the efficacy of crizotinib is thought to be specific to ALK inhibition, crizotinib also acts via the generation of superoxide and induction of apoptosis [191].

Resistance against angiogenesis inhibitor bevacizumab was also reported [192]. VEGF binds not only to its tyrosine kinase receptors (VEGFR), which can also interact to Neuropilin-1 (NP1) and Neuropilin-2 (NP2) [193]. Co-expression of NP1 and NP2 in NSCLC tissues is significantly correlated with tumour progression and bad prognosis [194]. Given that bevacizumab blocks VEGF-A, NP1 and NP2 under other stimuli may still increase the effects of VEGFR-1 and VEGFR-2, promoting angiogenesis and activating alternative pathways.

4. Metabolic Remodelling in Lung Cancer in Response to Oxidative/Alkylating Treatment

As mentioned above, cisplatin interacts with reducing equivalents, such as GSH and DNA, accounting for increased ROS and DNA damage, which leads to apoptosis [195]. Different studies suggest that metabolic remodelling, in cisplatin-resistant lung cancer cells, involved redox buffering to abrogate cisplatin effect [196–199]. Those lung cancer cells display higher levels of ROS, in part related to the low levels of intracellular thioredoxin [198], but also due to the high levels of GSH and GCL-C (glutamate cysteine ligase catalytic subunit), the first enzyme acting on the synthesis of GSH [195], possibly to counteract the high ROS levels induced by cisplatin [200]. Several studies showed that cisplatin-resistant cells are vulnerable to rapidly ROS-inducing agents. Indeed, a study reported that cisplatin-resistant lung cancer cells were more sensitive to elesclomol, an agent known to exponentially augment ROS [197] inside the cell in such a fast way that cancer cells do not have means of adapting. The xCT cyst(e)ine/glutamate pump, which supplies cells with cysteine essential for GSH production, is upregulated in cisplatin-resistant cells, as they are more sensitive to the xCT inhibitor riluzole as compared to their parental nonresistant cells [197]. Additionally, cisplatin-resistant lung cancer cells have lower rates of glycolysis and rather rely on OXPHOS [201]. Lower levels of hexokinase 1 (HK1) and 2 (HK2), the enzymes that catalyse the first step of glycolysis [202], were also observed in these lung cancer cells, in agreement with the fact that cisplatin exposure decreases HK expression [201,203]. Cisplatin-resistant lung cancer cells also showed decreased levels of glycolysis and lactate production in comparison to the sensitive parental cell lines [197] indicating a lower glycolytic activity or an increased OXPHOS. Under normal growth conditions, cisplatin-resistant lung cancer cells are not sensitive to glucose starvation, however, under hypoxic conditions, these cells are more vulnerable for 2-deoxyglucose (2-DG) treatment, a competitive inhibitor of HK, as compared to the parental cells [196]. Since cells depend on glycolysis for their energy production in the absence of oxygen, the reduced levels of HK in cisplatin-resistant cells probably makes them more vulnerable for 2-DG [201]. Greater rates of OXPHOS and mitochondrial activity, as well as a higher dependence on glutamine are observed in cisplatin-resistant lung cancer cells to compensate the lower glycolytic activity [197,198,201]. Furthermore, to fuel the TCA-cycle, β-oxidation of fatty acids has been reported in cisplatin-resistant lung cancer cells [201]. A recent study showed that cisplatin-resistant lung adenocarcinoma cells have higher MMP and intracellular ATP levels than the nonresistant cells, which also confer them increased aggressiveness [204].

5. New Therapies, a Comprehensive Adjustment of Therapy to the Metabolic Remodelling

Anticancer drug resistance is often linked to metabolic alterations and these may be targeted to overcome this issue (Figure 2). Riluzole, a FDA-approved drug for the treatment of amyotrophic lateral sclerosis [205], interferes with glutamate flux and blocks metabotropic glutamate receptors (GRM) signalling. This drug blocked proliferation of melanoma cells expressing GRM in vitro, in vivo and in a phase-0 trial, making riluzole a promising drug to treat melanoma [206,207]. In cisplatin-resistant lung cancer cells, treatment of riluzole disrupted the oxidative defense by significantly reducing glutamate release which, in turn, suppressed GSH levels, resulting in higher ROS accumulation. Riluzole treatment increased ROS by suppressing lactate dehydrogenase A (LDHA) and NAD$^+$ levels and blocked the cystine/glutamate pump, leading to cell death in cisplatin-resistant cells [197]. Therefore, using riluzole as an antitumour agent against cisplatin resistance in lung cancer should be further explored.

The sugar analogue 2-DG has been shown to be selectively cytotoxic to several tumour cell lines when cultured under anaerobic and/or hypoxic conditions [208,209] and to reduce resistance to cisplatin in an in vivo xenograft model of lung cancer [201]. Another clinical trial reported that 2-DG in combination with docetaxel was well tolerated [213]. Under hypoxia, 2-DG and 2-fluorodeoxyglucose (2-FDG) treatment inhibited glycolysis, and thus lactate production, and also induced higher cell death in cisplatin-resistant cells with low levels of HK2, as compared to their respective parental cells [201]. Hence, targeting metabolic pathways using glycolytic inhibitors, such as 2-FDG or 2-DG, to kill cisplatin-resistant lung cancer cells under anaerobic/hypoxic conditions can be an interesting therapeutic approach.

The ability to repair cisplatin-DNA adducts appears to be involved in the development of cisplatin resistance. The nucleotide excision-repair (NER) pathway and the *ERCC1* gene have been pointed out as attractive molecular targets to increase the cytotoxic effects of platinum compounds and overcome their resistance [214]. Metformin has been used for more than 50 years for the treatment of type 2 diabetes mellitus [215] and several studies showed that it has anticancer properties, improving the prognosis of patients with multiple cancers and decreasing the risk of cancer development [216,217]. A study demonstrated that metformin enhanced the sensitivity to a combined treatment of cisplatin and ionizing radiation in in vitro NSCLC models, with a greater effect in cells that are less sensitive to cisplatin [212]. These authors also showed a significant reduction in the expression of *ERCC1* after metformin treatment, pointing a possible involvement of the NER pathway in the radio-enhancement effect of the combined cisplatin and metformin treatment. Moreover, metformin was found to reverse resistance to TKIs and ALK inhibitors in lung cancer [211]. Another study using metformin showed that it increased the sensitivity of carboplatin-resistant NSCLC cells to carboplatin treatment in in vitro and in vivo models [210]. Metformin treatment decreased the expression of pyruvate kinase muscle isozyme M2 (PKM2), the enzyme that catalyses the final step in glycolysis, and consequently inhibited partially glucose metabolism and reduced ATP levels in carboplatin-resistant NSCLC cells [210].

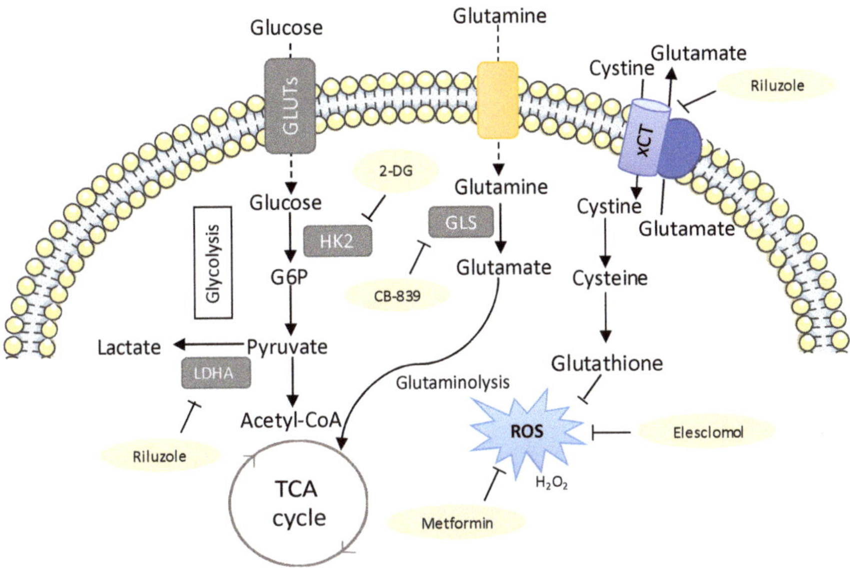

Figure 2. New metabolic therapies targeting drug-resistant lung cancer. The metabolic inhibitors that can be used to target drug-resistant cancers are shown in yellow. The sugar analogue 2-deoxyglucose (2-DG, competitive inhibitor of Hexokinase (HK), has been shown to be selectively cytotoxic to several tumour cell lines under anaerobic and/or hypoxic conditions [208,209] and to reduce resistance to cisplatin in lung cancer cells and an in vivo xenograft model of lung cancer [201]. Thus, 2-DG has the potential of being used clinically as an adjuvant to the classical chemotherapeutic compounds, such as cisplatin. Riluzole interferes with glutamate flux and was shown to increase reactive oxygen species (ROS) by suppressing lactate dehydrogenase A (LDHA) and NAD+ [197]. Glutamate stimulates the glutamine import and glutaminolysis, sustaining the import of cyst(e)ine, which has been related to increased therapy resistance in different cancer models [51], mostly due to its role in glutathione synthesis [52,54]. Riluzole interferes with system xCT-cystine/glutamate antiporter, resulting in decreased GSH levels [197]. Thus, using riluzole as an antitumour agent against cisplatin resistance in lung cancer patients could be further explored. Treatment with CB-389, a glutaminase inhibitor generated a metabolic crisis in EGFR mutant NSCLC cells, resulting in cell death and in rapid tumour regression in mouse NSCLC xenografts [64]. Metformin, an antidiabetic drug, was observed to increase the sensitivity of carboplatin-resistant NSCLC cells to carboplatin treatment in vitro and in vivo [210] and also reversed resistance to tyrosine kinase inhibitors (TKIs) and ALK inhibitors in lung cancer [211]. Another study demonstrated that metformin enhanced the sensitivity to a combined treatment of cisplatin and ionizing radiation in H460 and A549 NSCLC cell lines, with a greater effect in the A549 cell line, which is less sensitized by cisplatin [212]. Elesclomol is another potential therapeutic agent, since it further increased ROS in cisplatin-resistant cells, pushing them beyond their tolerance limit, which ultimately leads to cell death [199].

Tumour cells have the ability to adapt their metabolism to different environments and stressful conditions, increasing adaptability and tumour response to therapies [13,15]. The control of the mitochondrial biogenesis can be a mean of adaption, which is preferentially regulated by the peroxisome proliferator-activated receptor gamma (PPAR-γ) transcriptional coactivator-1 alpha (PGC-1α) [218]. During the OXPHOS process, protons are pumped into the mitochondrial inner membrane potential

(MIMP), which is finally dissipated through Complex V, generating ATP [219]. In a study using cisplatin-resistant NSCLC cells, two out of three cell lines showed stable changes towards an augmented OXPHOS function and decreased glycolysis [220]. However, the three cell lines responded in a similar way increasing ROS, MIMP and mitochondrial mass as an early response to cisplatin treatment. The authors also observed a decrease in AIF (proapoptosis) and an increase in Bcl2 (anti-apoptosis), indicating that this mechanism does not replace other classical mechanisms of cisplatin resistance [220]. A stable increase of PGC-1α, is seen in cells with increased OXPHOS activity. Treatment with the mitochondrial inhibitors metformin or rotenone (inhibitors of the complex I of the OXPHOS system) reduces the viability of the cell lines proportionally to their OXPHOS requirements [220]. This study provides new insights into cisplatin resistance mechanism in NSCLC cells which may lead to the design of new therapeutic approaches targeting mitochondria.

In NSCLC, Yuan et al., suggested that PKM2 knockdown could serve as a chemosensitizer to docetaxel, leading to the inhibition of cell viability, cell cycle arrest at G2/M phase and apoptosis [221]. These results further suggest that application of targeting the PKM2 has the potential to be a therapeutic strategy for NSCLC and provides one possible way to improve the chemotherapy effect of docetaxel. The use of NRF2-targeting agents to overcome this chemoresistance has been studied extensively. Stable knockdown of either NRF2 or KEAP1 in NSCLC cells resulted in sensitization to chemotherapeutic drugs. In particular, silencing of KEAPI augmented the expression of PPARγ and genes associated with differentiation [222]. Another study using a mouse model of mutant KrasG12D-induced lung cancer showed that suppressing the NRF2 pathway with the chemical inhibitor brusatol enhanced the antitumour efficacy of cisplatin and reduced the tumour burden as well as improving survival [223].

Although long-term gefitinib treatment can provide effective action against its primary target (aberrant EGFR activity), secondary effects result in high generation of ROS [177]. In a study using lung adenocarcinoma cells, gefitinib treatment, in the presence of a ROS scavenger, provided a partial rescue of mitochondrial aberrations. In addition, the withdrawal of gefitinib from a priori resistant clones correlated with normalized expression of EMT genes. These findings suggest that antioxidants potentially provide therapeutic benefits by attenuating TKI-induced ROS and EMT [177].

In a recent study by Apicella et al., lactate metabolism was found to be involved in the resistance to MET and EGFR TKIs (JNJ-605, crizotinib and erlotinib, respectively), in which patient-derived NSCLC showed upregulated glycolytic metabolism, with high release of lactate [224]. Lactate can act as a signaling molecule which instructs cancer associated fibroblasts (CAFs) to produce hepatocyte growth factor (HGF), whose secretion activates MET signaling in cancer cells, overcoming TKI inhibitory effects [31,224]. The pharmacological inhibition of LDH, MCT4 and MCT1 was sufficient to abrogate the in vivo resistance, making these inhibitors a new therapeutic approach that simultaneously targets lactate metabolism and oncogenes to overcome targeted therapy resistance [224].

6. Conclusions

Improved understanding of both cellular metabolism and resistance mechanisms at the molecular level promotes new opportunities to combine chemotherapeutic agents with targeted agents, which may be a promising strategy to overcome chemoresistance and to increase the effectiveness of therapy for lung cancer patients. A concomitant challenge is to find the exact killing profile and adapt it to the metabolic drift in which cancer cells continuously undergo. The course of metabolic adaptation to stressful conditions, such as drug exposure, brings together specialized tumour cell characteristics that are specific to the tumour and the individual. In the future, precision medicine will have to combine metabolic monitoring and evolution data in order to adapt clinical regimens and take advantage of the weaknesses of cancer for therapeutic purposes.

Funding: The research was funded by iNOVA4Health—UID/Multi/04462, a program financially supported by Fundação para a Ciência e Tecnologia/Ministério da Educação e Ciência (FCT-MCTES), through national funds and co-funded by FEDER under the PT2020 Partnership Agreement.

Conflicts of Interest: The authors declare no conflict of interest.

References

1. Cairns, R.A.; Mak, T.W. The current state of cancer metabolism. *Nat. Rev. Cancer* **2016**, *16*, 613–614. [CrossRef]
2. De Berardinis, R.J.; Chandel, N.S. Fundamentals of cancer metabolism. *Sci. Adv.* **2016**, *2*, e1600200. [CrossRef] [PubMed]
3. Warburg, O.; Minami, S. Versuche an Überlebendem Carcinomgewebe. *Klin. Wochenschr.* **1923**, *2*, 776–777. [CrossRef]
4. Dang, C.V. Links between metabolism and cancer. *Genes. Dev.* **2012**, *26*, 877–890. [CrossRef] [PubMed]
5. Koppenol, W.H.; Bounds, P.L.; Dang, C.V. Otto Warburg's contributions to current concepts of cancer metabolism. *Nat. Rev. Cancer* **2011**, *11*, 325–337. [CrossRef]
6. Kroemer, G.; Pouyssegur, J. Tumor Cell Metabolism: Cancer's Achilles' Heel. *Cancer Cell* **2008**, *13*, 472–482. [CrossRef]
7. Zheng, J. Energy metabolism of cancer: Glycolysis versus oxidative phosphorylation (review). *Oncol. Lett.* **2012**, *4*, 1151–1157. [CrossRef]
8. Gentric, G.; Mieulet, V.; Mechta-Grigoriou, F. Heterogeneity in Cancer Metabolism: New Concepts in an Old Field. *Antioxid. Redox Signal* **2017**, *26*, 462–485. [CrossRef]
9. Bost, F.; Decoux-Poullot, A.-G.; Tanti, J.F.; Clavel, S. Energy disruptors: Rising stars in anticancer therapy? *Oncogenesis* **2016**, *5*, e188. [CrossRef]
10. Vander Heiden, M.G.; Cantley, L.C.; Thompson, C.B. Understanding the Warburg effect: The metabolic requirements of cell proliferation. *Science* **2009**, *324*, 1029–1033. [CrossRef]
11. Pavlova, N.N.; Thompson, C.B. The Emerging Hallmarks of Cancer Metabolism. *Cell Metab.* **2016**, *23*, 27–47. [CrossRef] [PubMed]
12. Altman, B.J.; Stine, Z.E.; Dang, C.V. Corrigendum: From Krebs to clinic: Glutamine metabolism to cancer therapy. *Nat. Rev. Cancer* **2016**, *16*, 773. [CrossRef] [PubMed]
13. Obre, E.; Rossignol, R. Emerging concepts in bioenergetics and cancer research: Metabolic flexibility, coupling, symbiosis, switch, oxidative tumors, metabolic remodeling, signaling and bioenergetic therapy. *Int. J. Biochem. Cell Biol.* **2015**, *59*, 167–181. [CrossRef] [PubMed]
14. Cocetta, V.; Ragazzi, E.; Montopoli, M. Mitochondrial Involvement in Cisplatin Resistance. *Int. J. Mol. Sci.* **2019**, *20*, 3384. [CrossRef] [PubMed]
15. Cruz-Bermúdez, A.; Vicente-Blanco, R.J.; Gonzalez-Vioque, E.; Provencio, M.; Fernández-Moreno, M.; Garesse, R. Spotlight on the relevance of mtDNA in cancer. *Clin. Transl. Oncol.* **2017**, *19*, 409–418. [CrossRef]
16. Okon, I.S.; Zou, M.H. Mitochondrial ROS and cancer drug resistance: Implications for therapy. *Pharmacol. Res.* **2015**, *100*, 170–174. [CrossRef]
17. Che, M.; Wang, R.; Li, X.; Wang, H.Y.; Zheng, X.F.S. Expanding roles of superoxide dismutases in cell regulation and cancer. *Drug Discov. Today* **2016**, *21*, 143–149. [CrossRef]
18. Bansal, A.; Celeste Simon, M. Glutathione metabolism in cancer progression and treatment resistance. *J. Cell Biol.* **2018**, *17*, 2291–2298. [CrossRef]
19. Gandin, V.; Fernandes, A.P. Metal- and semimetal-containing inhibitors of thioredoxin reductase as anticancer agents. *Molecules* **2015**, *20*, 12732–12756. [CrossRef]
20. Torre, L.A.; Siegel, R.L.; Jemal, A. Lung cancer statistics. *Adv. Exp. Med. Biol.* **2016**, *893*, 1–19. [CrossRef]
21. Sher, T.; Dy, G.K.; Adjei, A.A. Small cell lung cancer. *Mayo Clin. Proc.* **2008**, *83*, 355–367. [CrossRef] [PubMed]
22. Denisenko, T.V.; Budkevich, I.N.; Zhivotovsky, B. Cell death-based treatment of lung adenocarcinoma. *Cell Death Dis.* **2018**, *9*, 117. [CrossRef] [PubMed]
23. Sellers, K.; Allen, T.D.; Bousamra, M.; Tan, J.L.; Méndez-Lucas, A.; Lin, W.; Bah, N.; Chernyavskaya, Y.; MacRae, J.I.; Higashi, R.M.; et al. Metabolic reprogramming and Notch activity distinguish between non-small cell lung cancer subtypes. *Br. J. Cancer* **2019**, *121*, 51–64. [CrossRef] [PubMed]
24. Hanahan, D.; Weinberg, R.A. Leading Edge Review Hallmarks of Cancer: The Next Generation. *Cell* **2011**, *144*, 646–674. [CrossRef] [PubMed]
25. Lopes-Coelho, F.; Gouveia-Fernandes, S.; Nunes, S.C.; Serpa, J. Metabolic Dynamics in Breast Cancer: Cooperation between Cancer and Stromal Breast Cancer Cells. *J. Clin. Breast Cancer Res.* **2017**, *1*, 1–7.
26. Reisz, J.A.; D'Alessandro, A. Measurement of metabolic fluxes using stable isotope tracers in whole animals and human patients. *Curr. Opin. Clin. Nutr. Metab. Care* **2017**, *20*, 366–374. [CrossRef]

27. Davidson, S.M.; Papagiannakopoulos, T.; Olenchock, B.A.; Heyman, J.E.; Keibler, M.A.; Luengo, A.; Bauer, M.R.; Jha, A.K.; O'Brien, J.P.; Pierce, K.A.; et al. Environment impacts the metabolic dependencies of ras-driven non-small cell lung cancer. *Cell Metab.* **2016**, *23*, 517–528. [CrossRef]

28. Hensley, C.T.; Faubert, B.; Yuan, Q.; Lev-Cohain, N.; Jin, E.; Kim, J.; Jiang, L.; Ko, B.; Skelton, R.; Loudat, L.; et al. Metabolic Heterogeneity in Human Lung Tumors. *Cell* **2016**. [CrossRef]

29. Faubert, B.; Li, K.Y.; Cai, L.; Hensley, C.T.; Kim, J.; Zacharias, L.G.; Yang, C.; Do, Q.N.; Doucette, S.; Burguete, D.; et al. Lactate Metabolism in Human Lung Tumors. *Cell* **2017**. [CrossRef]

30. Fan, T.W.M.; Lane, A.N.; Higashi, R.M.; Farag, M.A.; Gao, H.; Bousamra, M.; Miller, D.M. Altered regulation of metabolic pathways in human lung cancer discerned by 13C stable isotope-resolved metabolomics (SIRM). *Mol. Cancer* **2009**. [CrossRef]

31. Vander Linden, C.; Corbet, C. Reconciling environment-mediated metabolic heterogeneity with the oncogene-driven cancer paradigm in precision oncology. *Semin. Cell. Dev. Biol.* **2019**. [CrossRef] [PubMed]

32. Buescher, J.M.; Antoniewicz, M.R.; Boros, L.G.; Burgess, S.C.; Brunengraber, H.; Clish, C.B.; DeBerardinis, R.J.; Feron, O.; Frezza, C.; Ghesquiere, B.; et al. A roadmap for interpreting 13 C metabolite labeling patterns from cells. *Curr. Opin. Biotechnol.* **2015**, *34*, 189–201. [CrossRef] [PubMed]

33. Sellers, K.; Fox, M.P.; Li, M.B.; Slone, S.P.; Higashi, R.M.; Miller, D.M.; Wang, Y.; Yan, J.; Yuneva, M.O.; Deshpande, R.; et al. Pyruvate carboxylase is critical for non-small-cell lung cancer proliferation. *J. Clin. Investig.* **2015**. [CrossRef] [PubMed]

34. Adekola, K.; Rosen, S.T.; Shanmugam, M. Glucose transporters in cancer metabolism. *Curr. Opin. Oncol.* **2012**, *24*, 650–654. [CrossRef]

35. Mathupala, S.P.; Rempel, A.; Pedersen, P.L. Glucose catabolism in cancer cells: Identification and characterization of a marked activation response of the type II hexokinase gene to hypoxic conditions. *J. Biol. Chem.* **2001**, *276*, 43407–43412. [CrossRef]

36. Sun, Q.; Chen, X.; Ma, J.; Peng, H.; Wang, F.; Zha, X.; Wang, Y.; Jing, Y.; Yang, H.; Chen, R.; et al. Mammalian target of rapamycin up-regulation of pyruvate kinase isoenzyme type M2 is critical for aerobic glycolysis and tumor growth. *Proc. Natl. Acad. Sci. USA* **2011**, *108*, 4129–4134. [CrossRef]

37. Wang, H.; Wang, L.; Zhang, Y.; Wang, J.; Deng, Y.; Lin, D. Inhibition of glycolytic enzyme hexokinase II (HK2) suppresses lung tumor growth. *Cancer Cell Int.* **2016**, *16*, 9. [CrossRef]

38. Makinoshima, H.; Takita, M.; Matsumoto, S.; Yagishita, A.; Owada, S.; Esumi, H.; Tsuchihara, K. Epidermal growth factor receptor (EGFR) signaling regulates global metabolic pathways in EGFR-mutated lung adenocarcinoma. *J. Biol. Chem.* **2014**, *289*, 20813–20823. [CrossRef]

39. Min, H.Y.; Lee, H.Y. Oncogene-driven metabolic alterations in cancer. *Biomol. Ther.* **2018**, *26*, 45–56. [CrossRef]

40. Zhang, J.; Song, F.; Zhao, X.; Jiang, H.; Wu, X.; Wang, B.; Zhou, M.; Tian, M.; Shi, B.; Wang, H.; et al. EGFR modulates monounsaturated fatty acid synthesis through phosphorylation of SCD1 in lung cancer. *Mol. Cancer.* **2017**. [CrossRef]

41. Martín-Bernabé, A.; Cortés, R.; Lehmann, S.G.; Steve, M.; Cascante, M.; Bourgoin-Voillard, S. Quantitative proteomic approach to understand metabolic adaptation in non-small cell lung cancer. *J. Proteome. Res.* **2014**, *13*, 4695–4704. [CrossRef] [PubMed]

42. Palsson-McDermott, E.M.; O'Neill, L.A.J. The Warburg effect then and now: From cancer to inflammatory diseases. *BioEssays* **2013**, *35*, 965–973. [CrossRef] [PubMed]

43. Choi, H.; Paeng, J.C.; Kim, D.W.; Lee, J.K.; Park, C.M.; Kang, K.W.; Chung, J.K.; Lee, D.S. Metabolic and metastatic characteristics of ALK-rearranged lung adenocarcinoma on FDG PET/CT. *Lung Cancer* **2013**, *79*, 242–247. [CrossRef] [PubMed]

44. Halestrap, A.P.; Price, N.T. The proton-linked monocarboxylate transporter (MCT) family: Structure, function and regulation. *Biochem. J.* **1999**, *343*, 281–299. [CrossRef] [PubMed]

45. Ke, Q.; Costa, M. Hypoxia-Inducible Factor-1 (HIF-1). *Mol. Pharmacol.* **2006**, *70*, 1469–1480. [CrossRef] [PubMed]

46. Papandreou, I.; Cairns, R.A.; Fontana, L.; Lim, A.L.; Denko, N.C. HIF-1 mediates adaptation to hypoxia by actively downregulating mitochondrial oxygen consumption. *Cell. Metab.* **2006**, *3*, 187–197. [CrossRef]

47. Hiyama, N.; Ando, T.; Maemura, K.; Sakatani, T.; Amano, Y.; Watanabe, K.; Kage, H.; Yatomi, Y.; Nagase, T.; Nakajima, J.; et al. Glutamate-cysteine ligase catalytic subunit is associated with cisplatin resistance in lung adenocarcinoma. *Jpn. J. Clin. Oncol.* **2018**, *48*, 303–307. [CrossRef]

48. Migita, T.; Narita, T.; Nomura, K.; Miyagi, E.; Inazuka, F.; Matsuura, M.; Ushijima, M.; Mashima, T.; Seimiya, H.; Satoh, Y.; et al. ATP citrate lyase: Activation and therapeutic implications in non-small cell lung cancer. *Cancer Res.* **2008**. [CrossRef]

49. Zhang, W.C.; Ng, S.C.; Yang, H.; Rai, A.; Umashankar, S.; Ma, S.; Soh, B.S.; Sun, L.L.; Tai, B.C.; Nga, M.E.; et al. Glycine decarboxylase activity drives non-small cell lung cancer tumor-initiating cells and tumorigenesis. *Cell* **2012**. [CrossRef]

50. Ji, X.; Qian, J.; Rahman, S.M.J.; Siska, P.J.; Zou, Y.; Harris, B.K.; Hoeksema, M.D.; Trenary, I.A.; Heidi, C.; Eisenberg, R.; et al. xCT (SLC7A11)-mediated metabolic reprogramming promotes non-small cell lung cancer progression. *Oncogene* **2018**, *37*, 5007–5019. [CrossRef]

51. Nunes, S.C.; Ramos, C.; Lopes-Coelho, F.; Sequeira, C.O.; Silva, F.; Gouveia-Fernandes, S.; Rodrigues, A.; Guimarães, A.; Silveira, M.; Abreu, S.; et al. Cysteine allows ovarian cancer cells to adapt to hypoxia and to escape from carboplatin cytotoxicity. *Sci. Rep.* **2018**, *8*, 9513. [CrossRef] [PubMed]

52. Colla, R.; Izzotti, A.; De Ciucis, C.; Fenoglio, D.; Ravera, S.; Speciale, A.; Ricciarelli, R.; Furfaro, A.L.; Pulliero, A.; Passalacqua, M.; et al. Glutathione-mediated antioxidant response and aerobic metabolism: Two crucial factors involved in determining the multi-drug resistance of high-risk neuroblastoma. *Oncotarget* **2016**, *7*, 70715–70737. [CrossRef] [PubMed]

53. Mallappa, S.; Neeli, P.K.; Karnewar, S.; Kotamraju, S. Doxorubicin induces prostate cancer drug resistance by upregulation of ABCG4 through GSH depletion and CREB activation: Relevance of statins in chemosensitization. *Mol. Carcinog.* **2019**. [CrossRef] [PubMed]

54. Nunes, S.C.; Serpa, J. Glutathione in ovarian cancer: A double-edged sword. *Int. J. Mol. Sci.* **2018**, *19*, 1882. [CrossRef] [PubMed]

55. Ciamporcero, E.; Daga, M.; Pizzimenti, S.; Roetto, A.; Dianzani, C.; Compagnone, A.; Palmieri, A.; Ullio, C.; Cangemi, L.; Pili, R.; et al. Crosstalk between Nrf2 and YAP contributes to maintaining the antioxidant potential and chemoresistance in bladder cancer. *Free Radic. Biol. Med.* **2018**, *115*, 447–457. [CrossRef]

56. Chen, P.-H.; Cai, L.; Kim, H.S.; Britt, R.; Xiao, G.; White, M.A.; Minna, J.D.; DeBerardinis, R.J. Metabolic diversity in human non-small cell lung cancer. *Cancer Metab.* **2014**. [CrossRef]

57. Vaupel, P. Hypoxia and Aggressive Tumor Phenotype: Implications for Therapy and Prognosis. *Oncologist* **2008**, *13*, 21–26. [CrossRef]

58. Zhou, J.; Schmid, T.; Schnitzer, S.; Brüne, B. Tumor hypoxia and cancer progression. *Cancer Lett.* **2006**, *237*, 10–21. [CrossRef]

59. Rankin, E.B.; Giaccia, A.J. Hypoxic control of metastasis. *Science* **2016**, *352*, 175–180. [CrossRef]

60. Giatromanolaki, A.; Koukourakis, M.I.; Sivridis, E.; Turley, H.; Talks, K.; Pezzella, F.; Gatter, K.C.; Harris, A.L. Relation of hypoxia inducible factor 1α and 2α in operable non-small cell lung cancer to angiogenic/molecular profile of tumours and survival. *Br. J. Cancer* **2001**, *85*, 881–890. [CrossRef]

61. Karetsi, E.; Ioannou, M.G.; Kerenidi, T.; Minas, M.; Molyvdas, P.A.; Gourgoulianis, K.I.; Paraskeva, E. Differential expression of hypoxia-inducible factor 1α in non-small cell lung cancer and small cell lung cancer. *Clinics* **2012**, *67*, 1373–1378. [CrossRef]

62. Pikor, L.A.; Ramnarine, V.R.; Lam, S.; Lam, W.L. Genetic alterations defining NSCLC subtypes and their therapeutic implications. *Lung Cancer* **2013**, *82*, 179–189. [CrossRef] [PubMed]

63. Kerr, E.M.; Martins, C.P. Metabolic rewiring in mutant Kras lung cancer. *FEBS J.* **2018**, *285*, 28–41. [CrossRef] [PubMed]

64. Momcilovic, M.; Bailey, S.T.; Lee, J.T.; Fishbein, M.C.; Magyar, C.; Braas, D.; Graeber, T.; Jackson, N.; Czernin, J.; Emberley, E.; et al. Targeted Inhibition of EGFR and Glutaminase Induces Metabolic Crisis in EGFR Mutant Lung Cancer. *Cell Rep.* **2017**, *18*, 601–610. [CrossRef] [PubMed]

65. Kimmelman, A.C. Metabolic dependencies in RAS-driven cancers. *Clin. Cancer Res.* **2015**, *21*, 1828–1834. [CrossRef] [PubMed]

66. Kawada, K.; Toda, K.; Sakai, Y. Targeting metabolic reprogramming in KRAS-driven cancers. *Int. J. Clin. Oncol.* **2017**, *22*, 651–659. [CrossRef]

67. Ying, H.; Kimmelman, A.C.; Lyssiotis, C.A.; Hua, S.; Chu, G.; Fletcher-Sananikone, E.; Locasale, J.W.; Son, J.; Zhang, H.; Coloff, J.L.; et al. Oncogenic kras maintains pancreatic tumors through regulation of anabolic glucose metabolism. *Cell* **2012**, *149*, 656–670. [CrossRef]

68. Son, J.; Lyssiotis, C.A.; Ying, H.; Wang, X.; Hua, S.; Ligorio, M.; Perera, R.M.; Ferrone, C.R.; Mullarky, E.; Shyh-Chang, N.; et al. Glutamine supports pancreatic cancer growth through a KRAS-regulated metabolic pathway. *Nature* **2013**, *496*, 101–105. [CrossRef]

69. Onetti, R.; Baulida, J.; Bassols, A. Increased glucose transport in ras-transformed fibroblasts: A possible role for N-glycosylation of GLUT1. *FEBS Lett.* **1997**, *407*, 267–270. [CrossRef]

70. Brunelli, L.; Caiola, E.; Marabese, M.; Broggini, M.; Pastorelli, R. Capturing the metabolomic diversity of KRAS mutants in non-small-cell lung cancer cells. *Oncotarget* **2014**. [CrossRef]

71. Kerr, E.M.; Gaude, E.; Turrell, F.K.; Frezza, C.; Martins, C.P. Mutant Kras copy number defines metabolic reprogramming and therapeutic susceptibilities. *Nature* **2016**, *531*, 110–113. [CrossRef] [PubMed]

72. Kim, W.Y.; Perera, S.; Zhou, B.; Carretero, J.; Jen, J.Y.; Heathcote, S.A.; Jackson, A.L.; Nikolinakos, P.; Ospina, B.; Naumov, G.; et al. HIF2α cooperates with RAS to promote lung tumorigenesis in mice. *J. Clin. Investig.* **2009**, *119*, 2160–2170. [CrossRef] [PubMed]

73. Herbst, R.S.; Morgensztern, D.; Boshoff, C. The biology and management of non-small cell lung cancer. *Nature* **2018**, *553*, 446–454. [CrossRef] [PubMed]

74. Hirsch, F.R.; Scagliotti, G.V.; Mulshine, J.L.; Kwon, R.; Curran, W.J.; Wu, Y.-L.; Paz-Ares, L. Lung cancer: Current therapies and new targeted treatments. *Lancet* **2017**, *389*, 299–311. [CrossRef]

75. Pao, W.; Hutchinson, K.E. Chipping away at the lung cancer genome. *Nat. Publ. Gr.* **2012**, *18*, 349–351. [CrossRef] [PubMed]

76. Mok, T.S.; Wu, Y.L.; Thongprasert, S.; Yang, C.H.; Chu, D.T.; Saijo, N.; Sunpaweravong, P.; Han, B.; Margono, B.; Ichinose, Y.; et al. Gefitinib or carboplatin-paclitaxel in pulmonary adenocarcinoma. *N. Engl. J. Med.* **2009**. [CrossRef]

77. Paz-Ares, L.; Tan, E.H.; O'Byrne, K.; Zhang, L.; Hirsh, V.; Boyer, M.; Yang, J.C.H.; Mok, T.; Lee, K.H.; Lu, S.; et al. Afatinib versus gefitinib in patients with EGFR mutation-positive advanced non-small-cell lung cancer: Overall survival data from the phase IIb LUX-Lung 7 trial. *Ann. Oncol.* **2017**. [CrossRef]

78. Soda, M.; Choi, Y.L.; Enomoto, M.; Takada, S.; Yamashita, Y.; Ishikawa, S.; Fujiwara, S.I.; Watanabe, H.; Kurashina, K.; Hatanaka, H.; et al. Identification of the transforming EML4-ALK fusion gene in non-small-cell lung cancer. *Nature* **2007**. [CrossRef]

79. Gainor, J.F.; Varghese, A.M.; Ou, S.H.I.; Kabraji, S.; Awad, M.M.; Katayama, R.; Pawlak, A.; Mino-Kenudson, M.; Yeap, B.Y.; Riely, G.J.; et al. ALK rearrangements are mutually exclusive with mutations in EGFR or KRAS: An analysis of 1,683 patients with non-small cell lung cancer. *Clin. Cancer Res.* **2013**, *19*, 4273–4281. [CrossRef]

80. Kwak, E.L.; Bang, Y.J.; Camidge, D.R.; Shaw, A.T.; Solomon, B.; Maki, R.G.; Ou, S.H.I.; Dezube, B.J.; Jänne, P.A.; Costa, D.B.; et al. Anaplastic lymphoma kinase inhibition in non-small-cell lung cancer. *N. Engl. J. Med.* **2010**, *28*, 1693–1703. [CrossRef]

81. Shaw, A.T.; Kim, D.W.; Nakagawa, K.; Seto, T.; Crinó, L.; Ahn, M.J.; De Pas, T.; Besse, B.; Solomon, B.J.; Blackhall, F.; et al. Crizotinib versus chemotherapy in advanced ALK-positive lung cancer. *N. Engl. J. Med.* **2013**. [CrossRef] [PubMed]

82. Shaw, A.T.; Kim, D.W.; Mehra, R.; Tan, D.S.W.; Felip, E.; Chow, L.Q.M.; Camidge, D.R.; Vansteenkiste, J.; Sharma, S.; De Pas, T.; et al. Ceritinib in ALK-rearranged non-small-cell lung cancer. *N. Engl. J. Med.* **2014**, *370*, 1189–1197. [CrossRef] [PubMed]

83. Facchinetti, F.; Rossi, G.; Bria, E.; Soria, J.; Besse, B.; Minari, R.; Friboulet, L.; Tiseo, M. Oncogene addiction in non-small cell lung cancer: Focus on ROS1 inhibition. *Cancer Treat. Rev.* **2017**, *55*, 83–95. [CrossRef] [PubMed]

84. Shaw, A.T.; Ou, S.H.I.; Bang, Y.J.; Camidge, D.R.; Solomon, B.J.; Salgia, R.; Riely, G.J.; Varella-Garcia, M.; Shapiro, G.I.; Costa, D.B.; et al. Crizotinib in ROS1-rearranged non-small-cell lung cancer. *N. Engl. J. Med.* **2014**, *371*, 1963–1971. [CrossRef]

85. Ye, M.; Zhang, X.; Li, N.; Zhang, Y.; Jing, P.; Chang, N.; Wu, J.; Ren, X.; Zhang, J. ALK and ROS1 as targeted therapy paradigms and clinical implications to overcome crizotinib resistance. *Oncotarget* **2016**, *7*, 12289–12304. [CrossRef]

86. Sadiq, A.A.; Salgia, R. MET As a possible target for non-small-cell lung cancer. *J. Clin. Oncol.* **2013**, *31*, 1089–1096. [CrossRef]

87. Camidge, D.R.; Ou, S.-H.I.; Shapiro, G.; Otterson, G.A.; Villaruz, L.C.; Villalona-Calero, M.; Iafrate, A.J.; Varella-Garcia, M.; Dacic, S.; Cardarella, S.; et al. Efficacy and safety of crizotinib in patients with advanced c-MET -amplified non-small cell lung cancer (NSCLC). *J. Clin. Oncol.* **2014**. [CrossRef]

88. Murray, S.; Dahabreh, I.J.; Linardou, H.; Manoloukos, M.; Bafaloukos, D.; Kosmidis, P. Somatic mutations of the tyrosine kinase domain of epidermal growth factor receptor and tyrosine kinase inhibitor response to TKIs in non-small cell lung cancer: An analytical database. *J. Thorac. Oncol.* **2008**. [CrossRef]

89. Yang, H.; Liang, S.-Q.; Schmid, R.A.; Peng, R.-W. New Horizons in KRAS-Mutant Lung Cancer: Dawn After Darkness. *Front. Oncol.* **2019**, *9*, 1–13. [CrossRef]

90. Blumenschein, G.R.J.; Smit, E.F.; Planchard, D.; Kim, D.W.; Cadranel, J.; De Pas, T.; Dunphy, F.; Udud, K.; Ahn, M.J.; Hanna, N.H.; et al. A randomized phase 2 study of the MEK1/MEK2 inhibitor trametinib (GSK1120212) compared with docetaxel in. *Ann. Oncol.* **2015**, *26*, 894–901. [CrossRef]

91. Sandler, A.; Gray, R.; Perry, M.C.; Brahmer, J.; Schiller, J.H.; Dowlati, A.; Lilenbaum, R.; Johnson, D.H. Paclitaxel-carboplatin alone or with bevacizumab for non-small-cell lung cancer. *N. Engl. J. Med.* **2006**, *355*, 2542–2550. [CrossRef] [PubMed]

92. Qin, A.; Coffey, D.G.; Warren, E.H.; Ramnath, N. Mechanisms of immune evasion and current status of checkpoint inhibitors in non-small cell lung cancer. *Cancer Med.* **2016**, *5*, 2567–2578. [CrossRef] [PubMed]

93. Iwai, Y.; Ishida, M.; Tanaka, Y.; Okazaki, T.; Honjo, T.; Minato, N. Involvement of PD-L1 on tumor cells in the escape from host immune system and tumor immunotherapy by PD-L1 blockade. *Proc. Natl. Acad. Sci. USA* **2002**. [CrossRef] [PubMed]

94. Brahmer, J.; Reckamp, K.L.; Baas, P.; Crinò, L.; Eberhardt, W.E.E.; Poddubskaya, E.; Antonia, S.; Pluzanski, A.; Vokes, E.E.; Holgado, E.; et al. Nivolumab versus docetaxel in advanced squamous-cell non-small-cell lung cancer. *N. Engl. J. Med.* **2015**, *373*, 123–135. [CrossRef] [PubMed]

95. Borghaei, H.; Paz-Ares, L.; Horn, L.; Spigel, D.R.; Steins, M.; Ready, N.E.; Chow, L.Q.; Vokes, E.E.; Felip, E.; Holgado, E.; et al. Nivolumab versus docetaxel in advanced nonsquamous non-small-cell lung cancer. *N. Engl. J. Med.* **2015**, *373*, 1627–1639. [CrossRef] [PubMed]

96. Curran, M.A.; Montalvo, W.; Yagita, H.; Allison, J.P. PD-1 and CTLA-4 combination blockade expands infiltrating T cells and reduces regulatory T and myeloid cells within B16 melanoma tumors. *Proc. Natl. Acad. Sci. USA* **2010**, *107*, 4275–4280. [CrossRef]

97. Hellmann, M.D.; Rizvi, N.A.; Goldman, J.W.; Gettinger, S.N.; Borghaei, H.; Brahmer, J.R.; Ready, N.E.; Gerber, D.E.; Chow, L.Q.; Juergens, R.A.; et al. Nivolumab plus ipilimumab as first-line treatment for advanced non-small-cell lung cancer (CheckMate 012): Results of an open-label, phase 1, multicohort study. *Lancet. Oncol.* **2017**, *18*, 31–41. [CrossRef]

98. Sharma, P.; Hu-Lieskovan, S.; Wargo, J.A.; Ribas, A. Primary, Adaptive, and Acquired Resistance to Cancer Immunotherapy. *Cell* **2017**, *168*, 707–723. [CrossRef]

99. Qu, J.; Zhang, Y.; Chen, X.; Yang, H.; Zhou, C.; Yang, N. Newly developed anti-angiogenic therapy in non-small cell lung cancer. *Oncotarget* **2018**, *9*, 10147–10163. [CrossRef]

100. Rizvi, N.A.; Antonia, S.J.; Shepherd, F.A.; Chow, L.Q.; Goldman, J.; Shen, Y.; Chen, A.C.; Gettinger, S. Nivolumab (Anti-PD-1; BMS-936558, ONO-4538) Maintenance as Monotherapy or in Combination With Bevacizumab (BEV) for Non-Small Cell Lung Cancer (NSCLC) Previously Treated With Chemotherapy. *Radiat. Oncol.* **2014**, *90*, S32. [CrossRef]

101. Shanker, M.; Willcutts, D.; Roth, J.A.; Ramesh, R. Drug resistance in lung cancer. *Lung Cancer Targets Ther.* **2010**, *1*, 23–36. [CrossRef]

102. Alfarouk, K.O.; Stock, C.M.; Taylor, S.; Walsh, M.; Muddathir, A.K.; Verduzco, D.; Bashir, A.H.H.; Mohammed, O.Y.; Elhassan, G.O.; Harguindey, S.; et al. Resistance to cancer chemotherapy: Failure in drug response from ADME to P-gp. *Cancer Cell Int.* **2015**, *15*, 71. [CrossRef] [PubMed]

103. Ambudkar, S.V.; Dey, S.; Hrycyna, C.A.; Ramachandra, M.; Pastan, I.; Gottesman, M.M. Biochemical, cellular, and pharmacological aspects of the multidrug transporter. *Annu. Rev. Pharmacol. Toxicol.* **1999**, *39*, 361–398. [CrossRef] [PubMed]

104. Eastman, A.; Schulte, N. Enhanced DNA Repair as a Mechanism of Resistance to cis-Diamminedichloroplatinum(II). *Biochemistry* **1988**. [CrossRef]

105. Green, S.K.; Frankel, A.; Kerbel, R.S. Adhesion-dependent multicellular drug resistance. *Anticancer. Drug Des.* **1999**, *14*, 153–168.

106. Morin, P.J. Drug resistance and the microenvironment: Nature and nurture. *Drug Resist. Updat.* **2003**. [CrossRef]

107. Shair, I.; He, W.; Yin, L. Understanding of human ATP binding cassette superfamily and novel multidrug resistance modulators to overcome MDR. *Biomed. Pharmacother.* **2018**, *100*, 335–348. [CrossRef]

108. Taylor, J.C.; Horvath, A.R.; Higgins, C.F.; Begley, G.S. The multidrug resistance P-glycoprotein. Oligomeric state and intramolecular interactions. *J. Biol. Chem.* **2001**, *276*, 36075–36078. [CrossRef]

109. Scheffer, G.L.; Schroeijers, A.B.; Izquierdo, M.A.; Wiemer, E.A.C.; Scheper, R.J. Lung resistance-related protein/major vault protein and vaults in multidrug-resistant cancer. *Curr. Opin. Oncol.* **2000**, *12*, 550–556. [CrossRef]

110. Damia, G.; D'Incalci, M. Mechanisms of resistance to alkylating agents. *Cytotechnology* **1998**, *27*, 165–173. [CrossRef]

111. Thirumoorthy, N.; Manisenthil Kumar, K.T.; Sundar, A.S.; Panayappan, L.; Chatterjee, M. Metallothionein: An overview. *World J. Gastroenterol.* **2007**, *21*, 993–996. [CrossRef] [PubMed]

112. Kelley, S.L.; Basu, A.; Teicher, B.A.; Hacker, M.P.; Hamer, D.H.; Lazo, J.S. Overexpression of metallothionein confers resistance to anticancer drugs. *Science* **1988**, *30*, 1813–1815. [CrossRef] [PubMed]

113. Theocharis, S.; Karkantaris, C.; Philipides, T.; Agapitos, E.; Gika, A.; Margeli, A.; Kittas, C.; Koutselinis, A. Expression of metallothionein in lung carcinoma: Correlation with histological type and grade. *Histopathology* **2002**, *40*, 143–151. [CrossRef]

114. Yang, M.; Chitambar, C.R. Role of oxidative stress in the induction of metallothionein-2A and heme oxygenase-1 gene expression by the antineoplastic agent gallium nitrate in human lymphoma cells. *Free Radic. Biol. Med.* **2008**, *45*, 763–772. [CrossRef] [PubMed]

115. Vandier, D.; Calvez, V.; Massade, L.; Gouyette, A.; Mickley, L.; Fojo, T.; Rixe, O. Transactivation of the metallothionein promoter in cisplatin-resistant cancer cells: A specific gene therapy strategy. *J. Natl. Cancer Inst.* **2000**, *92*, 642–647. [CrossRef] [PubMed]

116. Hao, X.Y.; Bergh, J.; Brodin, O.; Heltman, U.; Mannervik, B. Acquired resistance to cisplatin and doxorubicin in a small cell lung cancer cell line is correlated to elevated expression of glutathione-linked detoxification enzymes. *Carcinogenesis* **1994**, *15*, 1167–1173. [CrossRef] [PubMed]

117. Tew, K.D. Glutathione-associated Enzymes in Anticancer Drug Resistance. *Cancer Res.* **1994**, *54*, 4313–4320. [CrossRef] [PubMed]

118. Tew, K.D.; Gaté, L. Glutathione S-transferases as emerging therapeutic targets. *Expert Opin. Ther. Targets* **2001**, *5*, 477–489. [CrossRef]

119. Sau, A.; Pellizzari Tregno, F.; Valentino, F.; Federici, G.; Caccuri, A.M. Glutathione transferases and development of new principles to overcome drug resistance. *Arch. Biochem. Biophys.* **2010**, *500*, 116–122. [CrossRef]

120. Awasthi, Y.C.; Singh, S.V.; Ahmad, H.; Moller, P.C. Immunocytochemical evidence for the expression of GST1, GST2, and GST3 gene loci for glutathione S-transferase in human lung. *Lung* **1987**, *165*, 323–332. [CrossRef]

121. Illio, C.D.; Boccio, G.D.; Aceto, A.; Casaccia, R.; Mucilli, F.; Federici, G. Elevation of glutathione transferase activity in human lung tumor. *Carcinogenesis* **1988**, *9*, 335–340. [CrossRef] [PubMed]

122. Meijerman, I.; Beijnen, J.H.; Schellens, J.H.M. Combined action and regulation of phase II enzymes and multidrug resistance proteins in multidrug resistance in cancer. *Cancer Treat. Rev.* **2008**, *34*, 505–520. [CrossRef]

123. Homolya, L.; Váradi, A.; Sarkadi, B. Multidrug resistance-associated proteins: Export pumps for conjugates with glutathione, glucuronate or sulfate. *BioFactors* **2003**, *17*, 103–114. [CrossRef] [PubMed]

124. Jain, N.; Lam, Y.M.; Pym, J.; Campling, B.G. Mechanisms of resistance of human small cell lung cancer lines selected in VP-16 and cisplatin. *Cancer* **1996**, *77*, 1797–1808. [CrossRef]

125. Hospers, G.A.P.; Meuer, C.; De Leij, L.; Uges, D.R.A.; Mulder, N.H.; De Vries, E.G.E. A study of human small-cell lung carcinoma (hSCLC) cell lines with different sensitivities to detect relevant mechanisms of cisplatin (CDDP) resistance. *Int. J. Cancer* **1990**, *46*, 138–144. [CrossRef] [PubMed]

126. Moritaka, T.; Kiura, K.; Ueoka, H.; Tabata, M.; Segawa, Y.; Shibayama, T.; Takigawa, N.; Ohnoshi, T.; Harada, M. Cisplatin-resistant human small cell lung cancer cell line shows collateral sensitivity to vinca alkaloids. *Anticancer Res.* **1998**, *18*, 927–933. [PubMed]

127. Zhan, M.C.; Liu, X.Y. Schedule-dependent reversion of cisplatin resistance by 5-fluorouracil in a cisplatin-resistant human lung adenocarcinoma cell line A549DDP. *Chin. Med. J. (Engl.)* **1999**, *112*, 336–339.

128. Chio, I.I.C.; Tuveson, D.A. ROS in Cancer: The Burning Question. *Trends. Mol. Med.* **2017**, *23*, 411–429. [CrossRef]

129. No, J.H.; Kim, Y.-B.; Song, Y.S. Targeting Nrf2 Signaling to Combat Chemoresistance. *J. Cancer Prev.* **2014**, *19*, 11–117. [CrossRef]

130. Zhang, P.; Singh, A.; Yegnasubramanian, S.; Esopi, D.; Kombairaju, P.; Bodas, M.; Wu, H.; Bova, S.; Biswal, S. Loss of kelch-like ECH-associated protein 1 function in prostate cancer cells causes chemoresistance and radioresistance and promotes tumor growth. *Mol. Cancer Ther.* **2010**, *9*, 336–346. [CrossRef]

131. Huang, Y.; Dai, Z.; Barbacioru, C.; Sadée, W. Cystine-glutamate transporter SLC7A11 in cancer chemosensitivity and chemoresistance. *Cancer Res.* **2005**, *65*, 7446–7454. [CrossRef] [PubMed]

132. Okuno, S.; Sato, H.; Kuriyama-Matsumura, K.; Tamba, M.; Wang, H.; Sohda, S.; Hamada, H.; Yoshikawa, H.; Kondo, T.; Bannai, S. Role of cystine transport in intracellular glutathione level and cisplatin resistance in human ovarian cancer cell lines. *Br. J. Cancer* **2003**, *88*, 951–956. [CrossRef] [PubMed]

133. Liu, Y.; Li, Q.; Zhou, L.; Xie, N.; Nice, E.C.; Zhang, H.; Huang, C.; Lei, Y. Cancer drug resistance: Redox resetting renders a way. *Oncotarget* **2016**, *7*, 42740–42761. [CrossRef] [PubMed]

134. Minotti, G.; Menna, P.; Salvatorelli, E.; Cairo, G.; Gianni, L. Anthracyclines: Molecular advances and pharmacologie developments in antitumor activity and cardiotoxicity. *Pharmacol. Rev.* **2004**, *56*, 185–229. [CrossRef]

135. Ilizarov, A.M.; Koo, H.C.; Kazzaz, J.A.; Mantell, L.L.; Li, Y.; Bhapat, R.; Pollack, S.; Horowitz, S.; Davis, J.M. Overexpression of manganese superoxide dismutase protects lung epithelial cells against oxidant injury. *Am. J. Respir. Cell Mol. Biol.* **2001**, *24*, 436–441. [CrossRef]

136. Kinnula, K.; Linnainmaa, K.; Raivio, K.O.; Kinnula, V.L. Endogenous antioxidant enzymes and glutathione S-transferase in protection of mesothelioma cells against hydrogen peroxide and epirubicin toxicity. *Br. J. Cancer* **1998**, *77*, 1097–1102. [CrossRef]

137. Sluiter, W.J.; Mulder, N.H.; Timmer-Bosscha, H.; Jan-Meersma, G.; de Vries, E.G.E. Relationship of Cellular Glutathione to the Cytotoxicity and Resistance of Seven Platinum Compounds. *Cancer Res.* **1992**, *52*, 6885–6889.

138. Alexandre, J.; Batteux, F.; Nicco, C.; Chéreau, C.; Laurent, A.; Guillevin, L.; Weill, B.; Goldwasser, F. Accumulation of hydrogen peroxide is an early and crucial step for paclitaxel-induced cancer cell death both in vitro and in vivo. *Int. J. Cancer* **2006**, *119*, 41–48. [CrossRef]

139. Llobet, D.; Eritja, N.; Encinas, M.; Sorolla, A.; Yeramian, A.; Schoenenberger, J.A.; Llombart-Cussac, A.; Marti, R.M.; Matias-Guiu, X.; Dolcet, X. Antioxidants block proteasome inhibitor function in endometrial carcinoma cells. *Anticancer Drugs* **2008**, *19*, 115–124. [CrossRef]

140. Bairati, I.; Meyer, F.; Gélinas, M.; Fortin, A.; Nabid, A.; Brochet, F.; Mercier, J.P.; Têtu, B.; Harel, F.; Abdous, B.; et al. Randomized trial of antioxidant vitamins to prevent acute adverse effects of radiation therapy in head and neck cancer patients. *J. Clin. Oncol.* **2005**, *23*, 5805–5813. [CrossRef]

141. Schwartz, P.M.; Moir, R.D.; Hyde, C.M.; Turek, P.J.; Handschumacher, R.E. Role of uridine phosphorylase in the anabolism of 5-fluorouracil. *Biochem. Pharmacol.* **1985**, *34*, 3585–3589. [CrossRef]

142. Malet-Martino, M. Clinical Studies of Three Oral Prodrugs of 5-Fluorouracil (Capecitabine, UFT, S-1): A Review. *Oncologist* **2002**, *7*, 288–323. [CrossRef] [PubMed]

143. Kosuri, K.V.; Wu, X.; Wang, L.; Villalona-Calero, M.A.; Otterson, G.A. An epigenetic mechanism for capecitabine resistance in mesothelioma. *Biochem. Biophys. Res. Commun.* **2010**, *15*, 1465–1470. [CrossRef] [PubMed]

144. O'Hagan, H.M.; Wang, W.; Sen, S.; DeStefano Shields, C.; Lee, S.S.; Zhang, Y.W.; Clements, E.G.; Cai, Y.; Van Neste, L.; Easwaran, H.; et al. Oxidative Damage Targets Complexes Containing DNA Methyltransferases, SIRT1, and Polycomb Members to Promoter CpG Islands. *Cancer Cell* **2011**, *20*, 606–619. [CrossRef] [PubMed]

145. Yueh, M.F.; Tukey, R.H. Nrf2-Keap1 Signaling pathway regulates human UGT1A1 expression in vitro and in transgenic UGT1 mice. *J. Biol. Chem.* **2007**, *282*, 8749–8758. [CrossRef]

146. Bélanger, A.S.; Tojcic, J.; Harvey, M.; Guillemette, C. Regulation of UGT1A1 and HNF1 transcription factor gene expression by DNA methylation in colon cancer cells. *BMC Mol. Biol.* **2010**, *11*, 9. [CrossRef] [PubMed]

147. Barabas, K.; Milner, R.; Lurie, D.; Adin, C. Cisplatin: A review of toxicities and therapeutic applications. *Vet. Comp. Oncol.* **2008**, *6*, 1–18. [CrossRef]

148. Gossage, L.; Madhusudan, S. Current status of excision repair cross complementing-group 1 (ERCC1) in cancer. *Cancer Treat. Rev.* **2007**, *33*, 565–577. [CrossRef]

149. Steffensen, K.D.; Waldstrøm, M.; Jakobsen, A. The relationship of platinum resistance and ERCC1 protein expression in epithelial ovarian cancer. *Int. J. Gynecol. Cancer* **2009**, *19*, 820–825. [CrossRef]

150. Lord, R.V.N.; Brabender, J.; Gandara, D.; Alberola, V.; Camps, C.; Domine, M.; Cardenal, F.; Sánchez, J.M.; Gumerlock, P.H.; Tarón, M.; et al. Low ERRC1 expression correlates with prolonged survival after cisplatin plus gemcitabine chemotherapy in non-small cell lung cancer. *Clin. Cancer Res.* **2002**, *8*, 2286–2291.

151. Rosell, R.; Lord, R.V.N.; Taron, M.; Reguart, N. DNA repair and cisplatin resistance in non-small-cell lung cancer. *Lung Cancer* **2002**, *38*, 217–227. [CrossRef]

152. Li, G.M. Mechanisms and functions of DNA mismatch repair. *Cell Res.* **2008**, *18*, 85–98. [CrossRef] [PubMed]

153. Topping, R.P.; Wilkinson, J.C.; Scarpinato, D. Mismatch repair protein deficiency compromises cisplatin-induced apoptotic signaling. *J. Biol. Chem.* **2009**, *284*, 14029–14039. [CrossRef] [PubMed]

154. Adhikari, S.; Choudhury, S.; Mitra, P.; Dubash, J.; Sajankila, S.; Roy, R. Targeting Base Excision Repair for Chemosensitization. *Anticancer Agents. Med. Chem.* **2008**, *8*, 351–357. [CrossRef]

155. Housman, G.; Byler, S.; Heerboth, S.; Lapinska, K.; Longacre, M.; Snyder, N.; Sarkar, S. Drug resistance in cancer: An overview. *Cancers (Basel)* **2014**, *6*, 1769–1792. [CrossRef]

156. Fennel, D.A. Bcl-2 as a target for overcoming chemoresistance in small-cell lung cancer. *Clin. Lung Cancer* **2003**, *4*, 307–313. [CrossRef]

157. Okouoyo, S.; Herzer, K.; Ucur, E.; Mattern, J.; Krammer, P.H.; Debatin, K.M.; Herr, I. Rescue of death receptor and mitochondrial apoptosis signaling in resistant human NSCLC in vivo. *Int. J. Cancer* **2004**, *108*, 580–587. [CrossRef]

158. Yang, L.; Mashima, T.; Sato, S.; Mochizuki, M.; Sakamoto, H.; Yamori, T.; Oh-hara, T.; Tsuruo, T. Predominant suppression of apoptosome by inhibitor of apoptosis protein in non-small cell lung cancer H460 cells: Therapeutic effect of a novel polyarginine-conjugated Smac peptide. *Cancer Res.* **2003**, *63*, 831–837.

159. Monzó, M.; Rosell, R.; Felip, E.; Astudillo, J.; Sánchez, J.J.; Maestre, J.; Martín, C.; Font, A.; Barnadas, A.; Abad, A. A novel anti-apoptosis gene: Re-expression of survivin messenger RNA as a prognosis marker in non-small-cell lung cancers. *J. Clin. Oncol.* **1999**, *17*, 2100–2104. [CrossRef]

160. Shrivastav, M.; De Haro, L.P.; Nickoloff, J.A. Regulation of DNA double-strand break repair pathway choice. *Cell Res.* **2008**, *18*, 134–147. [CrossRef]

161. Srivastava, R.K.; Sasaki, C.Y.; Hardwick, J.M.; Longo, D.L. Bcl-2-mediated drug resistance: Inhibition of apoptosis by blocking nuclear factor of activated T lymphocytes (NFAT)-induced fas ligand transcription. *J. Exp. Med.* **1999**, *190*, 253–265. [CrossRef] [PubMed]

162. Facchinetti, F.; Proto, C.; Minari, R.; Garassino, M.; Tiseo, M. Mechanisms of Resistance to Target Therapies in Non-Small Cell Lung Cancer. *Handb. Exp. Pharmacol.* **2018**, *249*, 63–89. [PubMed]

163. Rosell, R.; Moran, T.; Queralt, C.; Porta, R.; Cardenal, F.; Camps, C.; Majem, M.; Lopez-Vivanco, G.; Isla, D.; Provencio, M.; et al. Screening for epidermal growth factor receptor mutations in lung cancer. *N. Engl. J. Med.* **2009**, 958–967. [CrossRef] [PubMed]

164. Yang, J.C.H.; Wu, Y.L.; Schuler, M.; Sebastian, M.; Popat, S.; Yamamoto, N.; Zhou, C.; Hu, C.P.; O'Byrne, K.; Feng, J.; et al. Afatinib versus cisplatin-based chemotherapy for EGFR mutation-positive lung adenocarcinoma (LUX-Lung 3 and LUX-Lung 6): Analysis of overall survival data from two randomised, phase 3 trials. *Lancet Oncol.* **2015**, 141–151. [CrossRef]

165. Lynch, T.J.; Bell, D.W.; Sordella, R.; Gurubhagavatula, S.; Okimoto, R.A.; Brannigan, B.W.; Harris, P.L.; Haserlat, S.M.; Supko, J.G.; Haluska, F.G.; et al. Activating Mutations in the Epidermal Growth Factor Receptor Underlying Responsiveness of Non-Small-Cell Lung Cancer to Gefitinib. *N. Engl. J. Med.* **2004**, *1*, S24–S31. [CrossRef]

166. Godin-Heymann, N.; Ulkus, L.; Brannigan, B.W.; McDermott, U.; Lamb, J.; Maheswaran, S.; Settleman, J.; Haber, D.A. The T790M "gatekeeper" mutation in EGFR mediates resistance to low concentrations of an irreversible EGFR inhibitor. *Mol. Cancer Ther.* **2008**. [CrossRef]

167. Sequist, L.V.; Waltman, B.A.; Dias-Santagata, D.; Digumarthy, S.; Turke, A.B.; Fidias, P.; Bergethon, K.; Shaw, A.T.; Gettinger, S.; Cosper, A.K.; et al. Genotypic and histological evolution of lung cancers acquiring resistance to EGFR inhibitors. *Sci. Transl. Med.* **2011**, 75ra26. [CrossRef]

168. Engelman, J.A.; Zejnullahu, K.; Mitsudomi, T.; Song, Y.; Hyland, C.; Park, J.O.; Lindeman, N.; Gale, C.-M.; Zhao, X.; Christensen, J.; et al. MET amplification leads to gefitinib resistance in lung cancer by activating ERBB3 signaling. *Science* **2007**, *316*, 1039–1043. [CrossRef]

169. Turke, A.B.; Zejnullahu, K.; Wu, Y.L.; Song, Y.; Dias-Santagata, D.; Lifshits, E.; Toschi, L.; Rogers, A.; Mok, T.; Sequist, L.; et al. Preexistence and Clonal Selection of MET Amplification in EGFR Mutant NSCLC. *Cancer Cell* **2010**, *17*, 77–88. [CrossRef]

170. Yu, H.A.; Arcila, M.E.; Rekhtman, N.; Sima, C.S.; Zakowski, M.F.; Pao, W.; Kris, M.G.; Miller, V.A.; Ladanyi, M.; Riely, G.J. Analysis of tumor specimens at the time of acquired resistance to EGFR-TKI therapy in 155 patients with EGFR-mutant lung cancers. *Clin. Cancer Res.* **2013**, 2240–2247. [CrossRef]

171. Takezawa, K.; Pirazzoli, V.; Arcila, M.E.; Nebhan, C.A.; Song, X.; de Stanchina, E.; Ohashi, K.; Janjigian, Y.Y.; Spitzler, P.J.; Melnick, M.A.; et al. HER2 amplification: A potential mechanism of acquired resistance to egfr inhibition in EGFR -mutant lung cancers that lack the second-site EGFR T790M mutation. *Cancer Discov.* **2012**, 922–933. [CrossRef] [PubMed]

172. Pao, W.; Wang, T.Y.; Riely, G.J.; Vincent, A.M.; Qiulu, P.; Marc, L.; Maureen, F.Z.; Robert, T.H.; Mark, G.K.; Harold, E.V. KRAS mutations and primary resistance of lung adenocarcinomas to gefitinib or erlotinib. *PLoS Med.* **2005**. [CrossRef] [PubMed]

173. Milella, M.; Falcone, I.; Conciatori, F.; Incani, U.C.; Curatolo, A.D.; Inzerilli, N.; Nuzzo, C.M.A.; Vaccaro, V.; Vari, S.; Cognetti, F.; et al. PTEN: Multiple functions in human malignant tumors. *Front. Oncol.* **2015**, *5*. [CrossRef] [PubMed]

174. Cantley, L.C.; Neel, B.G. New insights into tumor suppression: PTEN suppresses tumor formation by restraining the phosphoinositide 3-kinase/AKT pathway. *Proc. Natl. Acad. Sci. USA* **1999**, *96*, 4240–4245. [CrossRef] [PubMed]

175. Sos, M.L.; Koker, M.; Weir, B.A.; Heynck, S.; Rabinovsky, R.; Zander, T.; Seeger, J.M.; Weiss, J.; Fischer, F.; Frommolt, P.; et al. PTEN loss contributes to erlotinib resistance in EGFR-mutant lung cancer by activation of akt and EGFR. *Cancer Res.* **2009**, *69*, 3256–3261. [CrossRef] [PubMed]

176. Thomson, S.; Buck, E.; Petti, F.; Griffin, G.; Brown, E.; Ramnarine, N.; Iwata, K.K.; Gibson, N.; Haley, J.D. Epithelial to mesenchymal transition is a determinant of sensitivity of non-small-cell lung carcinoma cell lines and xenografts to epidermal growth factor receptor inhibition. *Cancer Res.* **2005**, 9455–9462. [CrossRef]

177. Okon, I.S.; Coughlan, K.A.; Zhang, M.; Wang, Q.; Zou, M.H. Gefitinib-mediated reactive oxygen specie (ROS) instigates mitochondrial dysfunction and drug resistance in lung cancer cells. *J Biol Chem* **2015**, *290*, 9101–9110. [CrossRef]

178. Haratani, K.; Hayashi, H.; Watanabe, S.; Kaneda, H.; Yoshida, T.; Takeda, M.; Shimizu, T.; Nakagawa, K. Two cases of EGFR mutationpositive lung adenocarcinoma that transformed into squamous cell carcinoma: Successful treatment of one case with rociletinib. *Ann. Oncol.* **2016**, *27*, 200–202. [CrossRef]

179. Minari, R.; Bordi, P.; Tiseo, M. Third-generation epidermal growth factor receptor-tyrosine kinase inhibitors in T790M-positive non-small cell lung cancer: Review on emerged mechanisms of resistance. *Transl. Lung Cancer Res.* **2016**, *5*, 695–708. [CrossRef]

180. Thress, K.S.; Paweletz, C.P.; Felip, E.; Cho, B.C.; Stetson, D.; Dougherty, B.; Lai, Z.; Markovets, A.; Vivancos, A.; Kuang, Y.; et al. Acquired EGFR C797S mutation mediates resistance to AZD9291 in non-small cell lung cancer harboring EGFR T790M. *Nat. Med.* **2015**, 560–562. [CrossRef]

181. Niederst, M.J.; Hu, H.; Mulvey, H.E.; Lockerman, E.L.; Garcia, A.R.; Piotrowska, Z.; Sequist, L.V.; Engelman, J.A. The allelic context of the C797S mutation acquired upon treatment with third-generation EGFR inhibitors impacts sensitivity to subsequent treatment strategies. *Clin. Cancer Res.* **2015**, 3924–3933. [CrossRef] [PubMed]

182. Yang, T.; Ng, W.H.; Chen, H.; Chomchopbun, K.; Huynh, T.H.; Go, M.L.; Kon, O.L. Mitochondrial-Targeting MET Kinase Inhibitor Kills Erlotinib-Resistant Lung Cancer Cells. *ACS Med. Chem. Lett.* **2016**, *7*, 807–812. [CrossRef] [PubMed]

183. Qian, X.; Li, J.; Ding, J.; Wang, Z.; Zhang, W.; Hu, G. Erlotinib activates mitochondrial death pathways related to the production of reactive oxygen species in the human non-small cell lung cancer cell line a549. *Clin. Exp. Pharmacol. Physiol.* **2009**, *35*, 487–494. [CrossRef] [PubMed]

184. Shan, F.; Shao, Z.; Jiang, S.; Cheng, Z. Erlotinib induces the human non–small-cell lung cancer cells apoptosis via activating ROS-dependent JNK pathways. *Cancer Med.* **2016**, *5*, 3166–3175. [CrossRef]

185. Choi, Y.L.; Soda, M.; Yamashita, Y.; Ueno, T.; Takashima, J.; Nakajima, T.; Yatabe, Y.; Takeuchi, K.; Hamada, T.; Haruta, H.; et al. EML4-ALK mutations in lung cancer that confer resistance to ALK inhibitors. *N. Engl. J. Med.* **2010**, 1734–1739. [CrossRef]

186. Gainor, J.F.; Dardaei, L.; Yoda, S.; Friboulet, L.; Leshchiner, I.; Katayama, R.; Dagogo-Jack, I.; Gadgeel, S.; Schultz, K.; Singh, M.; et al. Molecular mechanisms of resistance to first- and second-generation ALK inhibitors in ALK -rearranged lung cancer. *Cancer Discov.* **2016**, 1118–1133. [CrossRef]

187. Mengoli, M.C.; Barbieri, F.; Bertolini, F.; Tiseo, M.; Rossi, G. K-RAS mutations indicating primary resistance to crizotinib in ALK-rearranged adenocarcinomas of the lung: Report of two cases and review of the literature. *Lung Cancer* **2016**, 55–58. [CrossRef]

188. Doebele, R.C.; Pilling, A.B.; Aisner, D.L.; Kutateladze, T.G.; Le, A.T.; Weickhardt, A.J.; Kondo, K.L.; Linderman, D.J.; Heasley, L.E.; Franklin, W.A.; et al. Mechanisms of resistance to crizotinib in patients with ALK gene rearranged non-small cell lung cancer. *Clin. Cancer Res.* **2012**. [CrossRef]

189. Cargnelutti, M.; Corso, S.; Pergolizzi, M.; Mévellec, L.; Aisner, D.L.; Dziadziuszko, R.; Varella-Garcia, M.; Comoglio, P.M.; Doebele, R.C.; Vialard, J.; et al. Activation of RAS family members confers resistance to ROS1 targeting drugs. *Oncotarget* **2015**, 5182–5194. [CrossRef]

190. Shaw, A.T.; Felip, E.; Bauer, T.M.; Besse, B.; Navarro, A.; Postel-Vinay, S.; Gainor, J.F.; Johnson, M.; Dietrich, J.; James, L.P.; et al. Lorlatinib in non-small-cell lung cancer with ALK or ROS1 rearrangement: An international, multicentre, open-label, single-arm first-in-man phase 1 trial. *Lancet Oncol.* **2017**, *18*, 1590–1599. [CrossRef]

191. Doherty, K.R.; Wappel, R.L.; Talbert, D.R.; Trusk, P.B.; Moran, D.M.; Kramer, J.W.; Brown, A.M.; Shell, S.A.; Bacus, S. Multi-parameter in vitro toxicity testing of crizotinib, sunitinib, erlotinib, and nilotinib in human cardiomyocytes. *Toxicol. Appl. Pharmacol.* **2013**, *272*, 245–255. [CrossRef] [PubMed]

192. Loges, S.; Schmidt, T.; Carmeliet, P. Mechanisms of resistance to anti-angiogenic therapy and development of third-generation anti-angiogenic drug candidates. *Genes Cancer* **2010**, *1*, 12–25. [CrossRef] [PubMed]

193. Niu, G.; Chen, X. Vascular Endothelial Growth Factor as an Anti-Angiogenic Target for Cancer Therapy. *Curr. Drug Targets* **2010**, *11*, 1000–1017. [CrossRef] [PubMed]

194. Kawakami, T.; Tokunaga, T.; Hatanaka, H.; Kijima, H.; Yamazaki, H.; Abe, Y.; Osamura, Y.; Inoue, H.; Ueyama, Y.; Nakamura, M. Neuropilin 1 and neuropilin 2 co-expression is significantly correlated with increased vascularity and poor prognosis in nonsmall cell lung carcinoma. *Cancer* **2002**, 2196–2201. [CrossRef] [PubMed]

195. Galluzzi, L.; Kepp, O.; Heiden, M.G.V.; Kroemer, G. Metabolic targets for cancer therapy. *Nat. Rev. Drug Discov.* **2013**, 829–846. [CrossRef] [PubMed]

196. Zaal, E.A.; Berkers, C.R. The influence of metabolism on drug response in cancer. *Front. Oncol.* **2018**, *8*, 1–15. [CrossRef]

197. Wangpaichitr, M.; Wu, C.; Li, Y.Y.; Nguyen, D.J.M.; Kandemir, H.; Shah, S.; Chen, S.; Feun, L.G.; Prince, J.S.; Kuo, M.T.; et al. Exploiting ROS and metabolic differences to kill cisplatin resistant lung cancer. *Oncotarget* **2017**, 49275–49292. [CrossRef]

198. Wangpaichitr, M.; Sullivan, E.J.; Theodoropoulos, G.; Wu, C.; You, M.; Feun, L.G.; Lampidis, T.J.; Kuo, M.T.; Savaraj, N. The relationship of thioredoxin-1 and cisplatin resistance: Its impact on ROS and oxidative metabolism in lung cancer cells. *Mol. Cancer Ther.* **2012**, 604–615. [CrossRef]

199. Wangpaichitr, M.; Wu, C.; You, M.; Maher, J.; Dinh, V.; Feun, L.; Savaraj, N. N'1,N'3-Dimethyl-N'1,N'3-bis(phenylcarbonothioyl) Propanedihydrazide (Elesclomol) Selectively Kills Cisplatin Resistant Lung Cancer Cells through Reactive Oxygen Species (ROS). *Cancers* **2009**, 23–38. [CrossRef]

200. Kuo, M.T.; Chen, H.H.W. Role of glutathione in the regulation of cisplatin resistance in cancer chemotherapy. *Met. Based Drugs* **2010**, 1–7. [CrossRef]

201. Sullivan, E.J.; Kurtoglu, M.; Brenneman, R.; Liu, H.; Lampidis, T.J. Targeting cisplatin-resistant human tumor cells with metabolic inhibitors. *Cancer Chemother. Pharmacol.* **2014**, 417–427. [CrossRef]

202. Fadaka, A.; Ajiboye, B.; Ojo, O.; Adewale, O.; Olayide, I.; Emuowhochere, R. Biology of glucose metabolization in cancer cells. *J. Oncol. Sci.* **2017**, *3*, 45–51. [CrossRef]

203. Leyton, J.; Latigo, J.R.; Perumal, M.; Dhaliwal, H.; He, Q.; Aboagye, E.O. Early detection of tumor response to chemotherapy by 3'-deoxy- 3'-[18F] fluorothymidine positron emission tomography: The effect of cisplatin on a fibrosarcoma tumor model in vivo. *Cancer Res.* **2005**, *65*, 4202–4210. [CrossRef] [PubMed]

204. Jeon, J.H.; Kim, D.K.; Shin, Y.; Kim, H.Y.; Song, B.; Lee, E.Y.; Kim, J.K.; You, H.J.; Cheong, H.; Shin, D.H.; et al. Migration and invasion of drug-resistant lung adenocarcinoma cells are dependent on mitochondrial activity. *Exp. Mol. Med.* **2016**, *48*, e277. [CrossRef] [PubMed]

205. Miller, R.G.; Mitchell, J.D.; Lyon, M.; Moore, D.H. Riluzole for amyotrophic lateral sclerosis (ALS)/motor neuron disease (MND). *Amyotroph. Lateral Scler. Other Mot Neuron Disords* **2003**, *4*, 191–206. [CrossRef]

206. Namkoong, J.; Shin, S.S.; Hwa, J.L.; Marín, Y.E.; Wall, B.A.; Goydos, J.S.; Chen, S. Metabotropic glutamate receptor 1 and glutamate signaling in human melanoma. *Cancer Res.* **2007**, 2298–2305. [CrossRef] [PubMed]

207. Yip, D.; Le, M.N.; Chan, J.L.K.; Lee, J.H.; Mehnert, J.A.; Yudd, A.; Kempf, J.; Shih, W.J.; Chen, S.; Goydos, J.S. A phase 0 trial of riluzole in patients with resectable stage III and IV melanoma. *Clin. Cancer Res.* **2009**, 3896–3902. [CrossRef] [PubMed]

208. Liu, H.; Savaraj, N.; Priebe, W.; Lampidis, T.J. Hypoxia increases tumor cell sensitivity to glycolytic inhibitors: A strategy for solid tumor therapy (Model C). *Biochem. Pharmacol.* **2002**, *64*, 1745–1751. [CrossRef]

209. Maher, J.C.; Krishan, A.; Lampidis, T.J. Greater cell cycle inhibition and cytotoxicity induced by 2-deoxy-D-glucose in tumor cells treated under hypoxic vs aerobic conditions. *Cancer Chemother. Pharmacol.* **2004**, *53*, 116–122. [CrossRef]

210. Liu, Y.; He, C.; Huang, X. Metformin partially reverses the carboplatin-resistance in NSCLC by inhibiting glucose metabolism. *Oncotarget* **2017**, *8*, 75206–75216. [CrossRef]

211. Li, L.; Han, R.; Xiao, H.; Lin, C.; Wang, Y.; Liu, H.; Li, K.; Chen, H.; Sun, F.; Yang, Z.; et al. Metformin sensitizes EGFR-TKI-Resistant human lung cancer cells in vitro and in vivo through inhibition of IL-6 signaling and EMT reversal. *Clin. Cancer Res.* **2014**, 2714–2726. [CrossRef] [PubMed]

212. Riaz, M.A.; Sak, A.; Erol, Y.B.; Groneberg, M.; Thomale, J.; Stuschke, M. Metformin enhances the radiosensitizing effect of cisplatin in non-small cell lung cancer cell lines with different cisplatin sensitivities. *Sci. Rep.* **2019**, *9*, 1–16. [CrossRef] [PubMed]

213. Raez, L.E.; Papadopoulos, K.; Ricart, A.D.; Chiorean, E.G.; Dipaola, R.S.; Stein, M.N.; Rocha Lima, C.M.; Schlesselman, J.J.; Tolba, K.; Langmuir, V.K.; et al. A phase i dose-escalation trial of 2-deoxy-d-glucose alone or combined with docetaxel in patients with advanced solid tumors. *Cancer Chemother. Pharmacol.* **2013**, *71*, 523–530. [CrossRef] [PubMed]

214. Zamble, D.B.; Lippard, S.J. Cisplatin and DNA repair in cancer chemotherapy. *Trends Biochem. Sci.* **1995**, 435–439. [CrossRef]

215. Pernicova, I.; Korbonits, M. Metformin-Mode of action and clinical implications for diabetes and cancer. *Nat. Rev. Endocrinol.* **2014**, *10*, 143–156. [CrossRef]

216. Snima, K.; Pillai, P.; Cherian, A.; Nair, S.; Lakshmanan, V.-K. Anti-diabetic Drug Metformin: Challenges and Perspectives for Cancer Therapy. *Curr. Cancer Drug Targets* **2014**, 727–736. [CrossRef]

217. Yousef, M.; Tsiani, E. Metformin in lung cancer: Review of in vitro and in vivo animal studies. *Cancers* **2017**. [CrossRef]

218. Handschin, C.; Spiegelman, B.M. Peroxisome proliferator-activated receptor γ coactivator 1 coactivators, energy homeostasis, and metabolism. *Endocr. Rev.* **2006**, *27*, 728–735. [CrossRef]

219. Sabharwal, S.S.; Schumacker, P.T. Mitochondrial ROS in cancer: Initiators, amplifiers or an Achilles' heel? *Nat. Rev. Cancer* **2014**, *14*, 709–721. [CrossRef]

220. Cruz-Bermúdez, A.; Laza-Briviesca, R.; Vicente-Blanco, R.J.; García-Grande, A.; Coronado, M.J.; Laine-Menéndez, S.; Palacios-Zambrano, S.; Moreno-Villa, M.R.; Ruiz-Valdepeñas, A.M.; Lendinez, C.; et al. Cisplatin resistance involves a metabolic reprogramming through ROS and PGC-1α in NSCLC which can be overcome by OXPHOS inhibition. *Free Radic. Biol. Med.* **2019**, *135*, 167–181. [CrossRef]

221. Yuan, S.; Qiao, T.; Zhuang, X.; Chen, W.; Xing, N.; Zhang, Q. Knockdown of the M2 Isoform of Pyruvate Kinase (PKM2) with shRNA Enhances the Effect of Docetaxel in Human NSCLC Cell Lines In Vitro. *Yonsei Med. J.* **2016**, *57*, 1312–1323. [CrossRef] [PubMed]

222. Zhan, L.; Zhang, H.; Zhang, Q.; Woods, C.G.; Chen, Y.; Xue, P.; Dong, J.; Tokar, E.J.; Xu, Y.; Hou, Y.; et al. Regulatory role of KEAP1 and NRF2 in PPARγ expression and chemoresistance in human non-small-cell lung carcinoma cells. *Free Radic. Biol. Med.* **2012**, *53*, 758–768. [CrossRef] [PubMed]

223. Tao, S.; Wang, S.; Moghaddam, S.J.; Ooi, A.; Chapman, E.; Wong, P.K.; Zhang, D.D. Oncogenic KRAS confers chemoresistance by upregulating NRF2. *Cancer Res.* **2014**, *74*, 7430–7441. [CrossRef] [PubMed]
224. Apicella, M.; Giannoni, E.; Fiore, S.; Chiarugi, P.; Giordano, S.; Corso, S. Increased Lactate Secretion by Cancer Cells Sustains Non-cell-autonomous Adaptive Resistance to MET and EGFR Targeted Therapies. *Cell Metab.* **2018**, *28*, 848–865. [CrossRef] [PubMed]

 *antioxidants*

Review

Modulation of Oxidative Stress by Ozone Therapy in the Prevention and Treatment of Chemotherapy-Induced Toxicity: Review and Prospects

Bernardino Clavo [1,2,3,4,5,*], Francisco Rodríguez-Esparragón [1], Delvys Rodríguez-Abreu [6], Gregorio Martínez-Sánchez [7], Pedro Llontop [8], David Aguiar-Bujanda [9], Leandro Fernández-Pérez [4] and Norberto Santana-Rodríguez [4,10,11]

1   Research Unit, Dr. Negrín University Hospital, 35019 Las Palmas de Gran Canaria, Spain; frodesp@gobiernodecanarias.org
2   Chronic Pain Unit, Dr. Negrín University Hospital, 35019 Las Palmas de Gran Canaria, Spain
3   Radiation Oncology Department, Dr. Negrín University Hospital, 35019 Las Palmas de Gran Canaria, Spain
4   Universitary Institute for Research in Biomedicine and Health (IUIBS), Molecular and Translational Pharmacology Group, University of Las Palmas de Gran Canaria, 35016 Las Palmas de Gran Canaria, Spain; leandrofco.fernandez@ulpgc.es (L.F.-P.); norbesanrod@gmail.com (N.S.-R.)
5   Spanish Group of Clinical Research in Radiation Oncology (GICOR), 28290 Madrid, Spain
6   Medical Oncology Department, Complejo Hospitalario Universitario Insular Materno-Infantil de Gran Canaria, University of Las Palmas de Gran Canaria, 35016 Las Palmas de Gran Canaria, Spain; delvysra@yahoo.com
7   Scientific Advisor, Freelance, 60126 Ancona, Italy; gregorcuba@yahoo.it
8   Experimental Medicine and Surgery Unit of Hospital Gregorio Marañon and the Health Research Institute of Hospital Gregorio Marañon (IiSGM), 28007 Madrid, Spain; perollonsa@gmail.com
9   Medical Oncology Department, Dr. Negrín University Hospital, 35019 Las Palmas de Gran Canaria, Spain; dagubuj@gobiernodecanarias.org
10 Thoracic Surgery, Department of Surgery, King Faisal Specialist Hospital and Research Center, 12713 Riyadh, Saudi Arabia
11 College of Medicine, Department of Surgery, Alfaisal University, 11533 Riyadh, Saudi Arabia
*   Correspondence: bernardinoclavo@gmail.com; Tel.: +34-928-449-278

Received: 29 October 2019; Accepted: 23 November 2019; Published: 26 November 2019

Abstract: (1) Background: Cancer is one of the leading causes of mortality worldwide. Radiotherapy and chemotherapy attempt to kill tumor cells by different mechanisms mediated by an intracellular increase of free radicals. However, free radicals can also increase in healthy cells and lead to oxidative stress, resulting in further damage to healthy tissues. Approaches to prevent or treat many of these side effects are limited. Ozone therapy can induce a controlled oxidative stress able to stimulate an adaptive antioxidant response in healthy tissue. This review describes the studies using ozone therapy to prevent and/or treat chemotherapy-induced toxicity, and how its effect is linked to a modification of free radicals and antioxidants. (2) Methods: This review encompasses a total of 13 peer-reviewed original articles (most of them with assessment of oxidative stress parameters) and some related works. It is mainly focused on four drugs: Cisplatin, Methotrexate, Doxorubicin, and Bleomycin. (3) Results: In experimental models and the few existing clinical studies, modulation of free radicals and antioxidants by ozone therapy was associated with decreased chemotherapy-induced toxicity. (4) Conclusions: The potential role of ozone therapy in the management of chemotherapy-induced toxicity merits further research. Randomized controlled trials are ongoing.

Keywords: antioxidants; bleomycin; cancer treatment; chemotherapy-induced toxicity; cisplatin; doxorubicin; free radicals; methotrexate; oxidative stress; ozone therapy

1. Introduction

Chemotherapy (CT) is one of the main treatments for cancer. Its efficacy has been growing because of new chemotherapy agents, the new combination regimens and an increasing multimodal approach. Many effects of chemotherapy depend on the increase of free radicals and reactive oxygen species (ROS) in cancer cells. However, they can also mediate chemotherapy-induced toxicity (CIT). For many drugs, the most frequent and major toxicities are cytopenias, nausea, vomiting and hair loss. The latter is usually reversible; for the management and/or prevention of cytopenias, platelet or hemoglobin transfusions are available, as well as erythropoietin-stimulating agents and colony-stimulating growth factors (CSGF). However, some other major toxicities can affect different organs and tissues, depending on the CT agent. Usually, this damage is mediated by ROS and high oxidative stress, and frequently, preventive and therapeutic approaches are limited.

Cellular ROS are generated in mitochondria by oxidative phosphorylation. ROS also participate as signaling molecules in cell physiological processes of proliferation and survival. Thus, oxidative stress reflects the imbalance due to an excess of ROS or oxidants that overcome the capability of cells to exert effective antioxidant responses. Excessive ROS production may arise from mitochondria dysfunction or by the interaction between normal or excessive mitochondrial production with exogenous sources. The superoxide anion ($O_2\bullet^-$) is a free radical produced by the single electron reduction of O_2. It is the first ROS directly produced from O_2 and the precursor of all other ROS. Spontaneous and superoxide dismutase (SOD)-dependent $O_2\bullet^-$ dismutation generates hydrogen peroxide (H_2O_2), which itself can undergo the Fenton reaction to generate the hydroxyl radical ($OH\bullet$) in the presence of transition metals, most commonly Fe^{2+}. Oxidative stress results in macromolecular damage. Lipid peroxidation generates direct products such as malondialdehyde (MDA), isoprostanes, and 4-hydroxynonenal. Protein oxidation can cause fragmentation at amino acid residues, formation of protein-protein cross-linkages, and oxidation of the protein backbone. Oxidative damage to DNA causes alterations in DNA bases. Also, MDA can react with DNA to form DNA adducts [1,2].

Ozone (O_3) is the triatomic allotrope form of oxygen which is much reactive (less stable) and more soluble (10 times) in water and plasma than the diatomic allotrope form (O_2). Its antioxidant potency is the third after fluorine and persulfate and it is higher than O_2 [3]. Ozone therapy consists in the medical use of a gas mixture of O_3/O_2, obtained from medical-grade oxygen using an ozone generator device and which has to be administered *in situ* because of the short half-life (at 20 °C the O_3 concentration is halved within 40 min, at 30 °C within 25 min) [3]. Typical clinical O_3 concentrations range from 10 to 60 μg/mL (μg of O_3 / mL of O_2) of a mixture O_3 (0.5–0.05%) and O_2 (95–99.5%) [4]. So, although more than 95% of the gas mixture is always oxygen, small variations in O_3 content change its potential effects.

A higher concentration of ozone (maximum 0.02 μg/mL) is beneficial, preventing damaging UV light from reaching the Earth's surface [5]. However, exposure by inhalation to prolonged ground-level ozone damages the respiratory system and extrapulmonary organs. In the same way, in humans, ozone can be dangerous or beneficial, depending on the route and organ/tissue of administration and on the concentration of exposition. It is becoming clear how the respiratory system—when undergoing a chronic oxidative stress—can release slowly, but steadily, a huge number of toxic compounds that are able to enter the circulation and cause serious damage [6]. Moreover, the potent antioxidant capacity of blood exposed to a small and precisely calculated dose of ozone only for a few minutes can modulate the endogenous antioxidant system and aids in the control of different pathological conditions [7].

This review is mainly focused on four drugs: Cisplatin, Methotrexate, Doxorubicin, and Bleomycin, which belong to different CT groups—alkylating agents, antimetabolites and antitumor antibiotics, respectively. These drugs can induce severe and dose-limiting toxicity, which has been reduced in experimental models when ozone has been administered as a preventive or therapeutic approach. Later, some related works supporting the previous studies will be summarized.

2. Chemotherapy-Induced Toxicity and Free Radicals

Mitochondria is one of the key contributors to cancer development and progression. Most of the $O_2\bullet^-$ generated under physiological conditions are efficiently converted into H_2O_2 by superoxide dismutase (SOD). Catalase, glutathione peroxidase (GSH-Px, eight isoforms), and peroxiredoxins (Prxs, six isoforms) can convert H_2O_2 to water and O_2, The $O_2\bullet^-$ to H_2O_2 reaction also occurs spontaneously. Small ROS concentrations are required as messengers and signals for appropriated cell regulation. Higher levels of ROS and free radicals are produced by chemotherapy and radiotherapy as the main action mechanism for killing cancer cells. However, most of the chronic CIT are also influenced by the perpetuation of a pro-oxidative status and inflammation. Frequently, the most used approaches for many CIT include symptomatic treatments, substances with antioxidant effect and anti-inflammatory drugs and corticosteroids, although sometimes with limited efficacy. We include a short review of the action mechanisms and toxicities of the four drugs (Cisplatin, Methotrexate, Doxorubicin, and Bleomycin) that have undergone studies to evaluate CIT-modulation by ozone therapy.

2.1. Cisplatin-Induced Toxicity

Cisplatin, cis- diamine-dichloro-platinum (CDDP) is one of the most used chemotherapy drugs because it is effective against different types of tumors. Cisplatin is an alkylating agent which is cell-cycle-phase nonspecific. It can bond to proteins, RNA and DNA, inhibiting DNA synthesis and cell cycle and it can also induce apoptosis. The most common cisplatin-induced toxicities are nausea, vomiting, myelosuppression, ion alterations, alopecia, sterility and others. However, among the most relevant and dose-limiting are ototoxicity, peripheral neuropathy and especially nephrotoxicity. Today, it is suggested that cisplatin-associated toxicities are mainly induced by free radicals' production, which will result in oxidative organ injury. The evidence is growing over the protective effects of antioxidants on cisplatin-induced adverse reactions, especially, nephrotoxicity [8–10]. The main route for cisplatin elimination is via the kidneys, and around one out of three to four patients treated with full doses of cisplatin could develop renal dysfunction; the percentage could be higher than 50% in children. This damage can be produced at several renal structures: blood vessels (with vasoconstriction a decrease in renal blood flow), glomeruli and mainly, in proximal tubular cells [9,10].

2.2. Methotrexate-Induced Toxicity

Methotrexate (MTX) acts as an antimetabolite, blocking the dihydrofolate reductase and inhibiting the formation of tetrahydrofolic acid (reduced folic acid). This way, MTX inhibits formation of thymidylate from deoxyuridylate and inhibits the synthesis of DNA. This action and the additional inhibition of RNA and synthesis of proteins prevents cells to enter in the S phase of cell cycle (MTX is a cell cycle-specific agent).

MTX is used against many different tumors and in some autoimmune diseases such as rheumatoid arthritis. Although MTX is safely administered to most patients, it can cause significant toxicity, especially with chronic or high-dose schemes. In addition to myelosuppression, the most relevant could be pneumonitis (especially in irradiated areas), enteritis, leukoencephalopathy (intrathecal combined with high dose systemic administration) and especially, hepatic and acute kidney injury which can happen in 2–12% of patients. Nephrotoxicity results from crystallization of methotrexate in the renal tubular lumen, leading to tubular toxicity. Acute kidney injury and other toxicities of high-dose MTX can lead to significant morbidity, treatment delays, and diminished renal function [11]. The effects of MTX in vivo may be mediated by reducing cell proliferation, increasing the rate of apoptosis of T cells, increasing endogenous adenosine release, altering the expression of cellular adhesion molecules, influencing production of cytokines, humoral responses and bone formation. Several reports indicate that the effects of MTX are influenced by genetic variants, specific dynamic processes and micro-environmental elements such as nucleotide deprivation or glutathione levels [12].

MTX-induced toxicity has been related to oxidative stress [13] and down-regulation of the nuclear factor erythroid 2-related factor 2 (Nrf2) and heme oxygenase-1 (HO-1) [14].

2.3. Doxorubicin-Induced Toxicity

Doxorubicin (DOX) is an anthracycline antitumor antibiotic used against a large number of tumors. It is cell-cycle-phase nonspecific by intercalation between DNA base pairs, and it blocks the action of topoisomerase-II and inhibits the DNA and RNA synthesis. Similarly to many other chemotherapy agents, DOX frequently produces myelosuppression, nausea, vomiting, and alopecia. However, two potential DOX-induced toxicities are at cutaneous and cardiac level. DOX is a vesicant agent and its extravasation can produce local ulceration and necrosis. On the other hand, a potential and characteristic DOX-induced toxicity is cardiomyopathy with congestive heart failure. This cardiotoxic effect is dose-limiting and cumulative-dose dependent, with a high risk increase at cumulative doses higher than 550 mg/m^2, or even lower (400 mg/m^2) in patients with previous thoracic irradiation, previous cardiopathy or in combination with other drugs. Oxidative stress remains the most probable mechanism for the DOX-induced cardiotoxic effect, mediated by the production of iron-complex and the subsequent generation of free radicals [15,16]. In selected patients, Dexrazoxane can be used to prevent/diminish DOX-induced cardiotoxicity, because Dexrazoxane is an iron chelator that decreases the DOX-iron binding and the subsequent free radical generation.

2.4. Bleomycin-Induced Toxicity

Bleomycin (BLM) is a redox-active drug with anticancer and other clinical applications. BLM is an effective agent against lymphomas, testicular and ovarian germ cell cancers and certain squamous carcinomas. The antineoplastic effect of BLM is thought to involve the production of single- and double-strand breaks in DNA (scission) by a complex of BLM, ferrous ions, and molecular oxygen. Bleomycin binds to DNA by intercalation of the dithiazole moiety between base pairs of DNA and by electrostatic interactions of the terminal amines. The reduction of molecular oxygen by ferrous ions chelated by BLM leads to hydrogen subtraction from the C3 and C4 carbons of deoxyribose, resulting in cleavage of the C3–C4 bond and liberation of a base with a DNA strand break. BLM is inactivated in vivo by the enzyme BLM hydrolase, a cytosolic aminopeptidase that has lower activity in the skin and lungs. Bleomycin is selectively toxic to cells in the M and G2 phases of the cell cycle, and generally more effective against actively dividing rather than resting cells [17]. Despite being one of the most effective broad-spectrum chemotherapeutic agents in the treatment of cancers, the clinical applications of BLM have been limited due to the side effect of causing lung fibrosis [18]. The risk of BLM-induced fibrosis is increased by the improvements in overall survival and in those patients with previous lung diseases or thoracic irradiation.

The mechanism of BML-induced lung injury is not entirely clear, but likely includes components of oxidative damage, relative deficiency of the deactivating enzyme BML hydrolase, genetic susceptibility, and elaboration of inflammatory cytokines. Oxidative damage to the lung appears important in the pathophysiology of lung injury, and antioxidants may ameliorate the process [19]. Systemic administration of antioxidant artemisitene strongly inhibits bleomycin-induced lung damage, through the activation of the Nrf2 signaling pathway [20].

3. Modulation of Oxidative Stress by Ozone Therapy

Local ozone applications can induce direct effects and modulation effects at the local level. However, when ozone therapy is applied with systemic intent (principally by autohemotherapy and by rectal insufflation), ozone does not enter into the blood circulation and it is not able to reach any specific target tissues. Ozone that is not removed by the antioxidants of the medium interacts with unsaturated fatty acids from cell membranes in intestinal mucosa (rectal administration) or blood cells (in the extra-corporeal blood–ozone mixture, during auto-hemotherapy) generating aldehyde and hydroxy-hydroperoxide (ozone-peroxide), which forms H_2O_2 and a second aldehyde—4-hydroxynonenal (4-HNE), which is one of the most relevant aldehydes. These substances act as second messengers and induce a further adaptive response from the body (with potential over regulation of antioxidant systems) in a hormetic dose–response relationship [21–23]. This is, the action mechanism of systemic ozone therapy is an "indirect" effect. Ozone does not follow the standard principles of Pharmacology: absorption, distribution, metabolism and excretion. Ozone "only acts" as a modulator or pro-drug and, by inducing secondary messengers, will enhance the subsequent adaptive responses. After this fast reaction (few seconds), ozone disappears. Ozone concentration and effects do not follow a linear relationship: very low concentrations could have no effect and very high concentrations can lead to contrary effects to those produced by lower/middle concentrations [24]. Mediators such as 4-HNE and H_2O_2 are among the most relevant secondary messengers induced by ozone during lung toxicity following airway inhalation [25,26] but also, in the course of the induction of beneficial effects during medical application [2,27]. Moreover, H_2O_2 can enter the cytoplasm of mononuclear cells and modulate nuclear factor kappa B (NF-κB). H_2O_2 emerges not as an inducer of NF-κB, but as an agent able to modulate the activation of the NF-κB pathway by other agents. This modulation is generic at the level of the whole pathway but specific at the level of the single gene. Therefore, H_2O_2 is a fine-tuning regulator of NF-κB-dependent processes, as exemplified by its dual regulation of inflammation [28]. Most likely, the therapeutic dose of O_3 blocks the NF-κB signal, reducing inflammation [29]. In contrast, a high dose of O_3 promotes inflammation by activation of the NF-κB pathway [30]. In addition, H_2O_2 can act as promotor of the Nrf2 pathway. The important role of Nrf2 induction by ozone in order to enhance the antioxidant systems has been described recently [31–33].

There is a broad consensus on the relevance of the induction of protective molecules during small but repeated oxidative stress [22,34]. The most relevant aldehyde produced by the reaction of O_3 is 4-HNE, which remains more stable than ROS [22,27]. 4-HNE is known to be quite reactive; it participates in multiple physiological processes as a nonclassical secondary messenger and readily forms covalent modifications of numerous targets [35]. 4-HNE is rapidly degraded by alcohol dehydrogenases, aldehyde dehydrogenase, and by glutathione-S-transferase. 4-HNE will form adducts with the thiol (-SH) and amino groups of Cys34 present in domain-I of albumin. This way, 4-HNE can send a signal of a transient oxidative stress to different tissues in the body and its effects depends on concentration as well as cell/tissue origin. This pathway can activate the synthesis of several substances such as: γ-glutamyl transferase, γ-glutamyl transpeptidase, HSP-70, HO-1, and antioxidant enzymes such as SOD, GSH-Px, catalase and glucose-6-phosphate dehydrogenase (G6PDH, a critical enzyme electron-donor during erythropoiesis in the bone marrow) and the Nrf2 pathway. In addition, these pluripotent effects of 4-HNE can be explained by its concentration-dependent interactions with the cytokine networks and complex cellular antioxidant systems also showing cell and tissue specificities [2,36]. As it happens with the potential actions of ozone, the potential actions of 4-HNE are very different at lower concentrations (regulation of proliferation and differentiation and enhancement of Nrf2 and antioxidant systems) than at high concentrations (induction of oxidative stress, apoptosis, and necrosis).

Experimental results demonstrated that ozone ex vivo or in vivo can activate Nrf2 [7,37]. This mechanism can explain the genomic target of ozone, which induces the proteomic response (protein synthesis, as antioxidant enzymes: e.g., HO-1, SOD, CAT), providing far better protection against the total body damaging effects from free radicals. In addition, a very recent manuscript demonstrates the role of ozone on casein kinase 2 (CK2) (another regulator of the Nrf2 activity through its phosphorylation) in multiple sclerosis patients [38]. However, the effects of ozone also involve the modulation (inhibition) of the NF-κB pathway. This pathway activates the release of pro-inflammatory cytokines such as: TNFα, INFγ, IL1β, IL6, IL8, as well as pro-inflammatory genes such as cyclooxygenase-2 (COX-2) and inducible nitric oxide synthase (iNOS) [39]. As a result, the dose adminstered in ozone therapy and its hormetic response have a crucial role to manage the equilibrium inflammation/pro-inflammation responses. Both Nrf2 and NF-κB regulation are coordinated in order to maintain redox homeostasis in healthy cells. However, during pathological conditions, this regulation is perturbed, offering an opportunity for therapeutic intervention [39,40]. The regulation of inflammation by NF-κB signaling as well as Nrf2 pathways separately is widely documented. Since both these major signaling pathways modulate inflammation, they may crosstalk to bring about coordinated inflammatory responses (Figure 1) [41,42].

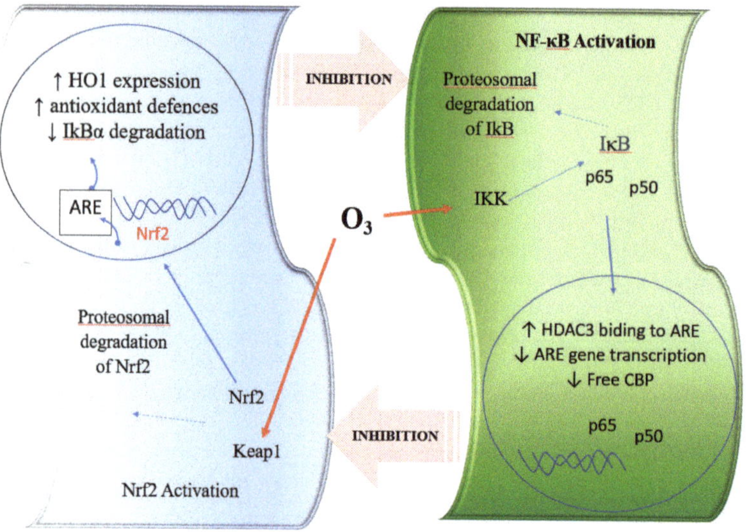

Figure 1. Representation of the interaction with the crosstalk between the nuclear factor erythroid 2-related factor 2 (Nrf2) and nuclear factor kappa B (NF-κB) pathways and the role of ozone. HO-1, haem-oxygenase-1; ARE, antioxidant response element; Keap1, Kelch-like ECH-associated protein 1; IKK: IκB kinase; CBP: CREB binding protein; HDAC3: histone deacetylase 3. Nrf2: nuclear erythroid 2 related factor 2; NF-κB: nuclear factor kappa light chain enhancer of B cell; LPS: lipopolysaccharide; O₃: ozone.

Preclinical studies indicated that ozone therapy could attenuate tubulointerstitial injury in rats with adenine-induced chronic kidney disease by mediating the modulation of Nrf2 and NF-κB [43]. In addition, clinical studies confirm this effect of O₃ modulating the balance Nrf2/NF-κB in patients with multiple sclerosis [38].

4. Ozone Therapy in Chemotherapy-Induced Toxicity

Because ozone can modulate oxidative stress, inflammation and ischemia/hypoxia, it could be expected to exert a beneficial effect in chronic CIT when those mechanisms are involved. Several experimental models and isolated clinical studies have demonstrated its benefit in the prevention and/or treatment of CIT by some chemotherapy drugs, especially Cisplatin, Methotrexate, Doxorubicin, and Bleomycin. Finally, we will describe some related studies that offer additional support to the protective effect of ozone against CIT.

4.1. Ozone and Cisplatin-Induced Toxicity

In the last 15 years, several experimental models have described the effects and potential action mechanisms of ozone for prevention (by ozone preconditioning) or for treatment (stablished alterations) of renal damage by cisplatin.

In 2004, Borrego et al. [44] described the effect of ozone preconditioning (ozone administration before cisplatin administration) to prevent cisplatin nephrotoxicity. Nine milliliters of ozone were administered, at different concentrations, by rectal insufflation: one session/day for 15 consecutive days before the day of intraperitoneal cisplatin injection. Rats were sacrificed 5 days after cisplatin injection. Regarding the control group without treatment, the groups with only O_2 or with only O_3 (without cisplatin) showed similar levels of serum creatinine (as a marker of renal damage), as well as renal levels of free radicals (measuring thiobarbituric acid-reactive substances—TBARS) and antioxidants (GSH, SOD, CAT, GSH-Px). Cisplatin group showed increased serum creatinine (four times) and TBARS (two times) and decrease of all antioxidants (between 15–40%). Regarding the cisplatin group, administration of cisplatin with O_2 or with low O_3 concentrations (10 µg/mL) did not show relevant changes and cisplatin with high O_3 concentrations (50 or 70 µg/mL) showed similar (or even worse) creatinine levels, with disappointing results in antioxidants levels. Cisplatin plus O_3/O_2 preconditioning at these higher concentrations (50 and 70 µg/mL) showed histopathological changes that were quite similar to those present with cisplatin alone. However, rats treated with cisplatin and with O_3 preconditioning at moderate concentrations (20 or 30 µg/mL) showed a relatively lower increase in creatinine levels (only two times) and TBARS, and a level of antioxidants similar or even higher than the levels of the control group. Patterns of change in levels of creatinine, free radicals and antioxidants were similar to those described in Figure 2. In the histopathological analysis, treatment with cisplatin alone showed intense tubular necrosis and cast formation in the lumen, whereas treatment with cisplatin O_3/O_2 preconditioning at 30 µg/mL showed no significant differences with non-treated rats.

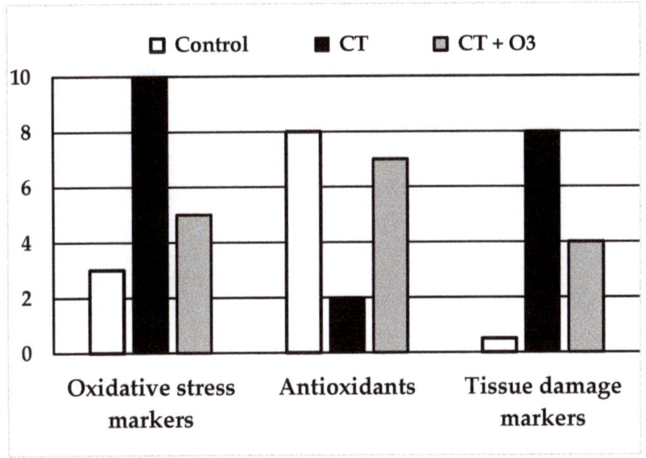

Figure 2. Schemes of results obtained in the experimental studies using systemic ozone therapy (rectal or intraperitoneal) using chemotherapy drugs. *(Left and Middle)*: "Oxidative stress markers" (MDA: malondialdehyde, TBARS: thiobarbituric acid-reactive substances) and "Tissue damage markers" (creatinine, pro-BNP: pro-brain natriuretic peptide) increased largely and significantly with chemotherapy. The increase was significantly lower in rats with chemotherapy + ozone therapy. *(Middle)*: levels of "Antioxidants" (GSH: glutathione, SOD: superoxide dismutase, CAT: catalase and GSH-GPx: glutathione peroxidase) decreased in chemotherapy group whereas those contents were closer to the control group in rats treated with chemotherapy + ozone therapy. All differences were statistically significant.

Also in 2004, this group studied the effect of ozone administration after cisplatin-induced acute nephrotoxicity [45]. In this study, cisplatin was administered before the ozone treatment. After that, O_3/O_2 was administered at different concentrations (10, 30 and 50 µg/mL) by rectal insufflation: one session/day for five consecutive days. Rats were sacrificed one day later. Cisplatin alone or cisplatin + oxygen showed similar levels of all parameters, that is: the addition of oxygen had no effect. In comparison with the control group (without cisplatin), the cisplatin group showed significant increases in creatinine (marker of renal damage) and TBARS. All cisplatin + ozone groups showed levels of parameters closer to those of the control group, and a statistically significant difference with cisplatin alone: lower increase in creatinine and TBARS, and lower decrease (or even increase) of antioxidants. Additionally, treatment with cisplatin alone showed severe and widespread tubular necrosis with dilation of proximal tubules and cast formation in the lumen, whereas treatment with cisplatin and further O_3/O_2 also showed tubular necrosis, but to a lesser extent. Therefore, in the previous work, this group described a preventive effect against cisplatin-induced damage in the kidneys by ozone preconditioning [44], whereas the current study showed partial recovery of already established damage by ozone treatment after cisplatin administration [45]. Patterns of change in levels of creatinine, free radicals and antioxidants were similar to those described in Figure 2.

Later, in 2006 [46], the same group, evaluated the renal expression pattern of Bax in rats treated with cisplatin without/with O_3/O_2 administration only at the optimal O_3/O_2 concentration of 30 µg/mL, following the two previous approaches: (1) with the prevention approach of the 1st study, with ozone preconditioning administered before the injection of cisplatin (by rectal insufflation, one session/day during 15 days), and (2) with the treatment approach of the 2nd study, after cisplatin injection, (by O_3/O_2 rectal insufflations, one session/day during 5 days). Bax protein expression plays a relevant role in the induction of apoptosis. As described years before [47], cisplatin-induced toxicity was also associated with increased expression of Bax protein, in cytoplasm and nucleus in this work [46]. Overall, in the immunohistochemical analysis, rats receiving cisplatin injection and O_3/O_2 insufflations at 30 µg/mL showed lower expression of Bax, both in the preventive and treatment approaches, although the latter (with only five O_3/O_2 sessions) showed a smaller decrease in Bax expression, which was more relevant in the cortex zone. As in previous studies, compared with the control group, the increase in creatinine levels was significantly lower in rats treated with cisplatin and ozone, and an even better effect was demonstrated in the preventive group (15 days of O_3/O_2 insufflations) compared to the treatment group (with only five O_3/O_2 sessions).

It has been described that high levels of ROS can decrease the expression of Bcl-2 and increase the expression of Bax, with a final reduction of the ratio Bcl-2/Bax with a proapoptotic effect, as occurs with cisplatin-induced damage, whereas low doses of ROS can activate cell survival signaling pathways such as Nrf2 and its downstream HO-1, which can potentially decrease cytotoxicity [48]. In this way, HO-1 expression has been effectively described as a modulator of cisplatin-induced renal toxicity and its increase as a potential approach for decreasing kidney injury [49,50]. As described in these works, rectal O_3/O_2 insufflation at appropriated concentrations enhances the antioxidant mechanisms in renal tissue, which can explain its effect to prevent o diminish cisplatin-induced renal damage. Further support was provided years later, when it was described that appropriated O_3/O_2 concentration (this is, a moderate ROS stimulus) induces Nrf2 as the mechanism for increasing HO-1 [27] and antioxidant mechanisms leading to decrease in oxidative stress and pro-inflammatory cytokines [7,37,38,51].

Finally, in 2016, Kocak et al. [52] published a different experimental work, evaluating the effect of O_3/O_2 in the management of already established cisplatin-induced ototoxicity. Rats were treated with intratympanic and rectal ozone one session/days for 7 days. All rats received intraperitoneal cisplatin (for 3 days) to produce ototoxicity. After 1 week, ototoxicity was confirmed by testing of distortion-product otoacoustic emissions. Then, the rats were randomized to the following: (1) no treatment (control group), (2) ozone by rectal insufflation or (3) "ozone by rectal insufflation + intratympanic ozone administration". Rectal and intratympanic insufflation were 2.3–3 mL of O_3/O_2 gas at concentration of 60 µg/mL. Ozone treatment was 1/day for 7 days. Rats were sacrificed after the 7th day. Compared with the control group, rats from both ozone groups showed statistical significance ($p < 0.05$): (1) better results in testing of distortion-product otoacoustic emissions (this is: partial recovery of audition), and (2) lower-outer hair cell damage in the histopathological examination score analysis of the inner ears. There were no differences observed between ozone groups. Therefore, it was concluded that rectal insufflation of ozone was effective in the treatment of cell damage in cisplatin-induced ototoxicity, and that the intratympanic administration of ozone had no additional advantage over the rectal administration. This study did not evaluate oxidative stress parameters.

Overall, the experimental models described above show that treatment with cisplatin was associated with a decrease in antioxidants, increase in free radicals and functional (creatinine) and histopathological damage in the kidneys and ears. However, the addition of ozone to the treatment was able to decrease all these alterations. These findings suggest a potential clinical benefit in the treatment and prevention of cisplatin-induced ototoxicity and nephrotoxicity, which are dose-limiting.

4.2. Ozone in Methotrexate-Induced Toxicity

In 2009, Kesik et al. [53] described the effect of ozone preconditioning to prevent abdominal injury by MTX, with assessment in liver, kidney and intestinal tissues. Ozone administration (total dose of 0.72 mg/kg) was by intraperitoneal route: one session/day for 15 consecutive days before the day of intraperitoneal MTX injection. Rats were sacrificed 5 days after MTX injection. They were evaluated in three groups: sham, MTX and MTX + ozone. Differences in free radicals and antioxidants among study groups were statistically significant and similar in all tissues: 1) Compared with sham, the MTX group showed an increase in MDA and decrease in SOD and GSH-Px compared with MTX alone; MTX + ozone showed decreased MDA and increased SOD and GSH-Px. Patterns of change were similar to those described in Figure 2. However, in this study, the histopathological scores for assessment of tissue damage were only statistically significant in ileum, which showed a lower damage score in the MTX + ozone group vs. MTX alone, that is: at histopathological level, the addition of ozone ameliorated intestinal damage at 5 days after MTX administration [53].

In 2015, Aslaner et al. published two articles with a similar methodology to evaluate the effect of "ozone preconditioning + ozone treatment" in MTX-induced nephrotoxicity [54] and hepatotoxicity [55]. The length of studies was 21 days. All groups received 5 mL of intraperitoneal administration of physiological saline (control and MTX groups) or O_3/O_2 (MTX + ozone groups). MTX and MTX + ozone groups received a single intraperitoneal administration of MTX at the 16th day. Additionally, the MTX + ozone groups received O_3/O_2 (at 25 µg/mL) intraperitoneally, one session/day for 15 consecutive days before the MTX injection and five additional days after the MTX injection. Rats were sacrificed at the 21st day of the study. Compared with control groups, the MTX groups showed a significant increase in serum levels of ALT, ST, TNF-α and IL-1β and tissue levels of MDA and myeloperoxidase (MPO), as well as a significant decrease in tissue levels of GSH. However, compared with MTX alone, the MTX + ozone groups showed significantly lower serum levels of ALT, ST, TNF-α and IL-1β and tissue levels of MDA and MPO, as well as significantly higher tissue levels of GSH [54,55]. Compared with the MTX groups, the MTX + ozone groups showed a lower histopathological damage score, with statistically significant differences in kidney tissue. Patterns of change in MDA and GSH levels were similar to those described in Figure 2.

In 2016, Leon Fernandez et al. [56], described the results of a randomized controlled trial (RCT) using MTX without/with concurrent ozone therapy in patients with rheumatoid arthritis. Sixty patients were randomized into two groups to: (1) standard treatment (MTX group), with MTX (12.5 mg intramuscular) 1/week + Ibuprofen + folic acid; or (2) standard treatment + ozone (MTX + ozone group), with 20 rectal insufflations, 1/day, 5 days/week for 4 weeks. The O_3/O_2 concentration and volume were progressively increased, in order to enhance the adaptive response: from 25 µg/mL for 100 mL the 1st week to 40 µg/mL for 200 mL the 4th week. Patients in the MTX group only received standard treatment. Patients in the MTZ + ozone group received the same standard treatment + ozone by 20 rectal insufflation, 1/day, 5 days/week for 4 weeks. The O_3/O_2 concentration and volume were progressively increased, in order to enhance the adaptive response: from 25 µg/mL for 100 mL the 1st week to 40 µg/mL for 200 mL the 4th week. Clinical parameters and biochemical markers of oxidative stress were evaluated before and after the treatment. The MTX group showed no differences in disease activity score or health assessment questionnaire-disability index, whereas the MTX + ozone group showed a significant and clinically relevant improvement in both parameters, as well as a more remarkable decrease in pain intensity, according to the visual analog scale (VAS). Compared with patients treated in the MTX group, at the end of the study, patients treated with concurrent ozone therapy showed significantly higher levels of antioxidants (SOD, CAT, GSH) and lower levels of oxidative stress markers such as advanced oxidation protein products (AOPP), nitric oxide (NO), total hydroperoxides (TH) and malondialdehyde (Figure 3).

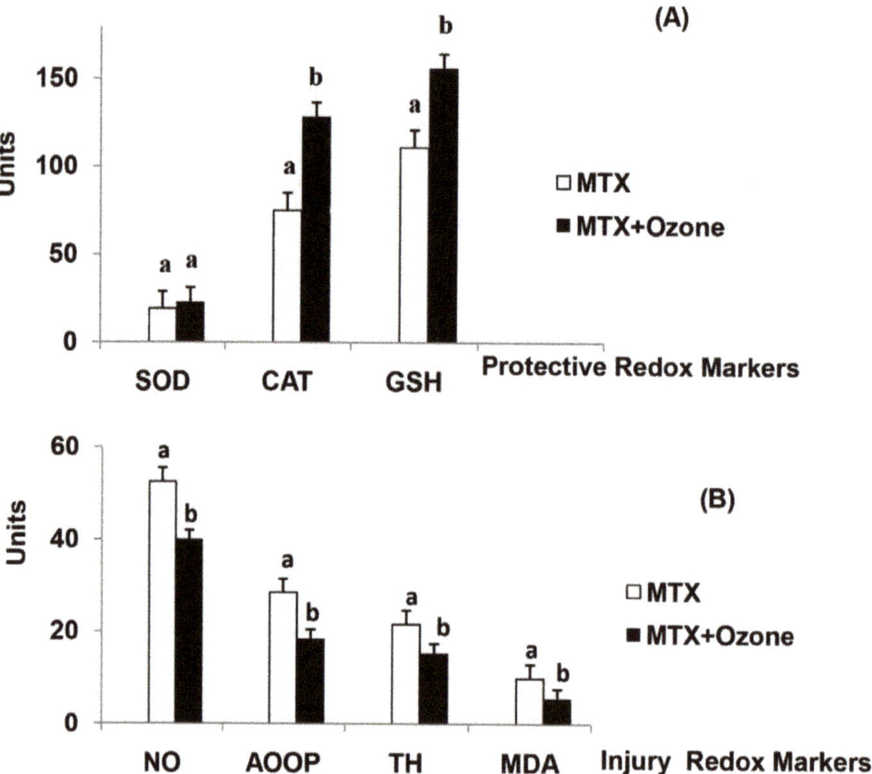

Figure 3. The redox status of patients with rheumatoid arthritis in (a): Methotrexate (MTX) and (b): "MTX + ozone" groups at the end of the study. (**A**) Protective redox markers, (**B**) Injury redox markers. The units of each marker are: SOD (superoxide dismutase, U/mL/min) and CAT (catalase, U/L/min) activities, GSH (reduced glutathione, μM), NO (nitric oxide, μM), AOPP (advanced oxidation protein products, μM), TH (total hydroperoxides, μM), MDA (malondialdehyde, μM). Data represent the mean $\pm$ S.E.M. of each group. Data analysis for each group was made by t-test. All differences between MTX vs. MTX + ozone groups were statistically significant, $p < 0.05$. From Ref. [56], with permission.

Overall, the works described above show that treatment with MTX was associated with a decrease in antioxidants, increase in free radicals and histopathological damage in kidney liver, and intestinal tissues. However, the addition of ozone to the treatment was able to decrease these alterations. These findings augur well for a potential clinical benefit of ozone in the treatment and prevention of MTX-induced toxicity in these issues, and they are further supported by the results in the only clinical trial published to date [57].

4.3. Ozone in Doxorubicin-Induced Toxicity

In 2004, Calunga et al. [58] described an experimental model of glomerulonephritis with a single DOX administration. After 10 weeks, rats were treated with O_3/O_2 rectal insufflation: one session/day for 15 days, at different concentrations. In this study, lower O_3/O_2 concentrations (15 μg/mL) showed better results than moderated concentrations (20 and 30 μg/mL) against the alterations induced by DOX on systolic arterial pressure, diuresis and proteinuria. However, this study did not evaluate the effect on oxidative stress or antioxidants.

In 2014, Delgado-Roche et al. [57] described that ozone preconditioning could prevent DOX-induced cardiotoxicity. Rats were assigned to four groups: (1) control (without DOX), (2) DOX alone, (3) DOX + oxygen, and (4) DOX + ozone. Intraperitoneal DOX was administered twice a week for 50 days. The O_3/O_2 administration was by rectal insufflation, at a volume of 6 mL, and concentrations of 50 μg/mL in the DOX + ozone group and 0 μg/mL (only oxygen) in the DOX + oxygen group. In both O_3/O_2 groups, 20 sessions, 1/day, were administered before the commencement of DOX injection. Rats were sacrificed after 50 days. There were no significant differences between the DOX group and DOX + oxygen group. Compared with the control group, both showed: (1) a decrease in antioxidants (CAT and SOD) and (2) an increase in free radicals (MDA, AOPP) and pro-brain natriuretic peptide (pro-BNP) as a marker of cardiac damage. However, the DOX + ozone group showed levels of pro-BNP, free radicals and antioxidants that were significantly closer to those of the control group. Patterns of change in levels of pro-BNP, free radicals and antioxidants were similar to those described in Figure 2. Additionally, histopathological analysis of the DOX group showed significant damage in heart tissue (subendocardial loss of muscular fibres, mild edema, and necrosis), whereas the DOX + ozone group only showed minor damage [57].

In 2016, Kesik et al. [59], described the effect of topical ozone application (ozonated olive oil) in the management of DOX-induced skin necrosis. This study assessed several topical treatments in an experimental model of skin necrosis induced by intradermal injection of Doxorubicin. The most relevant groups in this study were: (1) control group (DOX without further treatment), (2) DOX + dimethyl sulfoxide (DMSO), and (3) DOX + ozonated olive oil. It was expected the maximum skin necrosis occurred on day 14 after injection, so this was when analysis was carried out. Biopsies from the necrotic areas at 14 days did not show significant differences in tissue levels of MDA, IL1β, SOD or GSH-Px. However, compared with the control group, TNFα was significantly lower in DMS and ozonated olive oil groups, with no statistically significant differences observed between the last two groups. The ozonated olive oil group was the only one that showed a statistically significant decrease in ulcer size and in percentage of change (decrease) in the histopathologic ulcer score. DMSO is an antioxidant agent usually used in the management of DOX-induced extravasation injury. In this study, the authors demonstrated that topical use of ozonated olive oil improved this damage at least as well as DMSO [59]. Figure 4 shows a related clinical experience in our institution during the management of a patient with skin necrosis secondary to Doxorubicin extravasation.

In 2017, Salem et al. [60] evaluated the cytoprotective effects of ozone (and rutin and their combination) on DOX-induced testicular toxicity. Intraperitoneal DOX was administered 3 times/week for 2 weeks since the commencement time-point. Since the same commencement time-point, all groups received rectal gas insufflation (5 mL): one session/day, 5 days/week for 3 weeks. Placebo and doxorubicin groups received insufflations with O_2 only. In the ozone group, the gas insufflation was at O_3/O_2 concentrations of 25 μg/mL the 1st week, and 50 μg/mL the 2nd and the 3rd weeks. The study was terminated 21 days after treatment began. When compared to placebo, the DOX group showed a significant decrease in sperm count, motility and viability, and a significant increase in abnormal morphology. All these alterations were significantly lower in the group with DOX + ozone. In serum, DOX showed a significant decrease in testosterone levels and significant increases in luteinising hormone (LH) and follicle-stimulating hormone (FSH), whereas in the DOX + ozone group, these alterations were significantly lower. In testicular tissue, the DOX group showed a significant and relevant increase in g-glutamyltransferase (GGT), alkaline phosphatase (ALP), acid phosphatase, C-reactive protein (CRP), brain monocyte chemotactic protein-1 (MCP-1), malondialdehyde (MDA) and nitric oxide (NO), whereas all these values in the DOX + ozone group were significantly lower and closer to those of the placebo group. On the other hand, total antioxidant capacity (TAC) was significantly decreased in the DOX group, whereas in the DOX + ozone, there was a lower decrease and levels were closer to the placebo group. Patterns of change in MDA and antioxidant capacity were similar to those described in Figure 2.

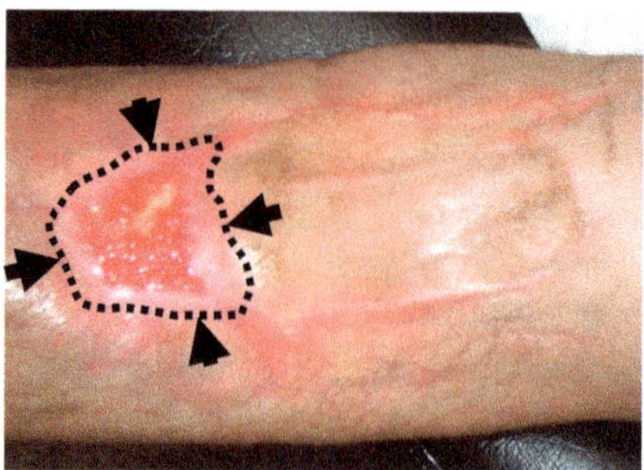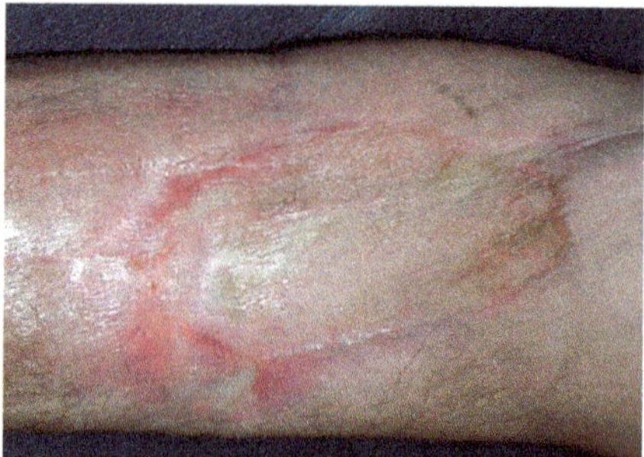

Figure 4. Topical ozone treatment in a patient with skin necrosis after Doxorubicin extravasation. A 61-year old patient under treatment for a stage IIIA multiple myeloma suffered a skin necrosis secondary to Doxorubicin (DOX) extravasation in the left elbow flexure. Because adverse evolution with conservative management, a muscle flap with a cutaneous graft was required (by the Department of Plastic Surgery). A second surgery was planned because of a loss of tissue in the distal area of the graft. (**Left**): Picture at the 9th session of local ozone therapy (wound size 25 × 15 mm). Black arrows and dotted lines show the limits of the wound at the commencement of ozone therapy (wound size 60 × 30 mm). (**Right**): Picture at the end of local ozone therapy, after 20 sessions. The planned second graft was avoided.

Finally, the recent work of Kamble et al. in 2018 merits mentioning. Using a different therapy (asiatic acid instead of ozone), they described that the activation of Nrf2 (as it is also induced by O_3/O_2 [7,27,37,38,51]) and the further enhancing of antioxidant systems can ameliorate DOX-induced toxicity in the heart, liver and kidneys [61].

Overall, the experimental models described above show that treatment with DOX was associated with a decrease in antioxidants, increase in free radicals and in functional and histopathological damage in the kidneys, heart, skin, and testicles. However, the addition of ozone to the treatment was able to ameliorate these alterations. These findings suggest a potential clinical benefit in the treatment and prevention of DOX-induced toxicity, and they are particularly relevant in DOX-induced cardiac-toxicity, which is dose-limiting.

4.4. Ozone in Bleomycin-Induced Toxicity

In 2015, Santana-Rodríguez et al. [62], showed preliminary results from an experimental model of Bleomycin-induced lung fibrosis. Twenty one Sprague-Dawley rats were randomized into four groups: (1) control, without intervention; (2) sham, with intratracheal administration of 500 µL saline; (3) BLM, with intratracheal administration of BLM; (4) BLM + ozone, treated as BLM group + O_3/O_2 rectal insufflation (20 mL/kg) before and after BLM administration. Administration of O_3/O_2 pre-BLM was 1/day for 15 days at increasing concentrations from 20 µg/mL to 50 µg/mL. After BLM administration, O_3/O_2 was administered at 50 µg/mL 3 times/week until sacrifice. Rats were sacrificed at 28 days after intratracheal administration of saline alone or with BLM. Lung fibrosis was assessed by the Ashcroft scale in a blinded histopathological analysis. Rats treated with BLM (with and without ozone) showed a significant and marked increase in lung fibrosis score. However, the fibrosis score was significantly lower in the BLM + ozone group in comparison with the BLM-alone group. Unfortunately, the levels of free radicals and antioxidants were not evaluated in this study [62].

4.5. Other Related Studies

There are few clinical works about the clinical effects of ozone in the management of CIT. They did not describe administered chemotherapy (or it was in a multidrug scheme) nor did they evaluate oxidative stress parameters. However, we consider that these are most relevant.

In 2008, a randomized study in children with chemo-induced mucositis showed that topical ozonated sunflower (Oleozon®, Centro Nacional de Investigaciones Científicas, La Habana, Cuba) leads to higher and faster mucositis recovery than conventional treatment with "Chlorhexidine + Nystatin" [63].

Borrelli, in 2012 [64], showed results from an RCT of 40 patients with advanced non-small lung cancer treated with (not specified) standard chemotherapy (control group) or standard chemotherapy and ozone (and viscum album injection). A concentration of O_3/O_2 of 30 µg/mL was administered by autohemotherapy once per week for 12 weeks. Compared with the control group, patients in the "chemotherapy and ozone group" showed a significant improvement in the Quality of Life Questionnaire QLQ-C30), lower ROS and higher biological antioxidant potential plasma values than baseline values.

Finally, a related topic to mention could be the chemotherapy-induced peripheral neuropathy (CIPN), which can happen in more than half of patients treated with platin compounds, taxanes, vincristine or bortezomib [65], and it can lead to dose-reduction or even interruption of chemotherapy. Once more, among the mechanisms associated to CIPN, the following have been described: (1) apoptosis induced by ROS and oxidative stress, (2) decrease in antioxidants as vitamin, E.; and (3) increase of proinflammatory cytokines (IL-1, IL-6, IL-8, TNFa) [65–67]. Different treatments have been evaluated, including several approaches with antioxidants: acetylcysteine, amifostine, glutathione, retinoic acid, or vitamin E. However, until now, preventive or therapeutic approaches are limited in number and efficacy [68,69]. In these clinical conditions, when non-proved or limited therapeutic options exist, some experts consider it reasonable to use treatment based on its mechanisms of action or its effects in related syndromes [69]. In this way, based on its mechanism of action and our clinical experience with ozone in neuropathic pain secondary to cancer treatments (*Personal Communication* [70]), a double-blinded RCT with ozone therapy in refractory peripheral neuropathy induced by chemotherapy is ongoing, which will include an extensive assessment of oxidative stress and proinflammatory parameters (EudraCT: 2019-000821-37).

5. Discussion and Prospects

Overall, in the experimental models described above, the administration of cisplatin, doxorubicin or methotrexate was associated with increased serum levels of tissue-damage markers (creatinine in renal injury, pro-BNP in cardiac injury) and increased tissue levels of free radicals (lipid peroxidation markers—TBARS, MDA). At the same time, these drugs decreased tissue levels of antioxidants (GSH, SOD, catalase, GSH-Px). When assessed, the addition of O_2 (O_3/O_2 = 0 µg/mL) to rats treated with these chemotherapy drugs did not show relevant changes in comparison with chemotherapy alone, that is: the addition of systemic O_2 did not induce a decrease of free radicals and did not increase antioxidant levels.

However, when the administration of cisplatin, doxorubicin or methotrexate in rats was associated with O_3 preconditioning or O_3 treatment at appropriate concentrations, the oxidative stress parameters were closer to the those from the control group, that is: (1) lower increase in serum levels of tissue-damage markers (creatinine in renal injury, pro-BNP in cardiac injury), (2) lower increase in tissue levels of free radicals (lipid peroxidation markers—TBARS, MDA), and lower decrease in the tissue levels of antioxidants (GSH, SOD, catalase, GSH-Px), 3) decreased damage in histopathologic analysis. Considering these effects, we know that oxidative preconditioning can induce an effect also described for other phenomena such as exercise or ischemic, thermal and chemical preconditioning. A common feature of all of these processes is that a repeated and "moderate-controlled" stress is able to protect against a prolonged and severe stress [44].

The results described with ozone in the experimental models detailed in this review augur well for a potential clinical benefit, and they are further supported by increased survival in the experimental model of doxorubicin plus ozone [57] or the results in the clinical trial of patients with arthritis treated with methotrexate with/without ozone therapy [56].

As with chemotherapy, chronic radiation-induced toxicity is also mediated by a local perpetuation of the ischemic process, proinflammatory and prooxidative status. Some experimental models have described the potential role of ozone therapy to protect/diminish toxicity at lung [71], liver or intestinal [72] levels. Unfortunately, once more, there are few related clinical studies, with the most remarkable being those regarding the use of ozone therapy during radiotherapy of prostate cancer to decrease local toxicity [73] or after radiotherapy to treat pelvic radiation-induced toxicity [70,74,75].

It is necessary to highlight that, as is usual in medicine, higher concentrations are not always better, as demonstrated by Borrego et al. [44], showing that cisplatin plus O_3/O_2 preconditioning at higher concentrations (50 and 70 µg/mL) showed histopathological changes that were quite similar to those present with cisplatin alone and disappointing results in biochemical parameters. Furthermore, the results were worse than those obtained with moderate concentrations (20 or 30 µg/mL) in rectal insufflations. That is to say, very high O_3/O_2 concentrations can induce free radical levels that are too high and exceed adaptive capacity, leading to worse results or even deleterious effects.

In the same way, the effects of 4-HNE, Nrf2 and NF-κB induced by ozone depend on its concentration and cell type or tissue. These three pathways interact in the redox processes and can show dual actions. If they lead to increased oxidative stress, they can induce initiation, promotion or progression of tumor cells, as well as treatment-induced toxicity; although an increase of oxidative stress is the foundation of chemotherapy and radiotherapy. On the other hand, an increase in antioxidants could be related with a lower risk of tumor initiation and treatment-induced toxicity; although, it could potentially protect cancer cells from cancer treatments. Therefore, their effects in cancer pathology and their potential modulation in cancer treatment are complex and not completely known [2,76,77].

At physiological (very low) doses, 4-HNE stimulates activity of the Nrf2 pathway as well as proliferation, differentiation, and apoptosis. However, low concentrations could protect cancer cells against further damage [2,77]. In opposition, studies have been described where 4-HNE correlates with tumor malignancy in astrocytomas and breast or liver carcinomas [2]; although, high levels of 4-HNE under oxidative stress conditions have been described to predispose cancer cells to apoptosis and enhance results of radio-chemo therapy in lung carcinomas [77,78]. The role of NF-κB and Nrf2 and their modulation in cancer pathology is also not clear. Coincident with the molecular cloning of NF-κB/RelA and identification of its kinship to the v-Rel oncogene, it was anticipated that NF-κB itself would be involved in cancer development. Oncogenic activating mutations in NF-κB genes are rare and have been identified only in some lymphoid malignancies, while most NF-κB activating mutations in lymphoid malignancies occur in upstream signaling components that feed into NF-κB. NF-κB activation is also prevalent in carcinomas, in which NF-κB activation is mainly driven by inflammatory cytokines within the tumor microenvironment. Importantly, however, in all malignancies, NF-κB acts in a cell-type-specific manner: activating within cancer cells genes involved in survival, proliferation, angiogenesis, expansion, and metastasis, as well as the enhancement of inflammation-promoting genes in the tumor microenvironment. Yet, the complex biological functions of NF-κB have made targeting it therapeutically a challenge [79,80]. Moreover, Nrf2 has also shown a dual action that can enhance resistance to cancer treatment as well as inhibit cancer initiation and development [76]. Nrf2 increase has been associated with malignant transformation and progression in colorectal carcinoma [81], limited the success of temozolomide and is implied to play a role in the drug resistance mechanism [82] in gastric cancer. Nrf2 expression is positively correlated with invasive gastric cancer, suggesting its utility as a predictive index for unfavorable prognosis [83]. However, controlled, oscillating activation of Nrf2 has also been related to the prevention of cancer initiation and development [76,84]. In conclusion, it seems that modification of the balance of Nrf2 or NF-κB is involved in regulation of cancer initiation/progression and the drug resistance mechanism.

As consequences, approaches that reestablish the equilibrium Nrf2/NF-κB should provide a potential benefit in oncology.

As commented for 4-HNE, NF-κB, and Nrf2, the use of "high-dose antioxidants" during chemotherapy to prevent toxicity is also controversial, because of the potential protective effect on tumor cells and prognostic impairment [85,86]. However, ozone does not lead to a high increase of one isolated substance or antioxidant. At appropriated concentrations, ozone will induce an adaptive response with an "overall potentiation" of the "endogenous antioxidant mechanisms", which are usually decreased in most tumor cells.

There is a rational support for a potential enhancing effect of the "chemotherapy + ozone" combination as we have described in a recent review, which merits further research [87]. However, this potential and controversial combination during cancer treatment should not be hugely relevant for patients in the following situations:

(1) Current or potential CIT leads to contraindication or dose-reduction chemotherapy. In some clinical conditions, the addition of ozone during chemotherapy in order to prevent/diminish CIT could open two additional interesting treatment-windows to explore: (a) to avoid/diminish chemotherapy dose-reduction when some kind of CIT is present, and (b) the potential possibility for exploring chemotherapy administration in current clinical contraindications (e.g., renal failure).

(2) Tumor cells are not present, e.g., in the treatment of CIT after cancer treatment. Based on the demonstrated modulation of oxidative stress by ozone, the complementary use of ozone as palliative or compassionate treatment for CIT could be supported when an effective or demonstrated treatment does not exist or does not work, as suggested by experts [69]. In this way, an RCT in refractory peripheral neuropathy induced by chemotherapy is ongoing (EudraCT: 2019-000821-37), with planned analysis of inflammatory and oxidative stress markers.

(3) Chemotherapy is used in the management of no-cancer disease. This is supported by the RCT in rheumatoid arthritis, where the addition of ozone therapy to MTX treatment improved the biochemical and clinical results [56].

There is no doubt that all the above-mentioned issues merit further research and RCT.

6. Conclusions

The relationship between free radicals and ROS vs. antioxidants is a complex balance that depends on their concentrations and cell/tissue type of action, and with a Janus effect—both sides of the balance can lead to beneficial or harmful effects. Increased oxidative stress is associated with cancer and CIT, although a further increase of oxidative stress in cancer cells is key in chemotherapy and radiotherapy actions. On the other hand, high antioxidant levels could be useful in the management of CIT, although we must be careful with the potential protective effect on cancer cells. Ozone therapy, by an initial "soft and controlled" oxidative stress induces an adaptive response of the tissues with a final increase of the "overall-endogenous antioxidant systems", which have been associated with protective and therapeutic effects in CIT in several experimental models and an RCT. The potential benefit of ozone in these clinical conditions merits further research.

Author Contributions: Conception and design of the review, B.C., F.R.-E., D.R.-A., G.M.-S., P.L., D.A.-B., L.F.-P. and N.S.-R.; writing—original draft preparation about ozone concepts and review of the literature, B.C., F.R.-E., G.M.-S., P.L. and N.S.-R.; writing—original draft preparation about molecular concepts, F.R.-E., G.M.-S. and L.F.-P.; writing—original draft preparation about chemotherapy-induced toxicity, B.C., G.M.-S., D.R.-A. and D.A.-B.; writing—review, editing and approval of the final version, B.C., F.R.-E., D.R.-A., G.M.-S., P.L., D.A.-B., L.F.-P. and N.S.-R.

Funding: The publication of this paper and the subsequent study (EudraCT number: 2019-000821-37) is supported by a grant (PI 19/00458) from the Instituto de Salud Carlos III (Spanish Ministry of Health and Social Affairs, Madrid, Spain), a grant (016/2019) from the Fundación DISA (Las Palmas, Spain) and by a grant (BF1-19-13) from the Fundación Española del Dolor (Spanish Pain Foundation, Madrid, Spain).

Acknowledgments: Some of the ozone therapy device (Ozonosan Alpha-plus®) used at the Negrín University Hospital were provided by Renate Viebahn (Hänsler GmbH, Iffezheim, Germany) and another one was bought by a grant (PI EC 07/90024) from the Instituto de Salud Carlos III (Spanish Ministry of Health and Social Affairs, Madrid, Spain). Some data from this review were presented as oral communication at the 2nd International Traditional and Complementary Medicine Congress (Istanbul, Turkey, April 2019) organized with technical sponsorship of the World Health Organization.

Conflicts of Interest: The authors declare no conflict of interest.

Abbreviations

4-HNE: 4-hydroxynonenal; AOPP: advanced oxidation protein products; BLM: Bleomycin; CAT: catalase; CDDP: Cisplatin; CIT: chemotherapy -induced toxicity; CSGF: colony-stimulating growth factors; CT: chemotherapy (CT); DOX: Doxorubicin; G6PDH: glucose-6-phosphate dehydrogenase; GSH: reduced glutathione; GSH-Px: glutathione peroxidase; H_2O_2: hydrogen peroxide; HO-1: heme oxygenase-1; IL: interleukin; MDA: malondialdehyde; MPO: myeloperoxidase; MTX: Methotrexate; MPO: myeloperoxidase; NO: nitric oxide; NF-κB: nuclear factor kappa B; Nrf2: nuclear factor erythroid 2-related factor; O_2: oxygen; O_2.-: superoxide anion; O_3: ozone; OH•: hydroxyl radical; Prxs: peroxiredoxins; RCT: randomized controlled trial; ROS: reactive oxygen species; SOD superoxide dismutase; TAC: total antioxidant capacity; TBARS: thiobarbituric acid-reactive substances; TH: total hydroperoxides; TNF-α: tumor necrosis factor-alpha

References

1. Zuo, L.; Prather, E.R.; Stetskiv, M.; Garrison, D.E.; Meade, J.R.; Peace, T.I.; Zhou, T. Inflammaging and Oxidative Stress in Human Diseases: From Molecular Mechanisms to Novel Treatments. *Int. J. Mol. Sci.* **2019**, *20*, 4472. [CrossRef] [PubMed]

2. Milkovic, L.; Cipak Gasparovic, A.; Zarkovic, N. Overview on major lipid peroxidation bioactive factor 4-hydroxynonenal as pluripotent growth-regulating factor. *Free Radic. Res.* **2015**, *49*, 850–860. [CrossRef]

3. Bocci, V. *Ozone A New Medical Drug*, 2nd ed.; Springer Publishing: Dordrecht, The Netherlands, 2011.

4. Schwartz-Tapia, A.; Martínez-Sánchez, G.; Sabah, F. *Madrid Declaration on Ozone Therapy*; ISCO3 (International Scientific Committee of Ozone Therapy): Madrid, Spain, 2015.

5. Andrady, A.; Aucamp, P.J.; Bais, A.; Ballare, C.L.; Bjorn, L.O.; Bornman, J.F.; Caldwell, M.; Cullen, A.P.; Erickson, D.J.; de Gruijl, F.R.; et al. Environmental effects of ozone depletion and its interactions with climate change: Progress report, 2008. *Photochem. Photobiol. Sci.* **2009**, *8*, 13–22.

6. Bocci, V. Is it true that ozone is always toxic? The end of a dogma. *Toxicol. Appl. Pharmacol.* **2006**, *216*, 493–504. [CrossRef] [PubMed]

7. Galie, M.; Covi, V.; Tabaracci, G.; Malatesta, M. The Role of Nrf2 in the Antioxidant Cellular Response to Medical Ozone Exposure. *Int. J. Mol. Sci.* **2019**, *20*, 4009. [CrossRef] [PubMed]

8. Hakiminia, B.; Goudarzi, A.; Moghaddas, A. Has vitamin E any shreds of evidence in cisplatin-induced toxicity. *J. Biochem. Mol. Toxicol.* **2019**, *33*, e22349. [CrossRef]

9. Pabla, N.; Dong, Z. Cisplatin nephrotoxicity: Mechanisms and renoprotective strategies. *Kidney Int.* **2008**, *73*, 994–1007. [CrossRef] [PubMed]

10. Dugbartey, G.J.; Peppone, L.J.; de Graaf, I.A. An integrative view of cisplatin-induced renal and cardiac toxicities: Molecular mechanisms, current treatment challenges and potential protective measures. *Toxicology* **2016**, *371*, 58–66. [CrossRef]

11. Howard, S.C.; McCormick, J.; Pui, C.H.; Buddington, R.K.; Harvey, R.D. Preventing and Managing Toxicities of High-Dose Methotrexate. *Oncologist* **2016**, *21*, 1471–1482. [CrossRef]

12. Wessels, J.A.; Huizinga, T.W.; Guchelaar, H.J. Recent insights in the pharmacological actions of methotrexate in the treatment of rheumatoid arthritis. *Rheumatology (Oxford)* **2008**, *47*, 249–255. [CrossRef]

13. Miyazono, Y.; Gao, F.; Horie, T. Oxidative stress contributes to methotrexate-induced small intestinal toxicity in rats. *Scand. J. Gastroenterol.* **2004**, *39*, 1119–1127. [CrossRef] [PubMed]

14. Mahmoud, A.M.; Hussein, O.E.; Hozayen, W.G.; Abd El-Twab, S.M. Methotrexate hepatotoxicity is associated with oxidative stress, and down-regulation of PPARgamma and Nrf2: Protective effect of 18beta-Glycyrrhetinic acid. *Chem. Biol. Interact.* **2017**, *270*, 59–72. [CrossRef]

15. Pugazhendhi, A.; Edison, T.; Velmurugan, B.K.; Jacob, J.A.; Karuppusamy, I. Toxicity of Doxorubicin (Dox) to different experimental organ systems. *Life Sci.* **2018**, *200*, 26–30. [CrossRef]

16. Luo, X.; Evrovsky, Y.; Cole, D.; Trines, J.; Benson, L.N.; Lehotay, D.C. Doxorubicin-induced acute changes in cytotoxic aldehydes, antioxidant status and cardiac function in the rat. *Biochim. Biophys. Acta* **1997**, *1360*, 45–52. [PubMed]

17. Sikic, B.I. Biochemical and cellular determinants of bleomycin cytotoxicity. *Cancer Surv.* **1986**, *5*, 81–91.

18. Yu, Z.; Yan, B.; Gao, L.; Dong, C.; Zhong, J.; DOrtenzio, M.; Nguyen, B.; Seong Lee, S.; Hu, X.; Liang, F. Targeted Delivery of Bleomycin: A Comprehensive Anticancer Review. *Curr. Cancer Drug Targets* **2016**, *16*, 509–521. [CrossRef] [PubMed]

19. Shariati, S.; Kalantar, H.; Pashmforoosh, M.; Mansouri, E.; Khodayar, M.J. Epicatechin protective effects on bleomycin-induced pulmonary oxidative stress and fibrosis in mice. *Biomed. Pharmacother.* **2019**, *114*, 108776. [CrossRef]

20. Chen, W.; Li, S.; Li, J.; Zhou, W.; Wu, S.; Xu, S.; Cui, K.; Zhang, D.D.; Liu, B. Artemisitene activates the Nrf2-dependent antioxidant response and protects against bleomycin-induced lung injury. *FASEB J.* **2016**, *30*, 2500–2510. [CrossRef]

21. Bocci, V.; Borrelli, E.; Travagli, V.; Zanardi, I. The ozone paradox: Ozone is a strong oxidant as well as a medical drug. *Med. Res. Rev.* **2009**, *29*, 646–682. [CrossRef]

22. Bocci, V.A.; Zanardi, I.; Travagli, V. Ozone acting on human blood yields a hormetic dose-response relationship. *J. Transl. Med.* **2011**, *9*, 66. [CrossRef]

23. Viebahn-Hansler, R.; Leon Fernandez, O.S.; Fahmy, Z. Ozone in Medicine: The Low-Dose Ozone Concept—Guidelines and Treatment Strategies. *Ozone-Sci. Eng.* **2012**, *34*, 408–424. [CrossRef]

24. Re, L.; Malcangi, G.; Martínez-Sánchez, G. Medical ozone is now ready for a scientific challenge: Current status and future perspectives. *J. Exp. Integr. Med.* **2012**, *2*, 193–196. [CrossRef]

25. Hamilton, R.F., Jr.; Li, L.; Eschenbacher, W.L.; Szweda, L.; Holian, A. Potential involvement of 4-hydroxynonenal in the response of human lung cells to ozone. *Am. J. Physiol.* **1998**, *274 (Pt 1)*, L8–L16. [CrossRef]

26. Pryor, W.A.; Squadrito, G.L.; Friedman, M. The cascade mechanism to explain ozone toxicity: The role of lipid ozonation products. *Free Radic. Biol. Med.* **1995**, *19*, 935–941. [CrossRef]

27. Pecorelli, A.; Bocci, V.; Acquaviva, A.; Belmonte, G.; Gardi, C.; Virgili, F.; Ciccoli, L.; Valacchi, G. NRF2 activation is involved in ozonated human serum upregulation of HO-1 in endothelial cells. *Toxicol. Appl. Pharmacol.* **2013**, *267*, 30–40. [CrossRef]

28. Oliveira-Marques, V.; Marinho, H.S.; Cyrne, L.; Antunes, F. Role of hydrogen peroxide in NF-kappaB activation: From inducer to modulator. *Antioxid. Redox Signal.* **2009**, *11*, 2223–2243. [CrossRef]

29. Huth, K.C.; Saugel, B.; Jakob, F.M.; Cappello, C.; Quirling, M.; Paschos, E.; Ern, K.; Hickel, R.; Brand, K. Effect of aqueous ozone on the NF-kappaB system. *J. Dent. Res.* **2007**, *86*, 451–456. [CrossRef]

30. Kafoury, R.M.; Hernandez, J.M.; Lasky, J.A.; Toscano, W.A., Jr.; Friedman, M. Activation of transcription factor IL-6 (NF-IL-6) and nuclear factor-kappaB (NF-kappaB) by lipid ozonation products is crucial to interleukin-8 gene expression in human airway epithelial cells. *Environ. Toxicol.* **2007**, *22*, 159–168. [CrossRef]

31. Wang, L.; Chen, Z.; Liu, Y.; Du, Y.; Liu, X. Ozone oxidative postconditioning inhibits oxidative stress and apoptosis in renal ischemia and reperfusion injury through inhibition of MAPK signaling pathway. *Drug Des. Devel. Ther.* **2018**, *12*, 1293–1301. [CrossRef]

32. Galie, M.; Costanzo, M.; Nodari, A.; Boschi, F.; Calderan, L.; Mannucci, S.; Covi, V.; Tabaracci, G.; Malatesta, M. Mild ozonisation activates antioxidant cell response by the Keap1/Nrf2 dependent pathway. *Free Radic. Biol. Med.* **2018**, *124*, 114–121. [CrossRef]

33. Siniscalco, D.; Trotta, M.C.; Brigida, A.L.; Maisto, R.; Luongo, M.; Ferraraccio, F.; D'Amico, M.; Di Filippo, C. Intraperitoneal Administration of Oxygen/Ozone to Rats Reduces the Pancreatic Damage Induced by Streptozotocin. *Biology (Basel)* **2018**, *7*, 10. [CrossRef] [PubMed]

34. Nakasone, M.; Nakaso, K.; Horikoshi, Y.; Hanaki, T.; Kitagawa, Y.; Takahashi, T.; Inagaki, Y.; Matsura, T. Preconditioning by Low Dose LPS Prevents Subsequent LPS-Induced Severe Liver Injury via Nrf2 Activation in Mice. *Yonago Acta Med.* **2016**, *59*, 223–231. [PubMed]

35. Ayala, A.; Munoz, M.F.; Arguelles, S. Lipid peroxidation: Production, metabolism, and signaling mechanisms of malondialdehyde and 4-hydroxy-2-nonenal. *Oxid. Med. Cell Longev.* **2014**, *2014*, 360438. [CrossRef] [PubMed]

36. Breitzig, M.; Bhimineni, C.; Lockey, R.; Kolliputi, N. 4-Hydroxy-2-nonenal: A critical target in oxidative stress? *Am. J. Physiol. Cell Physiol.* **2016**, *311*, C537–C543. [CrossRef] [PubMed]

37. Re, L.; Martínez-Sánchez, G.; Bordicchia, M.; Malcangi, G.; Pocognoli, A.; Angel Morales-Segura, M.; Rothchild, J.; Rojas, A. Is ozone pre-conditioning effect linked to Nrf2/EpRE activation pathway in vivo? A preliminary result. *Eur. J. Pharmacol.* **2014**, *742*, 158–162. [CrossRef]
38. Delgado-Roche, L.; Riera-Romo, M.; Mesta, F.; Hernandez-Matos, Y.; Barrios, J.M.; Martinez-Sanchez, G.; Al-Dalaien, S.M. Medical ozone promotes Nrf2 phosphorylation reducing oxidative stress and pro-inflammatory cytokines in multiple sclerosis patients. *Eur. J. Pharmacol.* **2017**, *811*, 148–154. [CrossRef]
39. Ahmed, S.M.; Luo, L.; Namani, A.; Wang, X.J.; Tang, X. Nrf2 signaling pathway: Pivotal roles in inflammation. *Biochim. Biophys. Acta Mol. Basis Dis.* **2017**, *1863*, 585–597. [CrossRef]
40. Ganesh Yerra, V.; Negi, G.; Sharma, S.S.; Kumar, A. Potential therapeutic effects of the simultaneous targeting of the Nrf2 and NF-kappaB pathways in diabetic neuropathy. *Redox Biol.* **2013**, *1*, 394–397. [CrossRef]
41. Mohan, S.; Gupta, D. Crosstalk of toll-like receptors signaling and Nrf2 pathway for regulation of inflammation. *Biomed. Pharmacother.* **2018**, *108*, 1866–1878. [CrossRef]
42. Wardyn, J.D.; Ponsford, A.H.; Sanderson, C.M. Dissecting molecular cross-talk between Nrf2 and NF-kappaB response pathways. *Biochem. Soc. Trans.* **2015**, *43*, 621–626. [CrossRef]
43. Yu, G.; Liu, X.; Chen, Z.; Chen, H.; Wang, L.; Wang, Z.; Qiu, T.; Weng, X. Ozone therapy could attenuate tubulointerstitial injury in adenine-induced CKD rats by mediating Nrf2 and NF-kappaB. *Iran. J. Basic Med. Sci.* **2016**, *19*, 1136–1143. [PubMed]
44. Borrego, A.; Zamora, Z.B.; Gonzalez, R.; Romay, C.; Menendez, S.; Hernandez, F.; Montero, T.; Rojas, E. Protection by ozone preconditioning is mediated by the antioxidant system in cisplatin-induced nephrotoxicity in rats. *Mediators Inflamm.* **2004**, *13*, 13–19. [CrossRef] [PubMed]
45. Gonzalez, R.; Borrego, A.; Zamora, Z.; Romay, C.; Hernandez, F.; Menendez, S.; Montero, T.; Rojas, E. Reversion by ozone treatment of acute nephrotoxicity induced by cisplatin in rats. *Mediators Inflamm.* **2004**, *13*, 307–312. [CrossRef] [PubMed]
46. Borrego, A.; Zamora, Z.B.; Gonzalez, R.; Romay, C.; Menendez, S.; Hernandez, F.; Berlanga, J.; Montero, T. Ozone/oxygen mixture modifies the subcellular redistribution of Bax protein in renal tissue from rats treated with cisplatin. *Arch. Med. Res.* **2006**, *37*, 717–722. [CrossRef] [PubMed]
47. Sheikh-Hamad, D.; Cacini, W.; Buckley, A.R.; Isaac, J.; Truong, L.D.; Tsao, C.C.; Kishore, B.K. Cellular and molecular studies on cisplatin-induced apoptotic cell death in rat kidney. *Arch. Toxicol.* **2004**, *78*, 147–155. [CrossRef] [PubMed]
48. Redza-Dutordoir, M.; Averill-Bates, D.A. Activation of apoptosis signalling pathways by reactive oxygen species. *Biochim. Biophys. Acta* **2016**, *1863*, 2977–2992. [CrossRef]
49. Shiraishi, F.; Curtis, L.M.; Truong, L.; Poss, K.; Visner, G.A.; Madsen, K.; Nick, H.S.; Agarwal, A. Heme oxygenase-1 gene ablation or expression modulates cisplatin-induced renal tubular apoptosis. *Am. J. Physiol. Renal. Physiol.* **2000**, *278*, F726–F736. [CrossRef]
50. Bolisetty, S.; Traylor, A.; Joseph, R.; Zarjou, A.; Agarwal, A. Proximal tubule-targeted heme oxygenase-1 in cisplatin-induced acute kidney injury. *Am. J. Physiol. Renal. Physiol.* **2016**, *310*, F385–F394. [CrossRef]
51. Bocci, V.; Valacchi, G. Nrf2 activation as target to implement therapeutic treatments. *Front. Chem.* **2015**, *3*, 4. [CrossRef]
52. Kocak, H.E.; Taskin, U.; Aydin, S.; Oktay, M.F.; Altinay, S.; Celik, D.S.; Yucebas, K.; Altas, B. Effects of ozone (O3) therapy on cisplatin-induced ototoxicity in rats. *Eur. Arch. Otorhinolaryngol.* **2016**, *273*, 4153–4159. [CrossRef]
53. Kesik, V.; Uysal, B.; Kurt, B.; Kismet, E.; Koseoglu, V. Ozone ameliorates methotrexate-induced intestinal injury in rats. *Cancer Biol. Ther.* **2009**, *8*, 1623–1628. [CrossRef] [PubMed]
54. Aslaner, A.; Cakir, T.; Celik, B.; Dogan, U.; Mayir, B.; Akyuz, C.; Polat, C.; Basturk, A.; Soyer, V.; Koc, S.; et al. Does intraperitoneal medical ozone preconditioning and treatment ameliorate the methotrexate induced nephrotoxicity in rats? *Int. J. Clin. Exp. Med.* **2015**, *8*, 13811–13817. [PubMed]
55. Aslaner, A.; Cakir, T.; Celik, B.; Dogan, U.; Akyuz, C.; Basturk, A.; Polat, C.; Gunduz, U.; Mayir, B.; Sehirli, A.O. The protective effect of intraperitoneal medical ozone preconditioning and treatment on hepatotoxicity induced by methotrexate. *Int. J. Clin. Exp. Med.* **2015**, *8*, 13303–13309. [PubMed]
56. Leon Fernandez, O.S.; Viebahn-Haensler, R.; Cabreja, G.L.; Espinosa, I.S.; Matos, Y.H.; Roche, L.D.; Santos, B.T.; Oru, G.T.; Polo Vega, J.C. Medical ozone increases methotrexate clinical response and improves cellular redox balance in patients with rheumatoid arthritis. *Eur. J. Pharmacol.* **2016**, *789*, 313–318. [CrossRef] [PubMed]

57. Delgado-Roche, L.; Hernandez-Matos, Y.; Medina, E.A.; Morejon, D.A.; Gonzalez, M.R.; Martinez-Sanchez, G. Ozone-Oxidative Preconditioning Prevents Doxorubicin-induced Cardiotoxicity in Sprague-Dawley Rats. *Sultan Qaboos Univ. Med. J.* **2014**, *14*, e342–e348.

58. Calunga Fernández, J.L.; Bello Ferro, M.; Chaple La Hoz, M.; Barber Gutierrez, E.; Menendez Cepero, S.; Merino, N. Ozonoterapia en la glomerulonefritis tóxica experimental por adriamicina. *Rev. Cubana Investig. Biomed.* **2004**, *23*, 139–143.

59. Kesik, V.; Yuksel, R.; Yigit, N.; Saldir, M.; Karabacak, E.; Erdem, G.; Babacan, O.; Gulgun, M.; Korkmazer, N.; Bayrak, Z. Ozone Ameliorates Doxorubicine-Induced Skin Necrosis - results from an animal model. *Int. J. Low Extrem. Wounds* **2016**, *15*, 248–254. [CrossRef]

60. Salem, E.A.; Salem, N.A.; Hellstrom, W.J. Therapeutic effect of ozone and rutin on adriamycin-induced testicular toxicity in an experimental rat model. *Andrologia* **2017**, *49*, e12603. [CrossRef]

61. Kamble, S.M.; Patil, C.R. Asiatic Acid Ameliorates Doxorubicin-Induced Cardiac and Hepato-Renal Toxicities with Nrf2 Transcriptional Factor Activation in Rats. *Cardiovasc Toxicol.* **2018**, *18*, 131–141. [CrossRef]

62. Santana-Rodriguez, N.; Santisteban, P.L.; Clavo, B.; Camacho, R.; Santana, C.; Fiuza, M.D. Protective effect of ozone in an experimental model of pulmonary fibrosis. *Br. J. Surg.* **2015**, *102*, 11.

63. Menendez Cepero, S.; Gonzalez Alvarez, R.; Ledea Lozano, O.E.; Hernandez Rosales, F.; Leon Fernandez, O.S.; Diaz Gomez, M. *Ozono, Aspectos Básicos y Aplicaciones Clínicas*; Centro de Investigaciones del Ozono: La Habana, Cuba, 2008.

64. Borrelli, E. Treatment of advanced non-small-cell lung cancer with oxygen ozone therapy and mistletoe: An integrative approach. *Eur. J. Integr. Med.* **2012**, *4*, 130.

65. Velasco, R.; Bruna, J. Chemotherapy-induced peripheral neuropathy: An unresolved issue. *Neurologia* **2010**, *25*, 116–131. [CrossRef]

66. Velasco, R.; Santos, C.; Soler, G.; Gil-Gil, M.; Pernas, S.; Galan, M.; Palmero, R.; Bruna, J. Serum micronutrients and prealbumin during development and recovery of chemotherapy-induced peripheral neuropathy. *J. Peripher. Nerv. Syst.* **2016**, *21*, 134–141. [CrossRef] [PubMed]

67. Starobova, H.; Vetter, I. Pathophysiology of Chemotherapy-Induced Peripheral Neuropathy. *Front. Mol. Neurosci.* **2017**, *10*, 174. [CrossRef] [PubMed]

68. Albers, J.W.; Chaudhry, V.; Cavaletti, G.; Donehower, R.C. Interventions for preventing neuropathy caused by cisplatin and related compounds. *Cochrane Database Syst. Rev.* **2014**, *3*, CD005228. [CrossRef]

69. Hershman, D.L.; Lacchetti, C.; Dworkin, R.H.; Lavoie Smith, E.M.; Bleeker, J.; Cavaletti, G.; Chauhan, C.; Gavin, P.; Lavino, A.; Lustberg, M.B.; et al. Prevention and management of chemotherapy-induced peripheral neuropathy in survivors of adult cancers: American Society of Clinical Oncology clinical practice guideline. *J. Clin. Oncol.* **2014**, *32*, 1941–1967. [CrossRef]

70. Clavo, B. Ozone therapy in the management of side effects related to cancer treatments. In Proceedings of the 2nd International Traditional and Complementary Medicine Congress (Organized with technical sponsorship of the World Health Organization), Istanbul, Turkey, 24 April 2019.

71. Bakkal, B.H.; Gultekin, F.A.; Guven, B.; Turkcu, U.O.; Bektas, S.; Can, M. Effect of ozone oxidative preconditioning in preventing early radiation-induced lung injury in rats. *Braz J. Med. Biol. Res.* **2013**, *46*, 789–796. [CrossRef]

72. Gultekin, F.A.; Bakkal, B.H.; Guven, B.; Tasdoven, I.; Bektas, S.; Can, M.; Comert, M. Effects of ozone oxidative preconditioning on radiation-induced organ damage in rats. *J. Radiat. Res.* **2013**, *54*, 36–44. [CrossRef]

73. Menendez, S.; Cepero, J.; Borrego, L.R. Ozone therapy in cancer treatment: State of the art. *Ozone Sci. Eng.* **2008**, *30*, 398–404. [CrossRef]

74. Clavo, B.; Ceballos, D.; Gutierrez, D.; Rovira, G.; Suarez, G.; Lopez, L.; Pinar, B.; Cabezon, A.; Morales, V.; Oliva, E.; et al. Long-term control of refractory hemorrhagic radiation proctitis with ozone therapy. *J. Pain Symptom Manag.* **2013**, *46*, 106–112. [CrossRef]

75. Clavo, B.; Santana-Rodriguez, N.; Llontop, P.; Gutierrez, D.; Ceballos, D.; Mendez, C.; Rovira, G.; Suarez, G.; Rey-Baltar, D.; Garcia-Cabrera, L.; et al. Ozone Therapy in the Management of Persistent Radiation-Induced Rectal Bleeding in Prostate Cancer Patients. *Evid. Based Complement. Alternat. Med.* **2015**, *2015*, 480369. [CrossRef] [PubMed]

76. Milkovic, L.; Zarkovic, N.; Saso, L. Controversy about pharmacological modulation of Nrf2 for cancer therapy. *Redox Biol.* **2017**, *12*, 727–732. [CrossRef] [PubMed]

77. Zhong, H.; Yin, H. Role of lipid peroxidation derived 4-hydroxynonenal (4-HNE) in cancer: Focusing on mitochondria. *Redox Biol.* **2015**, *4*, 193–199. [CrossRef] [PubMed]

78. Allen, B.G.; Bhatia, S.K.; Buatti, J.M.; Brandt, K.E.; Lindholm, K.E.; Button, A.M.; Szweda, L.I.; Smith, B.J.; Spitz, D.R.; Fath, M.A. Ketogenic diets enhance oxidative stress and radio-chemo-therapy responses in lung cancer xenografts. *Clin. Cancer Res.* **2013**, *19*, 3905–3913. [CrossRef]

79. DiDonato, J.A.; Mercurio, F.; Karin, M. NF-kappaB and the link between inflammation and cancer. *Immunol. Rev.* **2012**, *246*, 379–400. [CrossRef]

80. Garg, A.; Aggarwal, B.B. Nuclear transcription factor-kappaB as a target for cancer drug development. *Leukemia* **2002**, *16*, 1053–1068. [CrossRef]

81. El-Deek, H.E.M.; Ahmed, A.M.; Mohammed, R.A.A. Aberration of Nrf2Bach1 pathway in colorectal carcinoma; role in carcinogenesis and tumor progression. *Ann. Diagn. Pathol.* **2019**, *38*, 138–144. [CrossRef]

82. Rocha, C.R.; Kajitani, G.S.; Quinet, A.; Fortunato, R.S.; Menck, C.F. NRF2 and glutathione are key resistance mediators to temozolomide in glioma and melanoma cells. *Oncotarget* **2016**, *7*, 48081–48092. [CrossRef]

83. Zheng, H.; Nong, Z.; Lu, G. Correlation Between Nuclear Factor E2-Related Factor 2 Expression and Gastric Cancer Progression. *Med. Sci. Monit.* **2015**, *21*, 2893–2899. [CrossRef]

84. Rojo de la Vega, M.; Chapman, E.; Zhang, D.D. NRF2 and the Hallmarks of Cancer. *Cancer Cell* **2018**, *34*, 21–43. [CrossRef]

85. Yasueda, A.; Urushima, H.; Ito, T. Efficacy and Interaction of Antioxidant Supplements as Adjuvant Therapy in Cancer Treatment: A Systematic Review. *Integr. Cancer Ther.* **2016**, *15*, 17–39. [CrossRef] [PubMed]

86. Conklin, K.A. Chemotherapy-associated oxidative stress: Impact on chemotherapeutic effectiveness. *Integr. Cancer Ther.* **2004**, *3*, 294–300. [CrossRef] [PubMed]

87. Clavo, B.; Santana-Rodriguez, N.; Llontop, P.; Gutierrez, D.; Suarez, G.; Lopez, L.; Rovira, G.; Martinez-Sanchez, G.; Gonzalez, E.; Jorge, I.J.; et al. Ozone Therapy as Adjuvant for Cancer Treatment: Is Further Research Warranted? *Evid. Based Complement. Alternat. Med.* **2018**, *2018*, 7931849. [CrossRef] [PubMed]

Review

Oxidative Stress and Cancer: Chemopreventive and Therapeutic Role of Triphala

Sahdeo Prasad * and Sanjay K. Srivastava *

Department of Immunotherapeutics and Biotechnology, and Center for Tumor Immunology and Targeted Cancer Therapy, Texas Tech University Health Sciences Center, Abilene, TX 79601, USA

* Correspondence: sahdeo.prasad@ttuhsc.edu or spbiotech@gmail.com (S.P.); sanjay.srivastava@ttuhsc.edu (S.K.S.); Tel.: +1-325-696-0464 (S.K.S.); Fax: +1-325-696-3875 (S.K.S.)

Received: 21 November 2019; Accepted: 8 January 2020; Published: 13 January 2020

Abstract: Oxidative stress, caused by the overproduction of free radicals, leads to the development of many chronic diseases including cancer. Free radicals are known to damage cellular biomolecules like lipids, proteins, and DNA that results in activation of multiple signaling pathways, growth factors, transcription factors, kinases, inflammatory and cell cycle regulatory molecules. Antioxidants, which are classified as exogenous and endogenous, are responsible for the removal of free radicals and consequently the reduction in oxidative stress-mediated diseases. Diet and medicinal herbs are the major source of antioxidants. Triphala, which is a traditional Ayurvedic formulation that has been used for centuries, has been shown to have immense potential to boost antioxidant activity. It scavenges free radicals, restores antioxidant enzymes and non-enzyme levels, and decreases lipid peroxidation. In addition, Triphala is revered as a chemopreventive, chemotherapeutic, immunomodulatory, and radioprotective agent. Accumulated evidence has revealed that Triphala modulates multiple cell signaling pathways including, ERK, MAPK, NF-κB, Akt, c-Myc, VEGFR, mTOR, tubulin, p53, cyclin D1, anti-apoptotic and pro-apoptotic proteins. The present review focuses on the comprehensive appraisal of Triphala in oxidative stress and cancer.

Keywords: oxidative stress; cancer; antioxidant; triphala; ayurveda; chemoprevention and chemotherapy

---

1. Introduction

Cancer is a major health problem worldwide and the second leading cause of death. According to the American Cancer Society, 1,762,450 new cancer cases and 606,880 cancer deaths are projected to occur in the United States in 2019 [1]. However, from 2006 to 2015, it has been observed that the incidence rate of some cancers are either stable or have declined by approximately 2%. Moreover, the overall cancer death rate also dropped from 1991 to 2016 by a total of 27% [1]. This decline in cancer incidence and death is due to the significant drop in smoking and an increase in advances for early cancer detection and screening [2]. However, cancer is still a major health issue that burdens high care cost and causes physical and emotional difficulties to cancer patients. Besides preventive measures, several therapeutic modalities like surgery, chemotherapy and radiotherapy have been developed. These are very effective treatment measures but are very expensive, cause serious side effects, and subsequently may lead patients to the development of resistance to the therapy [3].

Several factors are associated with the causation of cancer. It may be caused by either external factors, internal factors or both. External factors include the consumption of tobacco and alcohol, exposure to hazardous chemicals, ionizing radiation, infectious organisms, and other lifestyle factors, whereas internal factors include inherited mutations, an imbalanced hormone level, and poor immune conditions [4]. These factors affect the incidence and mortality of cancer by modifying cellular systems of the organism. Internal factors such as hereditary mutation are not modifiable. Therefore, in order to control the incidence of cancer, external factors such as lifestyle and environmental factors need to be modified. This can be achieved through the cessation of smoking, minimal use of alcohol, increased consumption of fruits, vegetables and whole grains, physical activity, avoidance of direct exposure to sunlight, minimal red meat consumption, proper vaccinations, and routine screening. It has been shown that adopting changes in lifestyle can reduce over 90% of cancer incidence [5].

2. Oxidative Stress and Cancer

Free radicals, which are reactive oxygen species (ROS) and reactive nitrogen species (RNS), are constantly produced by biological systems. The antioxidants present in cells safely interact with the free radicals and neutralize them, thus establishing balance in the body. Oxidative stress occurs when there is an imbalance between the generation of free radicals and antioxidant defenses [6]. Free radicals are highly reactive and unstable molecules produced naturally as a byproduct of metabolism (oxidative phosphorylation), or by exposure to environmental factors. ROS, which include superoxide anion (O_2^-), hydrogen peroxide (H_2O_2), and hydroxyl radicals (OH•), are produced by the mitochondrial respiratory chain during oxidative metabolism through the one-electron reduction of molecular oxygen (O_2) [7]. It is known that complexes I and II of mitochondria produce ROS only into the matrix, while complex III produces ROS on both sides of the mitochondrial inner membrane [8,9]. However, RNS which includes a nitric oxide radical (NO•), peroxynitrite ($ONOO^-$), and a nitrogen dioxide radical (NO_2•), are produced via the enzymatic activity of inducible nitric oxide synthase 2 (NOS2) and NADPH oxidase [10,11].

Oxidative stress is a crucial factor in the development of chronic diseases including cancer. Low levels of free radicals are implicated in many fundamental cellular processes such as immune defense, cellular proliferation and differentiation, activation of important signaling pathways and against pathogens. However, chronic and excessive amounts of ROS/RNS induce oxidative stress and cause deleterious effects to the cells [12]. They can induce oxidative damage to genetic materials, lipids, and proteins and further carcinogenesis and tumor progression. Increased oxidative stress also results in dysregulation of various cellular processes through modulation of signaling molecules, production of antioxidant enzymes and non-enzymes, cell growth, and chronic inflammation, which play major roles in the incidence of chronic diseases such as cancer [6].

Antioxidant systems are thus required to counteract oxidative stress and overcome cellular damage for the prevention of oxidative stress-mediated diseases like cancer. The cellular antioxidants are regulated by the transcription factor nuclear factor erythroid 2-related factor 2 (NRF2). Under an unstressed cellular condition NRF2 remains inactive by forming a complex with KEAP1 in the cytoplasm. NRF2 undergoes ubiquitination and further proteasomal degradation by KEAP1 through the Cullin 3 (CUL3) based E3 ligase [13]. However, disruption in binding of NRF2 to KEAP1 leads to the nuclear translocation of NRF2, where it regulates the basal and inducible expression of several genes that contain antioxidant response elements (AREs) [14,15]. NRF2 not only regulates redox homeostasis [16] but also the other physiology of cells [15]. NRF2 is reported to prevent chemical and radiation-induced carcinogenesis by quenching ROS or managing oxidative damage. However, since the last decade, a 'dark side' of NRF2 has also been described [17]. Some studies have shown that in cancer cells NRF2 activation promotes cancer progression [18,19], metastasis [20], and causes resistance to therapeutic agents [21]. These studies explain that the activation of NRF2 prevents carcinogenesis but may facilitate tumor growth and metastasis in cancer cells. Thus, activation of NRF2 is beneficial for the prevention of carcinogenesis but may not be beneficial for cancer treatment.

The antioxidants can be produced by cells endogenously or can be supplied to the cells through food and/or supplements exogenously [22]. Because of the limitation in endogenous production of antioxidant by cells, exogenous supplement of antioxidants can satisfy the requirement and thereby reduce oxidative stress-mediated cellular damage and carcinogenesis. Plant products are one of the major source of antioxidants, which have little or mild toxicity, are abundantly available, and have high efficacy and cost effectiveness [6,23–25]. Traditional medicines, which utilize a variety of medicinal plants, are the inherent source of antioxidants. Because of their disease curing capability, traditional medicines have been used for centuries against a variety of ailments [26]. Like other traditional medicines, Ayurvedic medicines consist of a single constituent or a mixture of different constituents of one or multiple medicinal plants. Ayurveda is an ancient Indian medical system and considered as one of the world's oldest holistic healing systems, which is thought to have been developed more than 5000 years ago in India [26,27].

Accumulated evidence suggests that Ayurvedic medicines exhibit antioxidant properties by neutralizing free radicals, quenching ROS, and lowering peroxides [6,28,29]. In a study, Ayurvedic medicine *Jeevaneeya Rasayana* (an ayurvedic polyherbal formulation) was found to increase the activities of antioxidant enzymes and the level of glutathione content in arthritic rats. This formulation also decreased the concentration of C-reactive protein, thiobarbituric acid reactive substance, and ceruloplasmin in arthritic rats [30]. Moreover, using an arthritic rat model, Ratheesh et al. [31] also found that the Ayurvedic formulation *Kerabala* increased antioxidant enzymes such as superoxide dismutase (SOD), catalase (CAT), glutathione peroxidase (GPx), and decreased the lipid peroxidation product. Another Ayurvedic formulation, *Amalakayas Rasayana*, also showed antioxidant activity by scavenging free radicals [32]. Mathew et al. [33] analyzed antioxidant activity of many Ayurvedic plant extracts and found all extracts had positive DPPH (2,2-diphenyl-1-picryl-hydrazyl-hydrate) free radical scavenging activity. These studies indicate that Ayurvedic plants or formulations have antioxidant and free radical scavenging activity that may explain its effect and justify its use as a medicine against oxidative stress-associated diseases such as cancer. A variety of Ayurvedic formulations have been described in Ayurveda, however in this article, we discuss an increasingly popular Ayurvedic formulation, Triphala (Figure 1).

Figure 1. Constituents of Triphala.

3. Triphala: A Formulation of Three Fruits

Triphala, as the name indicates in the Sanskrit language (tri = three and phala = fruits), is a herbal formulation consisting of the dried powdered fruits of three plants, *Terminalia chebula*

(Haritaki), *Terminalia belerica* (Bibhitaki), and *Phyllanthus emblica* or *Emblica officinalis* (Amalaki or the Indian gooseberry) (Figure 1). Although the Triphala formulation generally consists of equal proportions of fruits from these plants, a modified formulation consisting of 1:2:4 parts of *T. chebula*, *T. belerica*, and *E. officinalis* are also used [34]. Chemical analysis of *T. chebula* extract shows that it contains many biologically active constituents like chebulin, ellagic acid, 2,4-chebulyl-D-glucopyranose, arjunglucoside I, arjungenin, chebulinic acid, gallic acid, ethyl gallate, punicalagin, terflavin A, terchebin, luteolin and tannic acid. However, the main chemical constituents of *T. bellerica* are tannins that mainly include β-sitosterol, gallic acid, ellagic acid, ethyl gallate, galloyl glucose and chebulaginic acid [35]. The fruit of *P. emblica* has been shown to be rich in quercetin, phyllaemblic compounds, gallic acid, tannins, flavonoids, pectin, and vitamin C [36] (Figure 2).

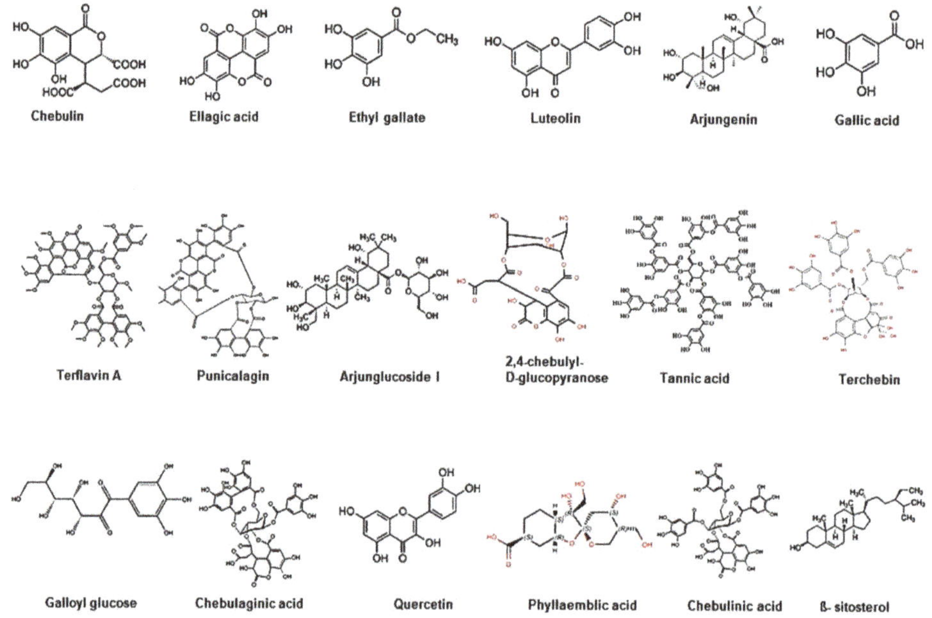

Figure 2. Chemical structure of bioactive components of Triphala.

A comparative study on *T. chebula*, *T. belerica* and *E. officinalis* (components of Triphala) have shown that they exhibit potent antioxidant activities. In a study, total antioxidant capacity was measured and results indicated that the *T. chebula* extract had a higher (4.52 ± 0.12) antioxidant capacity compared to *T. belerica* (1.01 ± 0.03) and *E. officinalis* (4.10 ± 0.17). The DPPH scavenging activity was found in the order of *T. chebula*, *T. belerica* and *E. officinalis* (1.73 ± 0.07 µg/mL, 1.45 ± 0.02 µg/mL & 1.43 ± 0.03 µg/mL). However, *T. chebula*, *T. belerica* and *E. officinalis* extracts showed a moderate effect on the scavenging singlet oxygen species with IC50 values of 424.50 ± 24.70 µg/mL, 233.12 ± 48.68 µg/mL and 490.42 ± 159.59 µg/mL, respectively [37]. These studies indicate that extracts of these fruits exhibit their antioxidative properties in the order of *T. chebula* > *E. officinalis* > *T. belerica*, which follow the order of their flavonoid contents [37].

Total phenolic, flavonoid and tannin contents were also analyzed in these fruits. The total phenolic content of *T. chebula* fruit extracts were varied from 867.2 to 1041.8 mg gallic acid/gm extract with the highest concentration of phenolic compounds in water extract followed by methanol and ethanol extracts. However, total triterpenoid content of the three extracts varied widely from 0.8 to 4.2 mg ursolic acid/gm extract with the lowest total triterpenoid content in water extract, whereas the methanol extract provided the highest triterpenoid content. The total tannin content of the three extracts varied from 33.9 to 40.3%/mg extract. The highest total tannin content was detected in the water extract followed by methanol and ethanol extracts [38]. However, *T. bellerica* fruits have shown flavonoids in ethanol and chloroform extracts but not in methanol extract. Additionally, triterpenoids and tannins were reported only in the ethanol extract [39]. The total phenolic content in *E. officinalis* has been reported to range from 188.8–237.0 mg gallic acid/gm. Nonetheless, the total flavonoid content ranged from 6.4–20.1 mg rutin/gm, whereas total tannin content ranged from 375.2–642.8 mg tannin/gm [40]. These studies indicate that *T. chebula* fruit extracts contain the highest amount of phytochemicals followed by *E. officinalis* and *T. chebula* fruits. Hazra et al. [37] also demonstrated that *T. chebula* fruits had the highest flavonoid content followed by *E. officinalis* and *T. belerica*. However, in the case of phenolic content, *E. officinalis* fruits had the highest, thereafter *T. belerica* and *T. chebula*. The variation in these phytochemical contents may depend on the geographical origin of these plants.

In Ayurveda, Tridosha defines three fundamental energies or principles (vata, pitta, and kapha) that govern the function of our bodies at the physical and emotional level [41]. Triphala is considered as a tridoshic rasayan that has the ability to balance and rejuvenate Tridosha, as well as promote health, immunity and longevity [42]. In Ayurvedic practice, Triphala is frequently used to treat digestion problems, poor food assimilation, constipation, and gastric acidity. Besides these, it is used in the treatment of many other diseases such as asthma, anemia, jaundice, fever cough, chronic ulcers, leucorrhoea, and pyorrhea. It is also recommended for use in the treatment of cardiovascular disorders, ophthalmic problems, liver dysfunction, inflammation, infection, obesity, anaemia, and fatigue [43]. Most people practicing Ayurvedic medicine consume Triphala as a 'health tonic'. Triphala improves blood circulation, reduces myocardial necrosis and serum cholesterol levels, and strengthens capillaries, which indicates its cardiotonic effects [44]. Recent studies showed that it had antioxidant, anti-inflammatory, antiaging, anti-mutagenic, anti-clastogenic and anticancer effects. Herewith an attempt was made to summarize the antioxidant and cancer preventive and therapeutic aspects of Triphala (Table 1).

Table 1. Antioxidant and chemotherapeutic effect of Triphala.

Effects	Studies	References
ROS scavenging	Eliminates X-radiation-induced ROS generation in HeLa cells.	[45]
	Quenches γ-radiation-induced free radicals.	[46]
	Scavenges free radicals comparable with ascorbic acid.	[47]
	Scavenges free radicals such as DPPH and superoxide.	[48]
Antioxidant enzymes	Increases expression of SOD-2 in HDF or HaCaT skin cells.	[49]
	Restores CAT, SOD, GST, GPx and GSH in bromobenzene treated rat kidney.	[50]
	Prevents peroxidative damage by increasing GSH and GST and decreasing LPO in DMH treated mouse liver.	[51]
	Restores GSH, CAT, SOD, GPx, and GST in cataract mouse model.	[52]
	Restores GSH content and decreases LPO in MTX-induced small intestinal damage in rats.	[34]
	Prevents noise-stress induced decrease in SOD, CAT, GPx, ascorbic acid, and increase in LPO in plasma and thymus tissues.	[53]
	Inhibits γ-radiation-induced lipid peroxidation in rat liver microsomes.	[48]
Radioprotective	Prevents γ-radiation-induced DNA damage in HeLa cells.	[45]
	Prevents DNA damage in blood leukocytes and splenocytes of mice exposed with whole body γ-radiation.	[46]
Chemopreventive	Reduces B(a)P-induced forestomach papillomagenesis in mice at a dose of 2.5% and 5% in diet.	[54]

Table 1. *Cont.*

Effects	Studies	References
Prooxidant	Increases ROS level and induces apoptosis in breast cancer MCF-7 and barcl-95 cells.	[55]
	Induces ROS and inhibits proliferation in MCF 7 and T47D breast cancer cells.	[56]
	Induces apoptosis and phosphorylation of p53 and ERK through ROS generation in Capan-2 cancer cells.	[57]
Therapeutic	Decreases survival and induces apoptosis in Capan-2 pancreatic cells cancer with an IC50 of 50 µg/mL.	[57]
	Inhibits gastric cancer cell proliferation and suppresses cell migration in vitro.	[58]
	Exerts anti-proliferative, apoptotic and anti-migratory effects in colon cancer cells.	[59]
	Inhibits proliferation of gynecological cancers cell with IC50 values of 98.28–101.23 µg/mL against SKOV-3, HeLa, and HEC-1B cells.	[60]
	Inhibits proliferation of HeLa, PANC-1, and MDA-MB-231 cells and suppresses the clonogenicity of HeLa cells.	[61]
	Inhibits proliferation of HCT116 and HCCSCs cells independent of p53 status.	[62]
	Inhibits colony formation and viability of breast cancer MCF-7 cells with wild type p53, which was more sensitive	[56]
	Induces cytotoxicity in Shionogi 115 and MCF-7 breast cancer cells and PC-3 and DU-145 prostate cancer cells.	[63]
	Oral administration at 50–100 mg/kg dose suppresses growth of Capan-2 pancreatic tumor-xenograft.	[57]
	Inhibits xenograft growth and metastasis of transplanted gastric carcinoma cells in vivo zebrafish xenograft model.	[58]
	Oral feeding to mice at 40 mg/kg inhibits barcl-95 tumor growth transplanted in nude mice.	[55]
Immunomodulatory	Stimulates neutrophil functions in the immunized rats and prevents stress-induced suppression in the neutrophil functions.	[64]
	Prevents the noise-stress induced changes in cell-mediated immune response in rats.	[53]
	Ameliorates functional and histological ovalbumin-induced bronchial hyperreactivity and increases CD4 counts in lung and spleen.	[65]
	Increases cytotoxic T cells and natural killer cells in healthy human volunteers.	[66]

CAT: Catalase, SOD: Superoxide dismutase, GPx: Glutathione peroxidase, GST: Glutathione-S-Transferase, GSH: Glutathione, ROS: Reactive oxygen species, DPPH: 2,2-diphenyl-1-picrylhydrazyl, DMH: Dimethylhydrazine, B(a)P: Benzo(a)pyrene, HCCSCs: Human colon cancer stem cells. MTX: Methotrexate, LPO: Lipid peroxidation.

4. Antioxidant Effects of Triphala

Triphala holds potential in restoring antioxidant levels and decreasing lipid peroxidation as shown in numerous in vitro, in vivo, and human studies (Figure 3). These antioxidant properties of Triphala are associated with the presence of polyphenols, vitamin C, and flavonoids. The active constituents of Triphala quench the ROS levels and reduce oxidative stress.

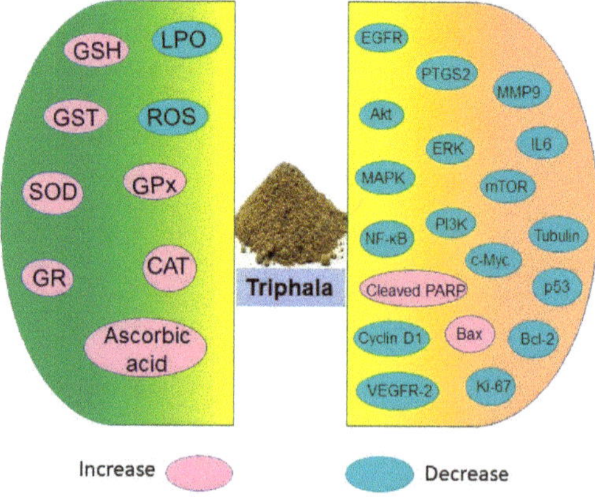

Figure 3. Antioxidative and chemoprotective molecules targeted by Triphala.

4.1. In Vitro Studies

Numerous in vitro studies have shown that Triphala has high antioxidant potential. In a study, both aqueous and methanolic extracts of Triphala were examined for antioxidant activities and were found to quench free radicals and induce SOD and CAT antioxidant enzymes. Triphala extract exhibited satisfactory free radical-scavenging activity that was comparable with ascorbic acid [47]. In HeLa cells, Triphala efficiently eliminates ROS levels generated by X-radiation and bleomycin and thus protects from X-radiation and bleomycin-mediated DNA strand breaks [45]. In another study, Triphala extract also inhibited radiation-induced lipid peroxidation in rat liver microsomes. The extracts were found to possess the ability to scavenge free radicals such as DPPH and superoxide. It was also found that Triphala extract had the ability to prevent gamma-radiation-induced strand break formation in plasmid DNA [48].

Triphala extract also inhibits H_2O_2-induced RBC haemolysis, nitric oxide production and shows high reducing power activity. As H_2O_2 induces cellular damage, it has been shown that pretreatment with Triphala rescues the human dermal fibroblast from H_2O_2-induced damage, inhibits cellular senescence, and protects DNA from damage [49]. Triphala has been found to increase glutathione (GSH) and decrease malondialdehyde levels in enucleated rat lenses. It can also restore the activities of antioxidant enzymes such as SOD, CAT, GPx, glutathione reductase (GR), and glutathione-S-transferase (GST) and improves selenite-induced cataract [52].

4.2. In Vivo Studies

Triphala is found to be effective in reducing oxidative stress in animal models. In a study, pretreatment with two doses (150 mg/kg and 300 mg/kg) of Triphala in a colitis rat model, it restored the antioxidant enzymes SOD and CAT, and decreased the malondialdehyde levels in the distal colon of rats. Triphala further relieved the rats from colitis, which could be attributed to its antioxidant activity [67]. In complete Freund's adjuvant-induced arthritic rat model, Triphala showed antioxidant properties. Administration of Triphala (100 mg/kg b wt, i.p.) restored the activities/levels of antioxidant (SOD ~75.6%, CAT ~62.7%, GPx ~55.8%, GST ~82.1%, and GSH ~72.7%), and decreased the lipid peroxidation in the paw tissues of arthritic rats [68]. A study on another monosodium urate crystal-induced arthritis model showed support for Triphala exhibiting antioxidant properties and decreasing inflammation. Oral treatment of Triphala (1 g/kg) inhibited paw volume and lipid peroxidation; however the antioxidant status was found to be increased in the plasma, liver, and spleen of monosodium urate crystal-induced mice when compared to control mice [69].

Triphala also exerts nephroprotective effects due to its antioxidant properties. In Wistar albino rats, bromobenzene treatment resulted in a decrease in the activities of antioxidant enzymes such as CAT, SOD, GST, and GPx as well as total reduced GSH in the kidney. Bromobenzene also increased lipid peroxidation in the kidney of animals. However, oral administration of two different doses (250 and 500 mg/kg) of Triphala in bromobenzene-treated rats restored antioxidant enzymes and decreased lipid peroxidation [50]. Thus, data indicates that Triphala has nephroprotective effects through its antioxidant nature. The antioxidative property of Triphala is also directed toward the protection of carcinogen-induced cellular damage. It has been shown to prevent 1,2-dimethylhydrazine dihydrochloride (DMH)-induced mouse liver damage by decreasing DMH-induced lipid peroxidation and increasing GSH and GST [51].

Administration of 25 mg/kg Triphala to the animals decreased nuclear cataract in the selenite-induced cataract model [52]. Sandhya et al. [46] have also shown that Triphala protects against radiation-induced oxidative damage in mice. They have found that 5 Gy radiation induces mortality in mice. However, oral treatment of Triphala (1 g/kg) reduced mortality by 60% in mice. This improvement was found to be associated with an increase in antioxidant enzymes such as SOD and protection from DNA damage in the intestine of mice exposed to irradiation. Triphala administration in animals also increased radiation tolerance, which was further mediated through its antioxidant activity and scavenging free radicals [70].

Noise-stress causes alterations in the antioxidant status and on the cell-mediated immune response. Thus, to determine the protective effect of Triphala, noise-stress (100 dB for 4 h/d/15 days) was employed on the rats and Triphala (1 g/kg/bw/48 days) was administered to the animals. Treatment with Triphala resulted in a decrease in noise-stress-induced lipid peroxidation and corticosterone level with concomitant increase of antioxidants in the plasma and tissues of rats. Thus, the study indicates that Triphala has preventive effects on noise-stress induced changes by increasing antioxidants as well as modulating the cell-mediated immune response in rats [53]. Besides these, equal (1:1:1) and unequal (1:2:4) formulations of Triphala have also been compared to determine their antioxidant and enteroprotective efficacy on methotrexate-induced small intestinal damage in rats. It has been observed that the unequal formulation of Triphala provides significantly more protection by restoring GSH than equal formulation of Triphala against methotrexate-induced damage in the rat intestine [34].

5. Prooxidant Nature of Triphala

Besides its antioxidant property, Triphala also exhibits prooxidant activity by inducing ROS production in cancer cells. Triphala has shown an insignificant level of ROS production in normal breast MCF-10F cells as well as in murine spleen and liver normal cells [55]. As increased levels of ROS in cancer cells causes lethality [71,72], the prooxidant nature of Triphala escalates the death of cancer cells and acts as an anticancer agent. In a study, it has been observed that Triphala inhibited proliferation and induced apoptosis in MCF-7 and T47D breast cancer cells through production of ROS. Further, it was observed that quenching ROS by antioxidants inhibited the anti-proliferative ability of Triphala suggesting its role in the induction of apoptosis through ROS production [56]. Triphala has also been shown to induce ROS generation in Capan-2 pancreatic cancer cells and further apoptosis. Triphala-induced ROS generation also led to phosphorylation of p53 and ERK in Capan-2 cells as pretreatment combined with the antioxidant N-acetylcysteine blocked Triphala-induced phosphorylation of these proteins [57]. Cancer cells are known to have high levels of ROS. Increasing ROS further crosses the threshold and forces the cell into apoptosis [72]. However, the level of ROS in normal cells is very low and not easy to raise to go over the threshold limit.

As a prooxidant, Triphala also produces a radiosensitizing action through oxidative damage, membrane alteration and damage to nucleic acids in various cancer cell lines. In tumor cell lines such as Ehrlich ascites (EAC), human cervical (HeLa), and breast (MCF-7) cells, treatment with Triphala induced a cytotoxic effect by initiating membrane oxidative damage and by triggering ROS generation by gamma radiation [73]. In contrast, Triphala showed protective effects against X-radiation and bleomycin in HeLa cells. However, it has also shown protective activity against ionizing radiation in mice [45].

6. Chemopreventive and Chemotherapeutic Effects

Extensive studies on Triphala have shown that it has preventive and therapeutic efficacy against malignancies such as breast, colon, pancreas, prostate, ovarian, cervical, endometrial, and lymphatic cancers as well as melanoma [58,60,61,63,74]. Although Triphala has been used for centuries against various ailments, recent in vitro, in vivo and human studies have demonstrated its safety and efficacy against multiple diseases including cancer. Experimental studies in the past decade have also shown that Triphala exhibits antineoplastic, radioprotective, and chemoprotective effects through modulation of multiple signaling molecules (Figure 3).

6.1. In Vitro Studies

Triphala has shown anticancer activities in various cancer cell lines. Using a cytotoxic assay, it was found that the aqueous extract of Triphala decreased the proliferation of breast and prostate cancer cells. Further chemical analysis showed that the extract was rich in polyphenol gallic acid, which was considered as a major factor in inducing cytotoxicity of cancer cells [63]. Sandhya et al. [55] also showed that the increasing concentrations of Triphala correspondingly decreased the viability of treated breast cancer MCF-7 cells. Besides cytotoxicity, Triphala treatment was found to induce apoptosis in MCF-7 and barcl-95 cells in vitro. Further mechanistic studies of Triphala on apoptosis and cytotoxicity were demonstrated by using single cell gel electrophoresis in breast cancer cells and found that it increased intracellular ROS and induced DNA damage, a characteristic of apoptosis. However, with similar concentrations of Triphala, it did not cause any cytotoxic effect or DNA damage on normal breast epithelial cells, MCF-10F, human peripheral blood mononuclear cells, mouse liver and spleen cells. This study indicated that Triphala was selectively cytotoxic to the cancer cells.

Further studies revealed the crucial role of p53 in Triphala-mediated apoptosis in breast cancer cells. It was observed that MCF-7 cells with wild type p53 were more sensitive to Triphala than p53 negative T47D breast cancer cells. Triphala-induced ROS generation plays a major role in apoptosis, since the addition of antioxidants inhibits the anti-proliferative ability of Triphala [56]. In contrast, it has been reported that the methanol extract of Triphala suppresses the proliferation of colon cancer HCT116 cells and human colon cancer stem cells (HCCSCs) independent of p53 status. This extract also induced p53-independent apoptosis in HCCSCs as indicated by elevated levels of cleaved PARP. It further suppressed c-Myc and cyclin D1, and induced apoptosis through elevation of Bax/Bcl-2 ratio. In addition, Triphala extract inhibited HCCSCs colony formation, a measure of CSCs self-renewal ability [62]. It is worth noting that Triphala scavenges ROS in normal cells thereby preventing oxidative damage, whereas it increases the ROS level and causes lethality in cancer cells.

Growth-inhibitory effects of Triphala were also evaluated in pancreatic cancer cells. It was observed that treatment with the aqueous extract of Triphala reduced the survival of pancreatic cancer Capan-2 cells. Triphala-mediated reduction in cell survival was correlated with the induction of apoptosis. Further it was shown that Triphala extract induced ROS production that led to phosphorylation of p53 and ERK in Capan-2 cells and then apoptosis, whereas antioxidant N-acetylcysteine (NAC) treatment blocked apoptosis [57]. In another study, Triphala inhibited the proliferation of multiple cancer cells such as HeLa (cervical adenocarcinoma), PANC-1 (pancreatic adenocarcinoma), and MDA-MB-231 (triple-negative breast carcinoma) cells and suppressed the clonogenicity of HeLa cells. The mechanism of the antiproliferative effect was mediated by disruption of secondary conformation of tubulin and inhibition of anilino naphthalene sulfonate binding to tubulin. Triphala acetylates cellular microtubules and stabilizes microtubule dynamics. In addition, Triphala interfered with the reassembly of microtubules. The microtubule interfering effects of Triphala lead to apoptotic cell death in cancer cells [61].

Triphala is also effective in suppressing gynecological cancer cell growth. Treatment with Triphala inhibited proliferation and induced apoptosis in SKOV-3, HeLa, and HEC-1B cells. The antiproliferative and proapoptotic activities were confirmed by cell cycle analysis and expression of Ki-67 protein. It was also found that Triphala decreased the expression of phospho-Akt, phospho-p44/42, and phospho-NF-κB p56 in these gynecological cancer cells, which indicated that MAPK/ERK, PI3K/Akt/mTOR, and NF-κB/p53 signaling pathways were the possible mechanism of Triphala-induced apoptosis [60]. Besides its anti-proliferative and apoptotic effects, Triphala suppressed cell migration of cancer cells in vitro thus indicating its anti-metastatic potential [58].

6.2. In Vivo Studies

Triphala also has cancer chemopreventive potential as shown in animal studies. In a study, Triphala (2.5%, supplemented in diet) significantly reduced benzo(a)pyrene [B(a)P] induced forestomach papillomagenesis in mice. It reduced tumor incidence by 77.77% in the short-term study and 66.66% in the long-term study. As it is a potent antioxidant, the chemopreventive effect of Triphala might be associated with an increased antioxidant status in animals [54]. Oral administration of Triphala (50–100 mg/kg) also suppresses the growth of Capan-2 pancreatic tumor-xenograft. It was found that reduction in tumor growth by Triphala in mice was due to increased apoptosis in the tumor cells, which was associated with increased activation of p53 and ERK [57]. Another study also revealed that tumor inhibitory effects of Triphala or its active constituents were through suppression of VEGF actions. Triphala and one of its active compounds, chebulinic acid, specifically inhibits VEGF-induced angiogenesis by suppressing VEGF receptor-2 (VEGFR-2) phosphorylation and thus reduces tumor growth and metastasis [75]. In a zebrafish xenograft model, administration of Triphala inhibited the growth and metastasis of transplanted gastric carcinoma cells. The antineoplastic effect of Triphala was analyzed by western blotting and results demonstrated that it inhibited phosphorylation of EGFR, Akt, and ERK [58].

7. Immunomodulatory Effect of Triphala

Triphala was shown to alter the immune system and act as an immunomodulatory agent. In a published study, the immunomodulatory activity of Triphala was assessed by testing the various functions of neutrophil-like adherence, phagocytosis and avidity index in albino rats. Upon Triphala administration, the avidity index was found to be increased in the animals. The neutrophil functions were also enhanced in the Triphala immunized group with a decrease in the corticosterone level [64]. Thus, Triphala appears to stimulate neutrophil functions in the immunized rats and prevent stress-induced suppression of neutrophil functions. Another study showed that the supplementation of Triphala prevented the noise-stress induced changes in the cell-mediated immune response in rats [53]. Immunostimulatory activity of Triphala was also evaluated in a phase I clinical study. Consumption of Triphala by healthy volunteers demonstrated significant immunostimulatory effects on cytotoxic T cells (CD3$^-$ CD8$^+$) and natural killer cells (CD16$^+$ CD56$^+$). However, Triphala did not change the cytokine level in volunteers [66]. The individual components of Triphala have also shown to exhibit immunomodulatory activity. The *T. chebula* fruit extract has illustrated an increase in spleen lymphocyte proliferation and enhanced the expression of cytokines such as IL-2, IL-10 and TNF-α in rats [76]. The methanolic extract of *T. bellerica* has affected the mouse immune system, specifically both the cellular and humoral immune response in vitro. This extract stimulated phagocytic activity and T-lymphocyte proliferation [77]. *E. officinalis* fruit extract exhibited immunostimulatory activity by its combined action on humoral and cell-mediated immune responses along with macrophages and phagocytes [78]. Thus, these studies indicate that Triphala and its three individual constituents have potential to stimulate immune systems.

8. Conclusions

Triphala has been used for centuries against various ailments in the Indian traditional medicine system. Studies in the recent past have indicated that Triphala has immense potential in the reduction of oxidative damage as well as in the prevention and treatment of cancer (Figure 4). Few studies indicated that antioxidants from dietary supplements may promote tumor growth and metastasis [79–81]. However, it is noteworthy that Triphala acts as an anticancer agent by exhibiting prooxidant effects in cancer cells. The dual nature of Triphala, acting as an antioxidant in normal cells and prooxidant in cancer cells, facilitates its function as both a chemopreventive and chemotherapeutic agent. Interestingly, Triphala has shown high efficacy and safety in humans as well as in experimental studies. However, most of the studies are done in animals and in vitro models. Clinical studies are required for its applicability as a chemopreventive, radioprotective, and chemotherapeutic agent. Three clinical trials on Triphala have been completed and another one is underway on different diseases like gut microbiome and skin (NCT03477825), gingivitis (NCT01898000), periodontal disease (NCT01900535) and stool microbiome and inflammation (NCT03907501). However, up to now, no clinical trial on Triphala in cancer has been done. Therefore, clinical studies to determine its efficacy in cancer patients are warranted. As 1:2:4 of Triphala has shown better enteroprotective effects over the conventional 1:1:1 combination, examining the effect of Triphala in different combination ratios is required with the hope that new formulations may exhibit better beneficial effects on oxidative stress-mediated chronic diseases.

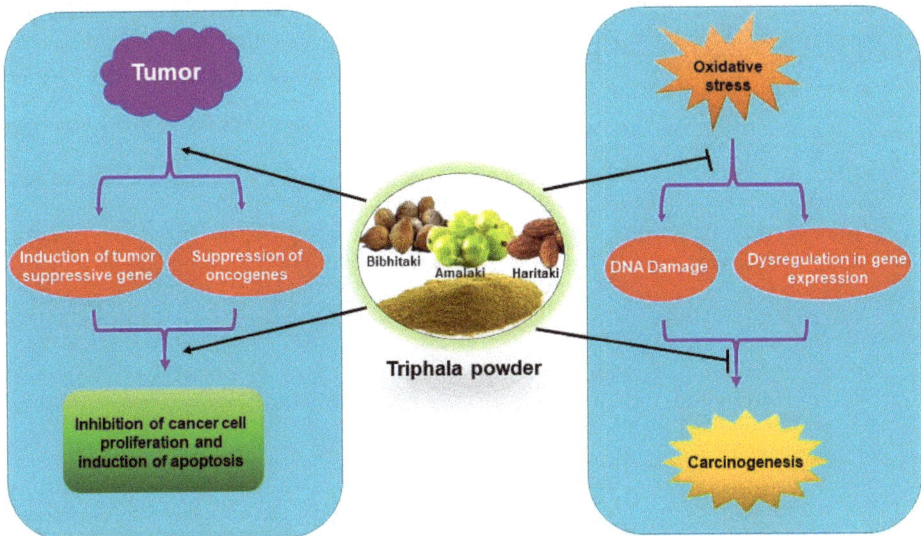

Figure 4. Antioxidative and chemopreventive/chemotherapeutic properties of Triphala.

Author Contributions: Conception and design of the review, S.P.; writing—original and final draft preparation, S.P.; editing, suggestions and supervision S.K.S. All authors have read and agreed to the published version of the manuscript.

Funding: This research received no external funding.

Acknowledgments: We thank Shyanne Page-Hefley from the Department of Pediatrics for carefully proofreading the manuscript.

Conflicts of Interest: The authors declare no conflict of interest.

References

1. Siegel, R.L.; Miller, K.D.; Jemal, A. Cancer statistics, 2019. *CA Cancer J. Clin.* **2019**, *69*, 7–34. [CrossRef] [PubMed]
2. Narod, S.A.; Iqbal, J.; Miller, A.B. Why have breast cancer mortality rates declined? *J. Cancer Policy* **2015**, *5*, 8–17. [CrossRef]
3. Huang, C.Y.; Ju, D.T.; Chang, C.F.; Reddy, P.M.; Velmurugan, B.K. A review on the effects of current chemotherapy drugs and natural agents in treating non-small cell lung cancer. *Biomedicine* **2017**, *7*, 12–23. [CrossRef] [PubMed]
4. Bouayed, J.; Bohn, T. Exogenous antioxidants-Double-edged swords in cellular redox state: Health beneficial effects at physiologic doses versus deleterious effects at high doses. *Oxid. Med. Cell. Longev.* **2010**, *3*, 228–237. [CrossRef] [PubMed]
5. Anand, P.; Kunnumakkara, A.B.; Sundaram, C.; Harikumar, K.B.; Tharakan, S.T.; Lai, O.S.; Sung, B.; Aggarwal, B.B. Cancer is a preventable disease that requires major lifestyle changes. *Pharm. Res.* **2008**, *25*, 2097–2116. [CrossRef]
6. Prasad, S.; Gupta, S.C.; Tyagi, A.K. Reactive oxygen species (ROS) and cancer: Role of antioxidative nutraceuticals. *Cancer Lett.* **2017**, *387*, 95–105. [CrossRef]
7. Hamanaka, R.B.; Chandel, N.S. Mitochondrial reactive oxygen species regulate cellular signaling and dictate biological outcomes. *Trends Biochem. Sci.* **2010**, *35*, 505–513. [CrossRef]
8. Turrens, J.F. Mitochondrial formation of reactive oxygen species. *J. Physiol.* **2003**, *552*, 335–344. [CrossRef]
9. Muller, F.L.; Liu, Y.H.; Van Remmen, H. Complex III releases superoxide to both sides of the inner mitochondrial membrane. *J. Biol. Chem.* **2004**, *279*, 49064–49073. [CrossRef]

10. Bedard, K.; Krause, K.H. The NOX family of ROS-generating NADPH oxidases: Physiology and pathophysiology. *Physiol. Rev.* **2007**, *87*, 245–313. [CrossRef]

11. Fang, F.C. Antimicrobial reactive oxygen and nitrogen species: Concepts and controversies. *Nat. Rev. Microbiol.* **2004**, *2*, 820–832. [CrossRef] [PubMed]

12. Di Meo, S.; Reed, T.T.; Venditti, P.; Victor, V.M. Role of ROS and RNS Sources in Physiological and Pathological Conditions. *Oxid. Med. Cell. Longev.* **2016**, *2016*, 1245049. [CrossRef] [PubMed]

13. Giudice, A.; Montella, M. Activation of the Nrf2-ARE signaling pathway: A promising strategy in cancer prevention. *Bioessays* **2006**, *28*, 169–181. [CrossRef] [PubMed]

14. Giudice, A.; Arra, C.; Turco, M.C. Review of molecular mechanisms involved in the activation of the Nrf2-ARE signaling pathway by chemopreventive agents. *Methods Mol. Biol.* **2010**, *647*, 37–74. [CrossRef] [PubMed]

15. Rojo de la Vega, M.; Chapman, E.; Zhang, D.D. NRF2 and the Hallmarks of Cancer. *Cancer Cell* **2018**, *34*, 21–43. [CrossRef] [PubMed]

16. Giudice, A.; Barbieri, A.; Bimonte, S.; Cascella, M.; Cuomo, A.; Crispo, A.; D'Arena, G.; Galdiero, M.; Della Pepa, M.E.; Botti, G.; et al. Dissecting the prevention of estrogen-dependent breast carcinogenesis through Nrf2-dependent and independent mechanisms. *Onco. Targets. Ther.* **2019**, *12*, 4937–4953. [CrossRef] [PubMed]

17. Wang, X.J.; Sun, Z.; Villeneuve, N.F.; Zhang, S.; Zhao, F.; Li, Y.; Chen, W.; Yi, X.; Zheng, W.; Wondrak, G.T.; et al. Nrf2 enhances resistance of cancer cells to chemotherapeutic drugs, the dark side of Nrf2. *Carcinogenesis* **2008**, *29*, 1235–1243. [CrossRef]

18. Satoh, H.; Moriguchi, T.; Takai, J.; Ebina, M.; Yamamoto, M. Nrf2 prevents initiation but accelerates progression through the Kras signaling pathway during lung carcinogenesis. *Cancer Res.* **2013**, *73*, 4158–4168. [CrossRef]

19. DeNicola, G.M.; Karreth, F.A.; Humpton, T.J.; Gopinathan, A.; Wei, C.; Frese, K.; Mangal, D.; Yu, K.H.; Yeo, C.J.; Calhoun, E.S.; et al. Oncogene-induced Nrf2 transcription promotes ROS detoxification and tumorigenesis. *Nature* **2011**, *475*, 106–109. [CrossRef]

20. Wang, H.; Liu, X.; Long, M.; Huang, Y.; Zhang, L.; Zhang, R.; Zheng, Y.; Liao, X.; Wang, Y.; Liao, Q.; et al. NRF2 activation by antioxidant antidiabetic agents accelerates tumor metastasis. *Sci. Transl. Med.* **2016**, *8*, 334–351. [CrossRef]

21. Choi, B.; Kwak, M. Shadows of NRF2 in cancer: Resistance to chemotherapy. *Curr. Opin. Toxicol.* **2016**, *1*, 20–28. [CrossRef]

22. Valko, M.; Rhodes, C.J.; Moncol, J.; Izakovic, M.; Mazur, M. Free radicals, metals and antioxidants in oxidative stress-induced cancer. *Chem. Biol. Interact.* **2006**, *160*, 1–40. [CrossRef] [PubMed]

23. Pramanik, K.C.; Boreddy, S.R.; Srivastava, S.K. Role of mitochondrial electron transport chain complexes in capsaicin mediated oxidative stress leading to apoptosis in pancreatic cancer cells. *PLoS ONE* **2011**, *6*, e20151. [CrossRef] [PubMed]

24. Prasad, S.; Gupta, S.C.; Tyagi, A.K.; Aggarwal, B.B. Curcumin, a component of golden spice: From bedside to bench and back. *Biotechnol. Adv.* **2014**, *32*, 1053–1064. [CrossRef]

25. Prasad, S.; Kalra, N.; Singh, M.; Shukla, Y. Protective effects of lupeol and mango extract against androgen induced oxidative stress in Swiss albino mice. *Asian J. Androl.* **2008**, *10*, 313–318. [CrossRef]

26. Aggarwal, B.B.; Prasad, S.; Reuter, S.; Kannappan, R.; Yadev, V.R.; Park, B.; Kim, J.H.; Gupta, S.C.; Phromnoi, K.; Sundaram, C.; et al. Identification of novel anti-inflammatory agents from Ayurvedic medicine for prevention of chronic diseases: "reverse pharmacology" and "bedside to bench" approach. *Curr. Drug Targets* **2011**, *12*, 1595–1653. [CrossRef]

27. Garodia, P.; Ichikawa, H.; Malani, N.; Sethi, G.; Aggarwal, B.B. From ancient medicine to modern medicine: ayurvedic concepts of health and their role in inflammation and cancer. *J. Soc. Integr. Oncol.* **2007**, *5*, 25–37. [CrossRef]

28. Shukla, S.D.; Bhatnagar, M.; Khurana, S. Critical evaluation of ayurvedic plants for stimulating intrinsic antioxidant response. *Front. Neurosci.* **2012**, *6*, 112. [CrossRef]

29. Sruthi, C.V.; Sindhu, A. A comparison of the antioxidant property of five Ayurvedic formulations commonly used in the management of vata vyadhis. *J. Ayurveda Integr. Med.* **2012**, *3*, 29–32. [CrossRef]

30. Shyni, G.L.; Ratheesh, M.; Sindhu, G.; Helen, A. Anti-inflammatory and antioxidant effects of Jeevaneeya Rasayana: An ayurvedic polyherbal formulation on acute and chronic models of inflammation. *Immunopharmacol. Immunotoxicol.* **2010**, *32*, 569–575. [CrossRef]

31. Ratheesh, M.; Sandya, S.; Pramod, C.; Asha, S.; Svenia, J.P.; Premlal, S.; GrishKumar, B. Anti-inflammatory and antioxidant effect of Kerabala: A value-added ayurvedic formulation from virgin coconut oil inhibits pathogenesis in adjuvant-induced arthritis. *Inflammopharmacology* **2017**, *25*, 41–53. [CrossRef] [PubMed]

32. Samarakoon, S.M.; Chandola, H.M.; Shukla, V.J. Evaluation of antioxidant potential of Amalakayas Rasayana: A polyherbal Ayurvedic formulation. *Int. J. Ayurveda Res.* **2011**, *2*, 23–28. [CrossRef] [PubMed]

33. Mathew, M.; Subramanian, S. In vitro screening for anti-cholinesterase and antioxidant activity of methanolic extracts of ayurvedic medicinal plants used for cognitive disorders. *PLoS ONE* **2014**, *9*, e86804. [CrossRef] [PubMed]

34. Nariya, M.; Shukla, V.; Jain, S.; Ravishankar, B. Comparison of enteroprotective efficacy of triphala formulations (Indian Herbal Drug) on methotrexate-induced small intestinal damage in rats. *Phytother. Res.* **2009**, *23*, 1092–1098. [CrossRef] [PubMed]

35. Kumar, N.; Khurana, S.M.P. Phytochemistry and medicinal potential of the Terminalia bellirica Roxb. (Bahera). *Indian J. Nat. Prod. Resour.* **2018**, *9*, 97–107.

36. Habib-ur-Rehman; Yasin, K.A.; Choudhary, M.A.; Khaliq, N.; Atta-ur-Rahman; Choudhary, M.I.; Malik, S. Studies on the chemical constituents of Phyllanthus emblica. *Nat. Prod. Res.* **2007**, *21*, 775–781. [CrossRef]

37. Hazra, B.; Sarkar, R.; Biswas, S.; Mandal, N. Comparative study of the antioxidant and reactive oxygen species scavenging properties in the extracts of the fruits of Terminalia chebula, Terminalia belerica and Emblica officinalis. *BMC Complement. Altern. Med.* **2010**, *10*, 20. [CrossRef]

38. Chang, C.L.; Lin, C.S. Phytochemical Composition, Antioxidant Activity, and Neuroprotective Effect of Terminalia chebula Retzius Extracts. *Evid.-Based Complement. Altern. Med.* **2012**, *2012*, 125247. [CrossRef]

39. Singh, M.P.; Gupta, A.; Sisodia, S.S. A Comparative Pharmacognostic Evaluation of Different Extracts of Terminalia bellerica Roxb. Fruit. *J. Res. Med. Dent. Sci.* **2018**, *6*, 213–218.

40. Poltanov, E.A.; Shikov, A.N.; Dorman, H.J.; Pozharitskaya, O.N.; Makarov, V.G.; Tikhonov, V.P.; Hiltunen, R. Chemical and antioxidant evaluation of Indian gooseberry (*Emblica officinalis* Gaertn., syn. *Phyllanthus emblica* L.) supplements. *Phytother. Res.* **2009**, *23*, 1309–1315. [CrossRef]

41. Shilpa, S.; Venkatesha Murthy, C.G. Understanding personality from Ayurvedic perspective for psychological assessment: A case. *AYU* **2011**, *32*, 12–19. [CrossRef] [PubMed]

42. Peterson, C.T.; Denniston, K.; Chopra, D. Therapeutic Uses of Triphala in Ayurvedic Medicine. *J. Altern. Complement. Med.* **2017**, *23*, 607–614. [CrossRef] [PubMed]

43. Sharma, R.K.; Dash, B. *Carka Samhita*; Chowkamba Sanskrit Series Office: Varanasi, India, 1998; Volume II.

44. Baliga, M.S. Triphala, Ayurvedic formulation for treating and preventing cancer: A review. *J. Altern. Complement. Med.* **2010**, *16*, 1301–1308. [CrossRef] [PubMed]

45. Takauji, Y.; Miki, K.; Mita, J.; Hossain, M.N.; Yamauchi, M.; Kioi, M.; Ayusawa, D.; Fujii, M. Triphala, a formulation of traditional Ayurvedic medicine, shows protective effect against X-radiation in HeLa cells. *J. Biosci.* **2016**, *41*, 569–575. [CrossRef]

46. Sandhya, T.; Lathika, K.M.; Pandey, B.N.; Bhilwade, H.N.; Chaubey, R.C.; Priyadarsini, K.I.; Mishra, K.P. Protection against radiation oxidative damage in mice by Triphala. *Mutat. Res.* **2006**, *609*, 17–25. [CrossRef]

47. Parveen, R.; Shamsi, T.N.; Singh, G.; Athar, T.; Fatima, S. Phytochemical analysis and in-vitro biochemical characterization of aqueous and methanolic extract of Triphala, a conventional herbal remedy. *Biotechnol. Rep.* **2018**, *17*, 126–136. [CrossRef]

48. Naik, G.H.; Priyadarsini, K.I.; Bhagirathi, R.G.; Mishra, B.; Mishra, K.P.; Banavalikar, M.M.; Mohan, H. In vitro antioxidant studies and free radical reactions of triphala, an ayurvedic formulation and its constituents. *Phytother. Res.* **2005**, *19*, 582–586. [CrossRef]

49. Varma, S.R.; Sivaprakasam, T.O.; Mishra, A.; Kumar, L.M.; Prakash, N.S.; Prabhu, S.; Ramakrishnan, S. Protective Effects of Triphala on Dermal Fibroblasts and Human Keratinocytes. *PLoS ONE* **2016**, *11*, e0145921. [CrossRef]

50. Baskaran, U.L.; Martin, S.J.; Mahaboobkhan, R.; Prince, S.E. Protective role of Triphala, an Indian traditional herbal formulation, against the nephrotoxic effects of bromobenzene in Wistar albino rats. *J. Integr. Med.* **2015**, *13*, 115–121. [CrossRef]

51. Sharma, A.; Sharma, K.K. Chemoprotective role of triphala against 1,2-dimethylhydrazine dihydrochloride induced carcinogenic damage to mouse liver. *Indian J. Clin. Biochem.* **2011**, *26*, 290–295. [CrossRef]

52. Gupta, S.K.; Kalaiselvan, V.; Srivastava, S.; Agrawal, S.S.; Saxena, R. Evaluation of anticataract potential of Triphala in selenite-induced cataract: In vitro and in vivo studies. *J. Ayurveda Integr. Med.* **2010**, *1*, 280–286. [CrossRef]

53. Srikumar, R.; Parthasarathy, N.J.; Manikandan, S.; Narayanan, G.S.; Sheeladevi, R. Effect of Triphala on oxidative stress and on cell-mediated immune response against noise stress in rats. *Mol. Cell Biochem.* **2006**, *283*, 67–74. [CrossRef] [PubMed]

54. Deep, G.; Dhiman, M.; Rao, A.R.; Kale, R.K. Chemopreventive potential of Triphala (a composite Indian drug) on benzo(a)pyrene induced forestomach tumorigenesis in murine tumor model system. *J. Exp. Clin. Cancer Res.* **2005**, *24*, 555–563. [PubMed]

55. Sandhya, T.; Lathika, K.M.; Pandey, B.N.; Mishra, K.P. Potential of traditional ayurvedic formulation, Triphala, as a novel anticancer drug. *Cancer Lett.* **2006**, *231*, 206–214. [CrossRef] [PubMed]

56. Sandhya, T.; Mishra, K.P. Cytotoxic response of breast cancer cell lines, MCF 7 and T 47 D to triphala and its modification by antioxidants. *Cancer Lett.* **2006**, *238*, 304–313. [CrossRef] [PubMed]

57. Shi, Y.; Sahu, R.P.; Srivastava, S.K. Triphala inhibits both in vitro and in vivo xenograft growth of pancreatic tumor cells by inducing apoptosis. *BMC Cancer* **2008**, *8*, 294. [CrossRef]

58. Tsering, J.; Hu, X. Triphala Suppresses Growth and Migration of Human Gastric Carcinoma Cells In Vitro and in a Zebrafish Xenograft Model. *Biomed. Res. Int.* **2018**, *2018*, 7046927. [CrossRef]

59. Wang, M.; Li, Y.; Hu, X. Chebulinic acid derived from triphala is a promising antitumour agent in human colorectal carcinoma cell lines. *BMC Complement. Altern. Med.* **2018**, *18*, 342. [CrossRef]

60. Zhao, Y.; Wang, M.; Tsering, J.; Li, H.; Li, S.; Li, Y.; Liu, Y.; Hu, X. An Integrated Study on the Antitumor Effect and Mechanism of Triphala Against Gynecological Cancers Based on Network Pharmacological Prediction and In Vitro Experimental Validation. *Integr. Cancer Ther.* **2018**, *17*, 894–901. [CrossRef]

61. Cheriyamundath, S.; Mahaddalkar, T.; Save, S.N.; Choudhary, S.; Hosur, R.V.; Lopus, M. Aqueous extract of Triphala inhibits cancer cell proliferation through perturbation of microtubule assembly dynamics. *Biomed. Pharmacother.* **2018**, *98*, 76–81. [CrossRef]

62. Vadde, R.; Radhakrishnan, S.; Reddivari, L.; Vanamala, J.K. Triphala Extract Suppresses Proliferation and Induces Apoptosis in Human Colon Cancer Stem Cells via Suppressing c-Myc/Cyclin D1 and Elevation of Bax/Bcl-2 Ratio. *Biomed. Res. Int.* **2015**, *2015*, 649263. [CrossRef] [PubMed]

63. Kaur, S.; Michael, H.; Arora, S.; Harkonen, P.L.; Kumar, S. The in vitro cytotoxic and apoptotic activity of Triphala–an Indian herbal drug. *J. Ethnopharmacol.* **2005**, *97*, 15–20. [CrossRef] [PubMed]

64. Srikumar, R.; Jeya Parthasarathy, N.; Sheela Devi, R. Immunomodulatory activity of triphala on neutrophil functions. *Biol. Pharm. Bull.* **2005**, *28*, 1398–1403. [CrossRef] [PubMed]

65. Horani, A.; Shoseyov, D.; Ginsburg, I.; Mruwat, R.; Doron, S.; Amer, J.; Safadi, R. Triphala (PADMA) extract alleviates bronchial hyperreactivity in a mouse model through liver and spleen immune modulation and increased anti-oxidative effects. *Ther. Adv. Respir. Dis.* **2012**, *6*, 199–210. [CrossRef] [PubMed]

66. Phetkate, P.; Kummalue, T.; Yaowalak, U.-P.; Kietinun, S. Significant increase in cytotoxic T lymphocytes and natural killer cells by triphala: A clinical phase I study. *Evid.-Based Complement. Altern. Med.* **2012**, *2012*, 239856. [CrossRef]

67. Rayudu, V.; Raju, A.B. Effect of Triphala on dextran sulphate sodium-induced colitis in rats. *AYU* **2014**, *35*, 333–338. [CrossRef]

68. Kalaiselvan, S.; Rasool, M.K. The anti-inflammatory effect of triphala in arthritic-induced rats. *Pharm. Biol.* **2015**, *53*, 51–60. [CrossRef]

69. Sabina, E.P.; Rasool, M. An in vivo and in vitro potential of Indian ayurvedic herbal formulation Triphala on experimental gouty arthritis in mice. *Vascul. Pharmacol.* **2008**, *48*, 14–20. [CrossRef]

70. Jagetia, G.C.; Malagi, K.J.; Baliga, M.S.; Venkatesh, P.; Veruva, R.R. Triphala, an ayurvedic rasayana drug, protects mice against radiation-induced lethality by free-radical scavenging. *J. Altern. Complement. Med.* **2004**, *10*, 971–978. [CrossRef]

71. Wang, J.; Yi, J. Cancer cell killing via ROS: to increase or decrease, that is the question. *Cancer Biol. Ther.* **2008**, *7*, 1875–1884. [CrossRef]

72. Wang, L.; Leite de Oliveira, R.; Huijberts, S.; Bosdriesz, E.; Pencheva, N.; Brunen, D.; Bosma, A.; Song, J.Y.; Zevenhoven, J.; Los-de Vries, G.T.; et al. An Acquired Vulnerability of Drug-Resistant Melanoma with Therapeutic Potential. *Cell* **2018**, *173*, 1413–1425. [CrossRef] [PubMed]

73. Girdhani, S.; Bhosle, S.M.; Thulsidas, S.A.; Kumar, A.; Mishra, K.P. Potential of radiosensitizing agents in cancer chemo-radiotherapy. *J. Cancer Res. Ther.* **2005**, *1*, 129–131. [CrossRef] [PubMed]

74. Birla, N.; Das, P.K. Phytochemical and anticarcinogenic evaluation of Triphala powder extract, against melanoma cell line induced skin cancer in rats. *Pharm. Biol. Eval.* **2016**, *3*, 366–370.

75. Lu, K.; Chakroborty, D.; Sarkar, C.; Lu, T.; Xie, Z.; Liu, Z.; Basu, S. Triphala and its active constituent chebulinic acid are natural inhibitors of vascular endothelial growth factor-a mediated angiogenesis. *PLoS ONE* **2012**, *7*, e43934. [CrossRef]

76. Aher, V.; Wahi, A. Immunomodulatory Activity of Alcohol Extract of Terminalia chebula Retz Combretaceae. *Trop. J. Pharm. Res.* **2011**, *10*, 567–575. [CrossRef]

77. Saraphanchotiwitthaya, A.; Sripalakit, P.; Ingkaninan, K. Effects of Terminalia bellerica Roxb. methanolic extract on mouse immune response in vitro. *Maejo Int. J. Sci. Technol.* **2008**, *2*, 400–407.

78. Suja, R.S.; Nair, A.M.C.; Sujith, S.; Preethy, J.; Deepa, A.K. Evaluation of immunomodulatory potential' of emblica officinalis fruit pulp extract in mice. *Indian J. Anim. Res.* **2009**, *43*, 103–106.

79. Neuhouser, M.L.; Barnett, M.J.; Kristal, A.R.; Ambrosone, C.B.; King, I.B.; Thornquist, M.; Goodman, G.G. Dietary supplement use and prostate cancer risk in the Carotene and Retinol Efficacy Trial. *Cancer Epidemiol. Biomark. Prev.* **2009**, *18*, 2202–2206. [CrossRef]

80. Goodman, G.E.; Thornquist, M.D.; Balmes, J.; Cullen, M.R.; Meyskens, F.L., Jr.; Omenn, G.S.; Valanis, B.; Williams, J.H., Jr. The Beta-Carotene and Retinol Efficacy Trial: incidence of lung cancer and cardiovascular disease mortality during 6-year follow-up after stopping beta-carotene and retinol supplements. *J. Natl. Cancer Inst.* **2004**, *96*, 1743–1750. [CrossRef]

81. Wiel, C.; Le Gal, K.; Ibrahim, M.X.; Jahangir, C.A.; Kashif, M.; Yao, H.; Ziegler, D.V.; Xu, X.; Ghosh, T.; Mondal, T.; et al. BACH1 Stabilization by Antioxidants Stimulates Lung Cancer Metastasis. *Cell* **2019**, *178*, 330–345. [CrossRef]

MDPI

St. Alban-Anlage 66

4052 Basel

Switzerland

Tel. +41 61 683 77 34

Fax +41 61 302 89 18

www.mdpi.com

Antioxidants Editorial Office

E-mail: antioxidants@mdpi.com

www.mdpi.com/journal/antioxidants